Agroecology

Konrad Martin • Joachim Sauerborn

Agroecology

Springer

Konrad Martin
Plant Production and Agroecology
University of Hohenheim
Stuttgart, Germany

Joachim Sauerborn
Plant Production and Agroecology
University of Hohenheim
Stuttgart, Germany

ISBN 978-94-007-5916-9 ISBN 978-94-007-5917-6 (eBook)
DOI 10.1007/978-94-007-5917-6
Springer Dordrecht Heidelberg New York London

Library of Congress Control Number: 2013931176

Printed on acid-free paper

Springer is part of Springer Science+Business Media (www.springer.com)

Preface

Approximately 10,000 years ago, humans began to engage in arable farming and, thus, to fundamentally modify parts of the natural landscape in a variety of vegetation zones around the world. The crop and livestock-production areas established since then are ecological systems with their own characteristics, because of regular human intervention. In total, these areas make up agroecosystems, whose structure and function are directed and affected by humans on the one hand but are, nevertheless, shaped by natural factors on the other.

This book deals with the patterns and processes of agroecosystems. It also tries to take a holistic view of the ecological relationships apparent from the interactions within agroecosystems—the agricultural activities of humans and their feedback in the natural world. Global aspects are considered throughout, because agriculture is a major cause of change in ecological systems and processes at different scales. The book's scope therefore covers agroecosystems of different regions and climate zones and the spatial and temporal dimensions of agriculture that affect ecological cycles, the environment, and the Earth's climate.

Finally, it is also our objective to illustrate the relationships between ecology as a basic science and the applied and production-oriented disciplines of agricultural science, for example crop production, plant protection, and livestock farming. It starts with the biotic and abiotic effects to which all species and their interactions in both natural and agricultural systems are subject. This involves examination not only of single factors but also their diverse direct and indirect effects on different system components. Building on ecological foundations, methods of managing and controlling agroecosystems to secure and increase production are discussed. These include such management measures as fertilization, irrigation, soil cultivation, selection of cultivated crop and livestock species, and management of pests, weeds and diseases. Such intervention also affects the environment. Not only does it affect natural habitats, their communities, and the global climate, it also affect the livelihoods of the growing human population in different ways.

With its interdisciplinary approach, this book is primarily directed at students of the agricultural sciences, biology, agro-biology, horticulture, and geography; it also provides relevant background information for interdisciplinary sciences, for example geoecology and landscape ecology, landscape management, and planning. The book also provides compact information for teachers and lecturers of different levels.

This book is an extended and updated version of the original German edition (Agrarökologie) published in 2006 by Eugen Ulmer Verlag, Stuttgart.

For their contributions to this edition, we wish to thank Mrs Anne Auffarth for preparation of all the graphs, Nicholas Mitchell for proof reading, and Christoph Allgaier for provision of additional illustrations. Unless otherwise noted, other illustrations used in the book are the work of Rolito Dumalang. We also wish to thank Mrs Maryse Walsh, Senior Publishing Editor, and Melanie van Overbeek, Senior Publishing Assistant, of the Agronomy and Life Sciences Unit of Springer Dordrecht, The Netherlands, for their helpful support and assistance in the preparation of the book manuscript.

Stuttgart-Hohenheim

Konrad Martin
Joachim Sauerborn

Contents

Chapter 1
Introduction

1.1 Basic Terms and Overview of Contents

The term "agroecology" is composed of three roots, derived from the Latin *agrarius* (pertaining to the field), the Greek *oikos* (the home or the household), and the Greek *logos* (study).

The term **ecology**, defined as *the science of the overall relations of organisms to both their living and non-living environment*, was first introduced by the German zoologist Ernst Haeckel (1866). Based on this, **agroecology** can be defined as *the science of the relationships of organisms in an environment purposely transformed by man for crop or livestock production*. However, the term agroecology is currently used with quite different meanings (Sect. 1.1.3).

The **organisms** include various **species** and **populations** (Box 1.1) of crops, wild plants, animals and microorganisms. Domesticated crops and animals originate from wild species and have subsequently evolved through the selection of particular traits by man (Sect. 2.2.1).

Box 1.1 Species and Population

Groups of individuals are members of the same **biological species** and form a genetic unit when they can produce offspring capable of reproduction. This applies to both the cultivated forms of plants (varieties) and animals (breeds) that were derived from wild species, although the common descent is often difficult to discern because of marked differences in their habit. Only when individuals no longer form a reproductive community, i.e. are isolated from each other in terms of reproduction, can they be regarded as separate species. A **population** is defined as a reproductive community that is composed of individuals of a species in a particular geographic area or in a particular habitat.

K. Martin and J. Sauerborn, *Agroecology*, DOI 10.1007/978-94-007-5917-6_1,

The **environment** consists of all factors affecting the living conditions of organisms. The different physical and chemical effects that originate from the non-living environment represent the **abiotic factors**. In terrestrial habitats, they essentially include the properties of the soil, specific geographic factors (e.g. topography and altitude), and climatic conditions. The latter primarily include factors related to insolation (thermal energy and light), precipitation, and the water balance.

The affects of **biotic factors** originate from organisms and can be exerted on other individuals of the same species (intraspecific), on individuals of a different species (interspecific), or on the abiotic environment (e.g. on specific soil properties). From the perspective of a species, the biotic environment essentially consists of other species, to which it can have different forms of relationship. These include feeding relationships, competition, and mutualism (Sect. 3.1).

Specific environmental factors are termed **resources**, defined as essential components of the environment that are used by the individuals and species. They can be of biotic or abiotic origin and differ by species, depending on their specific requirements. Basically, three categories of resource are of importance for all species: food, energy, and specific spatial structures. Important resources for plants include light (energy source), carbon dioxide, mineral nutrients, and, for species that form roots, a suitable soil structure to anchor and support the plant. Animals draw their food and energy from organic compounds that originate from either living organisms or their dead remains. Specific bacteria use chemical compounds (e.g. hydrogen sulfide or methane) as energy sources. Most living organisms also require oxygen for respiration. Spatial resources for animals can include shelter or hiding places and suitable locations to raise their offspring (e.g. nesting sites).

The **ecological niche** of a species comprises the total range of abiotic and biotic factors under which a species or population exists, maintaining itself by reproduction and reaching a specific density of individuals per spatial unit (abundance).

Organisms obtain their food directly from the natural environment in which they live and reproduce. There are few but notable exceptions to this general fact. One exception is the leaf-cutter ants (*Atta* and *Acromyrmex* species) in tropical America, which harvest leaf pieces from plants and carry them to their subterranean nests. There, the leaf material is chewed to a pulp and then used as a substrate for cultivation of fungi which, in turn, serve as food for the ant colony. Humans are also able to produce their own food by cultivation of crops. However, this strategy has not been used by humans during all their 150,000 years of existence as the so-called modern human (*Homo sapiens sapiens*) but was, instead, only adopted approximately 10,000 years ago. Therefore, important aspects of agroecology also concern the ecology of humans. This not only refers to the question of why and how there was this change in the use of natural resources (Chap. 2) but also to its future development, which is essentially determined by the growing world population. These questions concern human food supply and changes in the global environment (Chap. 8).

1.1.1 Ecosystems

Network structures involving organisms and their environment, in which a variety of abiotic and biotic factors are at work, are called **ecosystems**. They form structural and functional units that are both spatially related and involved in the exchange of materials, energy, and organisms, thus affecting each other. This characterization also applies to **agroecosystems** which differ from natural ecosystems in that they are managed by humans.

The most important patterns and processes describing natural and agricultural ecosystems include the relationships between species, species diversity, temporal changes of communities, and flows of materials and energy (Chap. 3).

There are no objective criteria with which to determine the spatial and functional boundaries of an ecosystem. However, on the basis of abiotic and biotic properties, some structural sections can be defined on the basis of different scales and used for description of patterns and for analysis of processes. For example, the Earth's surface can be divided into different ecological zones on the basis of climate, which broadly corresponds to specific vegetation and soil types (Chap. 7). Within the climate and vegetation zones, different landscapes and landscape elements continuing down to individual biotopes and habitats (Box 1.2) can be defined. Criteria for this include the respective temperature and precipitation conditions, geological and hydrological characteristics, topography and altitude, and the plant and animal communities affected by these factors. Corresponding to this, energy and material flows can be examined on different scales, ranging from global material cycles (Sect. 3.7) to small-scale translocations in the soil (Sect. 4.3).

Box 1.2 Biotope and Habitat

Originally, a **biotope** was understood to be the environment of an ecological community (biocoenosis; Sect. 3.2) characterized by abiotic factors whereby the biotope and the biocoenosis together form the ecosystem. However, in the broader, and currently most common form, the term is used mostly without separation of abiotic and biotic properties. In this case, the term generally refers to specific structures within a landscape (e.g. ponds, meadows, fields, woodlands etc.).

The term **habitat** was defined to characterize the space inhabited by a specific species. Today the term is also used for communities of species and, similar to biotope, includes abiotic and biotic characteristics.

Overall, biotopes, habitats and ecosystems do not represent hierarchically ordered and separable structures. They can be defined as spatial units only by more detailed characterization (e.g. on the basis of the vegetation).

1.1.2 Agroecosystems

Agroecosystems are ecosystems established for the production of useful plants and animals. They differ from natural ecosystems in that they are shaped by humans whose regular intervention manipulates the composition of their organisms and their function. They are characterized by regular cultivation measures, primarily sowing, harvest, and soil cultivation, but also by other types of management, for example mowing, grazing, and burning.

A classification of the different kinds of agroecosystem is presented in Sect. 2.4. The crops used in the different systems can be classified on the basis of their use:

- **Food crops** that serve as part of the human diet. These include cereals, root and tuber crops, legumes, oil crops, leaf vegetables, and fruit.
- **Beverages and stimulants, medicinal and spice plants**. Examples are coffee (*Coffea* species), tea (*Camellia sinensis*), chamomile (*Matricaria recuitita*), lavender (*Lavandula augustifolia*), and pepper (*Piper nigrum*).
- **Fodder plants** for animal production. These include a wide range of species and production types, depending on their use (Chap. 6). Food for livestock can be produced from different types of grassland, for example regularly mown hay meadows or pastures for livestock grazing. Field forage crops include a variety of cereals identical to human food plants, legumes, and other herbaceous species, for example fodder beet (*Beta vulgaris*).
- **Raw materials**, which can serve in the production of fibres and fuels. Fibre plants include cotton (*Gossypium* species), hemp (*Cannabis sativa*), and abaca (*Musa textilis*). For fuel, crops such as rapeseed (biodiesel), and alcohol (ethanol) are used; the latter is produced from a variety of crops, for example sugar cane (*Saccharum officinarum*).

Many crop species can be grown for different purposes and different uses, and so can be assigned to more than one of the categories listed.

One approach to description and analysis of agroecosystems is to focus on the crop plant. From the perspective of the plant, the agroecosystem is the environment where its growth, development, and physiological processes are directly or indirectly affected (Fig. 1.1). Each site is characterized by its own natural conditions, to which the crop plant is exposed (Chap. 4). Some of these conditions can be modified by the farmer to create the most suitable conditions for crop development. Specific measures include, e.g., fertilization and irrigation, and control of organisms, for example pests and weeds, that negatively effect the crop plants or cause yield losses (Chap. 5).

Another approach addresses analysis of matter and energy flows in agroecosystems at different spatial and temporal scales. Understanding and integration of ecological processes in farm production and land management can be used to improve resource use efficiency and to reduce external inputs. Overall, consideration of ecological principles, for example nutrient cycling, can contribute to sustainable crop and food production by minimizing both use of non-renewable

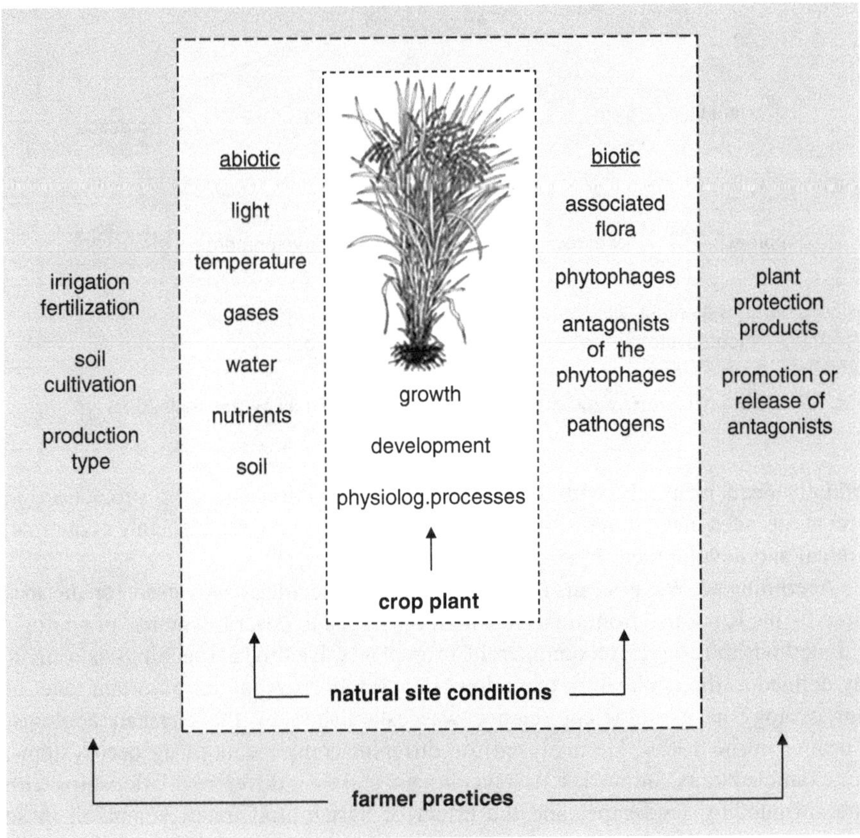

Fig. 1.1 The agroecosystem as the environment of the crop

inputs and detrimental effects on the environment. Concepts to achieve this include the combination and alternation of crops to enhance nutrient use efficiency, in particular the use of legumes for nitrogen fixation. In addition, incorporation of livestock in the system will increase the range of crops, including grassland, and provide nutrients in manure which can be used to fertilize annual crops and forage land (Chap. 6).

1.1.3 Agroecology: History and Concepts

Other than the generally accepted concept of ecology, the term "agroecology" is not clearly defined and can have quite different meanings. Wezel et al. (2009) reviewed the use and history of the term and found that it can imply a scientific discipline, an agricultural practice, or a social or political movement. Although agroecology

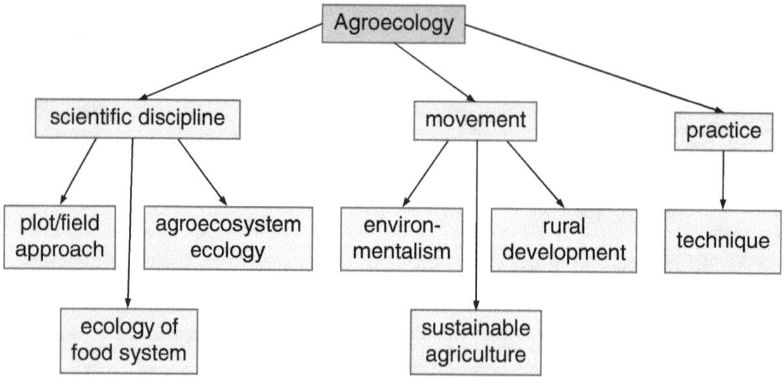

Fig. 1.2 Diversity of current meanings of agroecology (Based on Wezel et al. 2009)

initially dealt primarily with aspects of crop production and crop protection, in recent decades new dimensions, for example environmental, social, economic, ethical and development issues have become relevant.

According to Wezel et al. (2009), the term agroecology was used for the first time by the Russian agronomist Bensin in the 1930s to describe the use of ecological methods in research on commercial crop plants. By this, agroecology is primarily defined as the application of ecology to agriculture. Another important tenet of agroecology as a biological science was established by the German zoologist Tischler in the 1950s. He analyzed the different components of agroecosystems, for example plants, animals, soils, and climate, and their different relationships with the surrounding landscape, and the effect of agricultural management on these components.

As a consequence of different adverse effects of industrialized agriculture on the environment and on human welfare, agroecology emerged as both a movement and a set of practices from the 1970s onwards (Fig. 1.2).

Instead of narrow focus on short-term yields and economic returns, researchers increasingly considered the environmental and social factors of food production. From an ecological perspective, agroecology became defined as "*the scientific basis of alternative agriculture*" (Altieri 1987) and as "*the application of ecological concepts and principles to the design and management of sustainable agroecosystems*" (Gliessman 1997). Finally, agroecoecology as a scientific discipline went through a major change, moving beyond field or agroecosystem scales toward a larger focus on the whole food system, leading to a new and broader definition of agroecology as "*the integrative study of the ecology of the entire food systems, encompassing ecological, economic and social dimensions, or more simply the ecology of food systems*" (Francis et al. 2003). Therefore, according to Wezel et al. (2009), there are actually three approaches to agroecology, different aspects of which are also considered in this book: (1) investigations at plot and field scales, (2) investigations at agroecosystem and farm scales, and (3) investigations covering the whole food system.

References

Altieri MA (1987) Agroecology: the scientific basis of alternative agriculture. Westview Press, Boulder

Francis C, Lieblein G, Gliessman S, Breland TA, Creamer N, Harwood SL, Helenius J, Rickerl D, Salvador R, Wiedenhoeft M, Simmons S, Allen P, Altieri M, Flora C, Poincelot R (2003) Agroecology: the ecology of food systems. J Sustain Agric 22:99–118

Gliessman SR (1997) Agroecology: ecological processes in sustainable agriculture. CRC Press, Boca Raton, 384 p

Haeckel E (1866) Generelle morphologie der organismen. Georg Reimer Verlag, Berlin

Wezel A, Bellon S, Doré T, Francis C, Vallod D, David C (2009) Agroecology as a science, a movement and a practice. A review. Agron Sustain Dev 29:503–515

Chapter 2
Origin and Development of Agriculture

2.1 The First Humans

The savannas of Africa are believed to be the cradle of mankind. All members of *Homo sapiens sapiens* living today descended from populations that lived there approximately 150,000–200,000 years ago. Just as two earlier representatives of the genus *Homo* (*H. erectus* and the ancestors of the *H. neanderthalensis*) had done before, groups of modern humans left their homeland to colonize other continents. This occurred between 50,000 and 70,000 years ago. The reasons for this migration, which only began 100,000 years after the emergence of *H. sapiens*, are subject to speculation. It is possible that the first migrations coincided with changes in climate that also affected resource availability and the living conditions of humans. However, little is known about the diet of early *H. sapiens* in Africa. He was already a hunter of wild animals, which placed him in competition with large cats and other predators. Humans not only had to defend their prey against these predators, but also to protect themselves. The hunt for large animals for example gazelles, buffalo, or even elephants was of little importance for early humans partially because of their still primitive weapons. Presumably, they used a broad range of food sources including plant products (roots, seeds, and fruit), small animals (e.g. small mammals, reptiles, insects), wild honey, and eggs from birds and reptiles. They probably also ate the remains of large animals, for example bone marrow, which was accessed with the help of stone tools (Fig. 2.1). With the help of fire, it was possible to roast meat and to cook many plant species that were otherwise inedible or even poisonous.

According to today's knowledge, the first groups of *H. sapiens* left Africa via the Middle East and had reached Australia by at least 40,000 years ago (Fig. 2.2).

Because a land bridge has never existed between Asia and Australia, this continent could only be settled with the help of boats or rafts. Approximately 35,000 years ago the first humans reached Europe where the Neanderthal (*H. neanderthalensis*) was already living. The first modern Europeans were the Cro-Magnons, named after the first location in which they were found in France. After spreading into Asia, humans

K. Martin and J. Sauerborn, *Agroecology*, DOI 10.1007/978-94-007-5917-6_2,

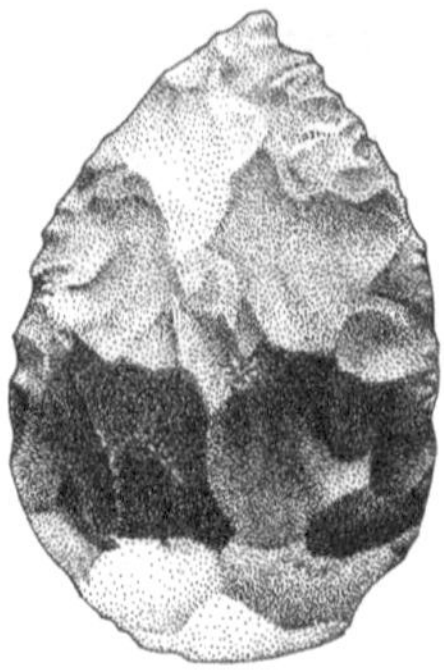

Fig. 2.1 Hand axe made of flint

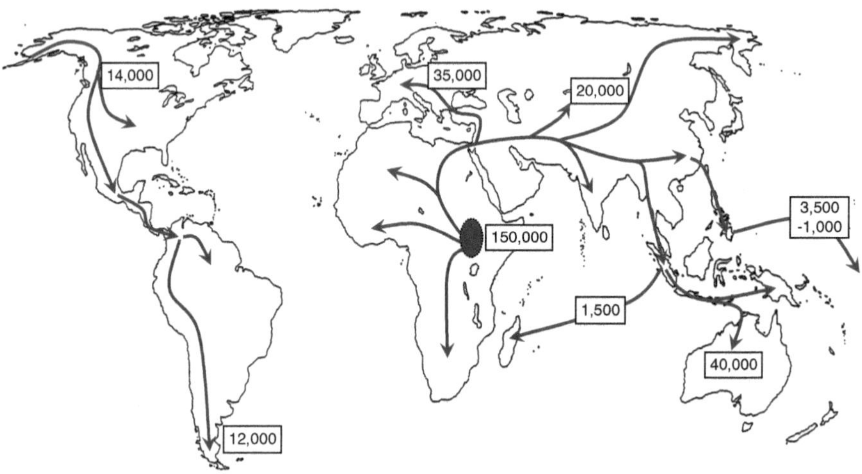

Fig. 2.2 The spread of *Homo sapiens* from Africa around the world (dates are years before present)

were able to settle the Americas via the Bering Strait, at least 14,000 years ago. During the ice age (i.e. the period before approx. 11,000 years ago), humans in Europe and North America were primarily hunters. With the development of effective weapons (Fig. 2.3) and other hunting techniques it became possible to successfully hunt large animals of the open steppe or tundra (for example mammoth, woolly rhinoceros, horse, Irish elk, giant deer, and European bison). Such species were not only of importance for food, but also provided fur and skin for the production of clothing and shelter. Use of plant resources may have been of minor importance because of the sparse vegetation in these areas during the ice age.

Humans reached the southern tip of South America within approximately 2,000 years of their initial colonization of North America. However, the history of

Fig. 2.3 Arrowhead made of flint

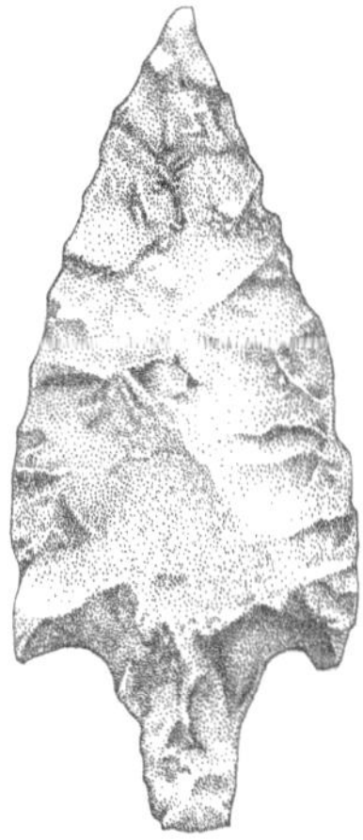

colonization of the Americas is subject to debate because there is also evidence of much earlier appearance of humans on this continent. The last large areas reached by humans, except for the polar regions, were the islands of the South Pacific (approx. 3,000–1,000 years ago) and Madagascar (approx. 1,500 years ago), both of which were colonized from Asia (Fig. 2.2).

2.2 The Origins of Agriculture

With the end of the last ice age, approximately 11,000 years ago, the climatic conditions fundamentally changed in many regions of the world. Temperature increases and altered precipitation patterns led to changes in vegetation. Forests became established in Central Europe and some other regions, replacing the treeless tundra or steppe that existed during the ice age. Distinct changes in the distribution of different vegetation zones also occurred in many regions of the tropics and subtropics. In addition, toward the end of the ice age most of the large animal species that existed in the ice age environments of Europe, North America, and parts of Asia became extinct. However, it is unlikely that changes in climate were the sole cause for the disappearance of this fauna. Their habitat did not completely disappear, but shifted toward the north, so migration into today's tundra regions was, in principle, possible and was achieved by a few species. Thus, musk oxen and reindeer still exist today, and the last dwarf mammoths became extinct 4,000 years ago on the Wrangel Islands in north-eastern Siberia. Another factor that probably contributed to the extinction of species among the large fauna of the ice age was hunting by humans.

In various regions of the world, the origin of agriculture dates back 11,000–10,000 years, during a phase in the climate that was accompanied by fundamental changes in the living conditions of organisms, including humans.

Why and how did the transition from hunting and gathering to farming occur? Numerous theories and models attempt to provide an answer to this. From these, two main groups of hypotheses can be distinguished:

1. According to the first hypothesis, agriculture is an innovation that enabled a way of life that is advantageous compared with the hunter-gatherer existence. Some groups of humans discovered the potential of producing plants in fields, whereby these earliest farmers not only acquired a secure source of food, but also became sedentary. This also initiated cultural progress and, overall, a higher standard of living. Such groups served as examples for the hunter-and-gatherer groups, which subsequently also began to practise agriculture.
2. According to the second hypothesis, a shortage of food resources (primarily the lack of wild animals for the hunters) was the precondition for the development of agriculture. Reasons for this include an increase in human population density in combination with decreases in big-game species because of overhunting. According to this view, the transition to agriculture was not a voluntary act, but rather occurred as a result of the need to find alternative sources of food. By no means does this have advantages over hunting and gathering, but is more labour and time-intensive and is, in addition, associated with the risk of crop failures and thus with hunger.

The emergence of agriculture was not a sudden event or a genius invention by individuals, but rather a far more gradual process. There is evidence from different regions of the world that the intended production of plants is generally associated with decreased use of, or a decline in, wild animal populations, which supports the second of the hypothesis stated above. In the earliest phases of crop production, the cultivated plants probably served as a kind of food reserve or alternative, in case of failure in hunting. Subsequently, crop production gained in importance as the hunted animals became increasingly scarce.

However, the development of agriculture in the different regions of the world in which it emerged (Sect. 2.2.2) did not always follow exactly the same pattern. There were, probably, corresponding to the given conditions, all imaginable kinds of transition between the nomadic hunter-and-gatherer existence and the sedentary farmer, in which both wild natural resources and crop production contributed to the food supply.

Complete dependence of humans on agriculture only emerged after wild animals and plants could no longer make a significant contribution to the food supply of the growing population.

2.2.1 Adapting Wild Plants and Animals for Agriculture

Both agricultural crop varieties and domestic livestock breeds originate from wild species, many of which were already gathered or hunted, respectively, by humans before the emergence of agriculture. In contrast with the wild forms, the cultivated plants and animals have altered characteristics that developed via selection (Box 2.1).

Box 2.1 Genetic Diversity and Selection

Individual differences are often already visible among members of the same population and species that reproduce by sexual reproduction. These are primarily morphological, physiological, and biochemical properties that together make up the **phenotype** of an individual. The phenotype is the sum of the interactions between the genes (which together represent the **genotype**) and the environment during the ontogenesis of an individual. A specific genotype can produce different phenotypes, depending on environmental conditions. For example, plants that are sufficiently supplied with water and nutrients form large individuals and produce more biomass than those that lack these resources. The range of variation of the phenotype, which the same genotype produces in relation to dominant environmental conditions, is termed the **reaction norm**. The second cause of phenotypic variation among members of a population are differences in their genotype. Individuals that are the product of sexual reproduction are never genetically identical, because each is a new and unique combination of the parental genes. The diversity of the genes of a population (the so-called **gene pool**) constantly changes because of the incidence of **mutations**. Mutations are random changes in the genotype of an individual which usually result from mistakes in the doubling of the genome during cell division (cytokinesis) and may be harmful, useful, or irrelevant to the individual.

On the basis of their unique genetic endowment, individuals of a population also have different probabilities of survival and reproduction (Fitness; Box 3.1), which means they are not all equally well-adapted to the given environmental conditions. This becomes particularly evident when the effect of individual factors changes drastically. This can include the effects of the abiotic environment (e.g. weather conditions) or effects that originate from other organisms (e.g. pathogens, predators, or competitors). Depending on the

(continued)

Box 2.1 (continued)

situation, some properties prove advantageous and others disadvantageous regarding survival under the given conditions. Individuals with unsuitable genotypes and phenotypes sometimes become eliminated from the population. This process is called **natural selection**.

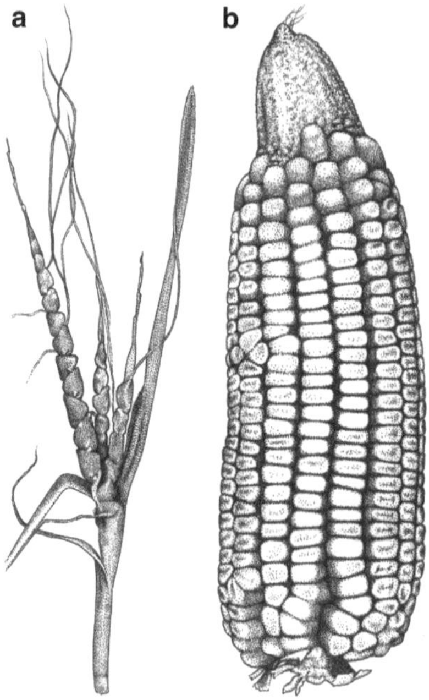

Fig. 2.4 Teosinte (**a**), a grass species from Central America, is regarded as the wild form of maize (**b**)

For cultivated species, however, selection is not based on natural processes alone, but also originates from humans, and probably happened unconsciously at the beginning. It can hardly be assumed that all wild seeds or fruits of a species were gathered randomly, instead the largest individuals were probably preferred. Of this yield, some was accidentally lost near the resting and settlement areas and thus inadvertently sown. In these plants, which were later harvested from such locations, the preferred characteristics already occurred more frequently than in the natural population. From here it was only a small step to conscious selection or selective breeding, through which, as a result of propagation and repeated selection of individuals, the desired traits were further improved. Over hundreds and thousands of years, crops and domestic livestock gradually emerged from wild ancestors via directed selection of the respective wild species (Fig. 2.4). Overall, conscious and unconscious changes in plant and animal species as a result of artificial selection, to make them more useful for humans than the original wild form, is called **domestication**.

2.2.1.1 Characteristics of Domesticated Crops

The altered characteristics of food crops are, primarily, the size of the utilized plant parts (for example seeds, fruit, leaves, or roots). For example, cabbage (*Brassica oleracea*) was variously selected for its leaves (cabbage and kale), stems (kohlrabi), flower shoots (broccoli and cauliflower), and buds (Brussels sprouts). Characteristics may also include altered concentrations of specific ingredients (especially secondary plant metabolites, Sect. 4.5.5.2) that determine edibility and taste. There are, in addition, other traits of importance for cultivation which became characteristic of domesticated crops. These include:

- **Dispersal mechanisms** of seeds of crop species (primarily cereals and legumes): the wild species have mechanisms that cause the release of ripe seeds from the plant and thus enable effective dispersal. The seed heads of wild grasses break apart ("shatter") at maturity to scatter the seed. The pods of legumes split explosively at maturity to scatter seed. These characteristics are undesirable in a crop plant because they lead to reduced yield when the seed falls off the plant before or during harvest. For many species, however, mutants appear that cannot drop their seeds. In mutated individuals of bean and pea, for example, the pod remains closed after ripening, and in the respective grass or cereal species, breaking of the spike is impeded. In the harvest of wild stands or in the field, these mutants are automatically preferred. Eventually, their characteristics establish themselves in the crop, which thus become entirely dependent on the farmer in its reproduction.
- **Synchronous germination**: The seeds of many annual plant species do not all germinate simultaneously in one season, but instead germinate over the course of several years. This mechanism prevents all the offspring of a population dying with the occurrence of unfavourable conditions (especially drought). When sown in fields, only individuals without this delay in germination emerge in the first year. They are harvested and utilized for a later sowing. Thus selection of plants that all germinate at once and ripen simultaneously occurs, and thus a higher yield is achieved. Lack of germination delay is, therefore, a characteristic of many domestic crops.

2.2.1.2 Characteristics of Domesticated Animals

Similar to useful plants, some wild animal species were domesticated as sources of food and clothing and/or for labour or transportation (called livestock or farm animals). In general, domestication of livestock species occurred somewhat later than crop domestication.

Because of the behaviour of animals, domestication of livestock is a more complex process than domestication of plants. Different from taming, animal domestication can be defined as a process which includes removal of the species from its natural habitat, adaptation to man and to a captive environment, control of its movements (keeping) and its food supply, and, finally, controlled breeding by practising artificial selection. Animals subject to less than complete mastery can be regarded as partially domesticated. Tameness, or lack of avoidance responses when approached by humans, is a desirable behavioural characteristic of captive animals, because it facilitates handling.

The transition from nature to captivity is accompanied by many changes in biological and physical environments. Providing animals with food and medical care, protecting them from predators, and assisting in the care of offspring are functions served by humans that may increase the genetic and phenotypic variability characterizing captive animal populations. Man's control including the selection of livestock for high fertility, docility, and early maturity often improves viability and reproductive success compared with their free-living counterparts (Price 1984).

The history of animal domestication shows that relatively few wild animal species have been successfully domesticated as farm animals. They are mainly large mammalian herbivores or omnivores and a small number of bird species (chicken, turkey, goose, duck), which are relatively convenient and economical to breed and to maintain in captivity. Diamond (2002) identified six main obstacles to domestication of wild mammal species:

- difficulty fulfilling specific food requirements;
- slow growth rate and long birth spacing;
- nasty disposition;
- reluctance to breed in captivity;
- lack of follow-the-leader dominance hierarchies; and
- tendency to panic in enclosures or when faced with predators.

For one or more of these reasons, such wild mammals as bears, elephants, antelopes, and gazelles were never domesticated. In contrast, behavioural characteristics that facilitate the domestication process include social organization in large groups of hierarchical structure, intensive mother–young interactions and precocial young, and low reactivity to man or sudden changes in the environment. Therefore all important livestock species, including poultry, are herd animals.

2.2.2 *Centres of Origin of Agriculture*

Worldwide, approximately 11 regions are believed to be centres of the origin of agriculture, identified as the locations in which native plant and some animal species were domesticated independently of each other (Fig. 2.5). In contrast, in other regions the origin of agriculture is based, at least in large part, on crops and livestock that were introduced to those regions and originally come from the centres of origin.

Fig. 2.5 Centres of origin of agriculture. **1** = Middle East (Fertile Crescent), **2a** = northern China, **2b** = southern China, **3** = Southeast Asia, **4a** = South American highlands, **4b** = South American lowlands, **5** = Central America, **6** = arid savannas of northern Africa, **7** = eastern North America, **8** = highlands of Ethiopia, **9** = humid savannas of West Africa (Based on Diamond 1998)

The most important regions in which plants and, sometimes, animals were domesticated are:

2.2.2.1 The Middle East

The Middle Eastern centre of origin, also known as the "Fertile Crescent", covers an area ranging from Jordan and Syria over the eastern part of Turkey to the valleys of the Euphrates and Tigris Rivers (Mesopotamia, today Iraq). The beginnings of agriculture in these regions date back at least 11,000 years. The oldest crop species that were domesticated there are the wheat species emmer (*Triticum dioccum*) and einkorn (*Triticum monococcum*; Fig. 2.6), barley (*Hordeum vulgare*), pea (*Pisum sativum*; Fig. 2.7), lentil (*Lens esculenta*), chickpea (*Cicer arietinum*), and flax (= linseed; *Linum usitatissimum*).

The first animal species domesticated in this region were sheep (*Ovis ammon*) and goat (*Capra hircus*), approximately 11,000 years ago, and subsequently pig (*Sus scrofa*) and cattle (*Bos taurus*). Agricultural economies reliant on a mixture of domesticated crops and livestock became established in this region approximately 9,500–9,000 years ago (Zeder 2008).

2.2.2.2 Northern and Southern China

It was, presumably, somewhat later than in the Middle East that the development of agriculture began in China, with at least two centres of origin. The first of these was the tropical/subtropical south, where rice (*Oryza sativa*) was domesticated

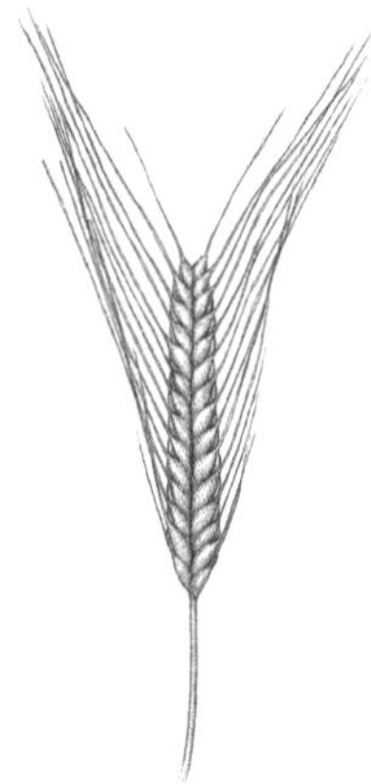

Fig. 2.6 Einkorn (*Triticum monococcum*)

Fig. 2.7 Pea (*Pisum sativum*)

approximately 10,000 years ago. The domestic chicken (*Gallus gallus*) is also assumed to originate from this region. The pig is also counted among the earliest livestock of China and was probably domesticated there independently of the Middle Eastern centre of domestication. The oldest crops of the cooler and dryer north include foxtail millet (*Setaria italica*), which was domesticated approximately 6,000 years ago, and soybean (*Glycine max*).

2.2.2.3 Southeast Asia

Banana (*Musa* species), sugar cane (*Saccharum officinarum*), taro (*Colocasia esculenta*; Fig. 2.8) and yam (*Dioscorea* species) originate from tropical Southeast

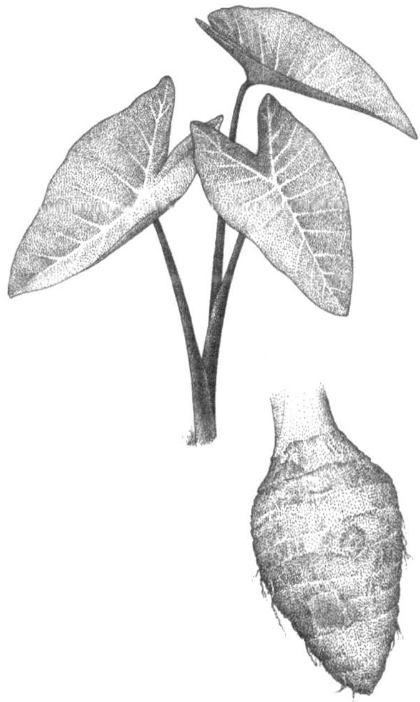

Fig. 2.8 Taro (*Colocasia esculenta*), habitus of the plant and rhizome

Asia. The agriculture of this region probably has its origins in the highlands of Papua New Guinea and began there, according to recent discoveries, approximately 10,000 years ago (Denham et al. 2003).

2.2.2.4 Tropical South America

Agriculture in South America also began at least 10,000 years ago, with evidence of the domestication of *Cucurbita* species (Piperno and Stothert 2003). In South America, depending on altitude, three regions can be identified from which specific crop species originate. The potato (*Solanum tuberuosum*) originates from the highlands of the Peruvian and Bolivian Andes. Peanut (*Arachis hypogaea*) and common bean (*Phaseolus vulgaris*) were domesticated in the mid-altitudes of the Andes. The tropical lowlands of South America were the centre of origin of squash and pumpkins (*Cucurbita*), peppers and chili (*Capsicum* species), pineapple (*Ananas comosus*), sweet potato (*Ipomoea batatas*), cassava (*Manihot esculenta*), avocado (*Persea americana*), and the cotton species *Gossypium barbadense*. The only animal species domesticated in South America were guanaco (*Llama guanicoe*; the wild form of llamas and alpacas) and guinea pig (*Cavia aperea*).

Fig. 2.9 Cotton (*Gossypium hirstum*), twig with open seed pod

2.2.2.5 Central America

In Mexico, the domestication of pumpkins or winter squash (*Cucurbita pepo*) dates back approximately 10,000 years (Smith 1997), and occurred independent of the process of domestication of the *Cucurbita* species in South America. Domestication of *Phaseolus* and *Capsicum* species, avocado and the cotton species *Gossypium hirstum* (Fig. 2.9) also occurred in Mexico independently of South America.

The most important crop species originating from Central America is maize (*Zea mays*). The potato only reached Central America thousands of years after its domestication in the Andes. This can be regarded as further evidence that the two regions developed independently of each other over long periods of time. The turkey (*Meleagris gallopavo*) was domesticated in Central America approximately 2,000 years ago.

2.2.2.6 North America

Approximately 4,000 years ago, an independent centre of origin of agricultural development emerged in eastern North America. There, several crops were cultivated that are of little importance today, for example sumpweed or marshelder (*Iva annua*) and pigweed or lambsquarters (*Chenopodium berlandieri*). The only important crop that originates from North America is the sunflower (*Helianthus annus*), which was probably grown as an oil crop.

Fig. 2.10 Coffee (*Coffea arabica*), twig with leaves and fruits

2.2.2.7 African Regions

In Africa, three climatically different regions that are believed to be centres of origin of agriculture have been identified. Agriculture developed in the **highlands of Ethiopia** approximately 6,000 years ago. Several domesticated species, for example coffee (*Coffea arabica*; Fig. 2.10), finger millet (*Eleusine coracana*), and the cereal species teff (*Eragrostis tef*), originate from this region.

In the **dry savannas of northern Africa**, sorghum (*Sorghum bicolor*), pearl millet (*Pennisetum glaucum*), and African rice (*Oryza glaberrima*) were domesticated. The **humid savannas of West Africa** are the centre of origin of oil palm (*Elaies guineensis*), cowpea (*Vigna unguiculata*), and the African yam (*Dioscorea* species).

2.2.3 Spread of Agriculture and Crops

In most regions in which agriculture developed, the cultivated plant species can be divided into four main groups, each of which meets specific nutritional requirements of humans. These are:

- starch-yielding crops (primarily cereals and tuber crops),
- crops rich in protein (primarily legumes),
- oil-producing crops, and
- fibre crops that primarily serve in the production of textiles.

In cases in which wild plants from these groups were not available for domestication, the respective crops could, under specific conditions, be acquired from other regions. The possibilities of this were primarily determined by

- the distance to another region and the existence of contact with the resident people, and
- the climatic suitability of the region for the production of crops that originate from other climate zones.

The agriculture that developed in the Middle East sooner or later spread into neighbouring regions. Recent evidence suggests that the expansion of domesticated species and agricultural economies across the Mediterranean was accomplished by several waves of colonists who established coastal farming enclaves around the Mediterranean Basin within approximately 3,000 years of the first farming activities. This process also involved the adoption of domesticated species and technology by indigenous populations and the local domestication of additional species. Likewise, agriculture spread in an easterly direction reaching the Indus Valley (today's Pakistan). The spread of agriculture to Central, Western, and Northern Europe proceeded over a period of thousands of years. Approximately 7,000 years ago, most Middle Eastern crops were known in Central Europe, but in some regions (e.g. coastal regions of the North and the Baltic Sea) they were only cultivated between 6,000 and 4,500 years ago. In some regions, for example northern Germany and the Alps, it was mainly animal husbandry that was adopted at that time. Already by pre-Roman times, crops of regions other than the Middle East had reached Central Europe. Examples include proso or common millet (*Panicum miliaceum*), which was an important crop well into the Middle Ages and probably originates from Central Asia, and the faba bean (*Vicia faba*), the wild form and centre of origin of which are not precisely known.

Among farm animals, the domestic chicken (*Gallus gallus*), originating from Southeast Asia, became known in Southern Europe approximately 3,500 years ago and in Central Europe 2,600 years ago. The first use of the domesticated horse (*Equus ferus caballus*) as a working animal in Europe dates back approximately 4,000 years.

Over the course of time, additional crop and animal species were domesticated, both in the original centres of agricultural development and in the regions of agricultural expansion. The wild forms of oats (*Avena sativa*; Fig. 2.11) were, presumably, the wild or animated oat (*Avena sterilis*) or the common wild oat (*A. fatua*), which reached Central Europe from the Middle East as weeds and were domesticated there approximately 4,000 years ago.

Other wild plants that became established in Europe were domesticated as late as the nineteenth century, and include lamb's lettuce (*Valerianella locusta*) and the witloof (*Cichorium intybus*), the domestic form of which is used as chicory.

Only several thousand years after the first food crops were domesticated did the domestic forms of most of the agriculturally utilized woody plants become established. These include many of the species typical of the Mediterranean region today, for example the grape vine (*Vitis vinifera*), olive tree (*Olea europaea*), fig tree (*Ficus carica*), and almond (*Prunus dulcis*), the wild forms of which originate from the Middle East to Central Asia, and the orange (*Citrus sinensis*), lemon (*C. limon*), peach (*Prunus persica*), and apricot (*P. armeniaca*), the centre of origin of which is China. The wild forms of the apple (*Malus* species) and pear (*Pyrus* species), and other fruit trees, originated in the deciduous forest zone of the mountainous regions of Central Asia.

After the re-discovery of America by Columbus in 1492, transcontinental exchange of domestic crops and livestock began. In Europe, Africa, and Asia,

Fig. 2.11 Oats (*Avena sativa*)

nothing was previously known about the plant species domesticated in America, for example maize, potato, peppers, tomato (*Solanum lycopersicum*; Fig. 2.12), cotton, and squash.

Over the course of hundreds of years, the most important crops and livestock species spread throughout the world, which means that today they are cultivated almost everywhere climatic conditions allow. In most regions, agriculture today is based on species that are exotic to the area. Europe, North America, and Australia are almost completely dependent on wheat and barley (from the Middle East), maize (from Central America), potatoes (from South America), and soybeans (from China). The same is true for Africa, where almost 80% of agricultural production is of species that originate from Central America (maize), South America (cassava, sweet potato), and Southeast Asia (banana).

The development of agriculture progressed independently in different regions of the world and was defined by gradual processes that sometimes stretched over thousands of years. These were essentially determined by the supply of prey for hunting, the availability of plant and animal species that could be domesticated, and the possibility of acquiring species from other regions.

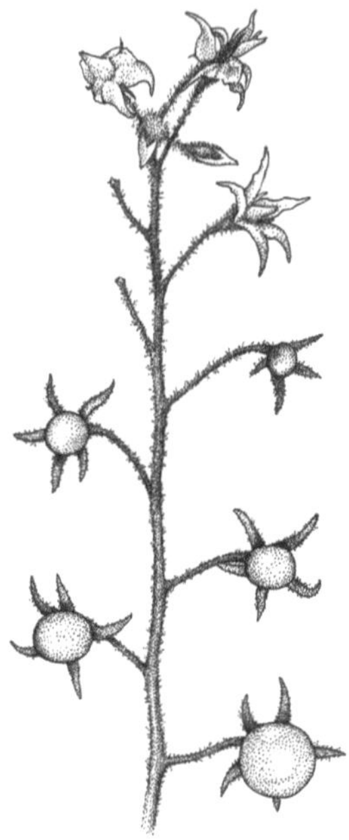

Fig. 2.12 Tomato (*Solanum lycopersicum*; synonym: *Lycopersicon esculentum*), inflorescence

2.3 Progress and Effects of Agriculture

The development of agriculture with the sedentary lifestyle initiated a chain of processes that strongly affected the living conditions of humans, their social and cultural contexts, and technical development. As a consequence of this, agricultural methods of production also changed, which was associated with a significant increase in yields over the course of time. Two developments that were dependent on each other provided the necessary conditions for this. First, sedentary living and the increased production of aliments beyond the needs of subsistence, led to an increase in population density and, second, agriculture enabled an increase in the size of settlements and division of labour. While farmers were responsible for the production of food, specialists for example blacksmiths, wagon makers, and carpenters could deal with other activities, which also included production of implements for agriculture. This altered social and material living conditions within societies and was the basis for the development of cities, states, and civilization in its entirety. Over the course of this development, the proportion of people working in agriculture declined. This enabled progress in technology and science, which contributed to the raising of living standards.

Worldwide, today, more than a billion farmers produce food for seven billion people. In Germany, the average farmer provides 137 people with food (Fig. 2.13). Today, the residents of industrial countries use no more than a few minutes per day for acquisition of food; for most people in developing countries this time is much longer.

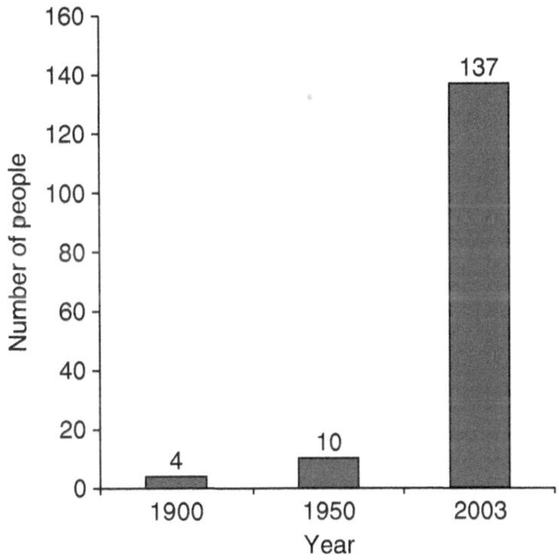

Fig. 2.13 Number of people supplied with food by a single farmer in Germany (Based on Bayerische Landesanstalt für Landwirtschaft 2000)

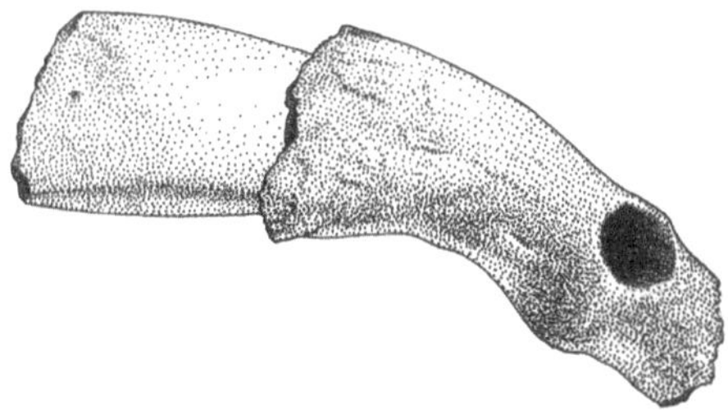

Fig. 2.14 Stone Age flat adze, used as a hoe or to fell small trees

2.3.1 Technical Development and Mechanization

With the discovery and invention of new tools which made it easier to cultivate the soil and to harvest crops, it was possible to increase agricultural production. In the Neolithic (approx. 7,000–4,000 years ago), the first implements for agriculture were made of wood and stone. At this time, digging sticks for sowing of plants, hoes to loosen the soil (Fig. 2.14), and stone scythes to harvest cereals already existed. The first primitive ploughs were used at least 5,000 years ago. They had a hook which was usually made of wood and served to create furrows in the field for the seed (ard or scratch plough; Fig. 2.15).

Later, from the Bronze Age (approx. 4,200–2,800 years ago), there is evidence of the first animal-drawn plough. Starting in the pre-Roman Iron Age (from 2,800 to 2,000 years ago), the ploughshare was reinforced with iron to prevent it wearing down as quickly as the purely wooden implement. In South, Central and North America, no animals that could be used for field work existed before the arrival of the Europeans in the sixteenth century.

With the Industrial Revolution, which originated in the middle of the eighteenth century in England, the development of modern agriculture began. Until then, agricultural implements had remained largely unchanged for hundreds of years

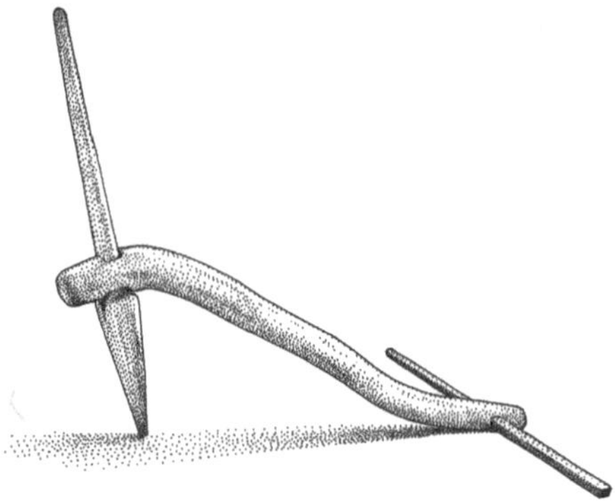

Fig. 2.15 Ard or scratch plough made of wood

and were constructed by the village blacksmith or wagon maker. At the beginning of the nineteenth century, improved ploughs were constructed that enabled more effective soil cultivation with less draught power. Further innovations included machines that made the sowing, harvesting, and threshing of cereals substantially easier and faster. The first German factory for agricultural implements and machines was founded in 1819 in Stuttgart-Hohenheim. Already by the 1860s, the first steam-powered ploughs were being used in Germany; however these could only replace the draught animals on large farms.

2.3.2 *Irrigation*

The availability of water determines the possibility of growing crops, and the security of the yield. In rain-fed agriculture, there is total dependence on the quantity and distribution of precipitation; this is insufficient to supply the plants over the entire vegetation period in some regions of the earth and/or in some years. The beginning and the duration of the rainy season is variable in many parts of the tropics and subtropics, which is why plant production there is associated with risk. The development of artificial irrigation systems, often associated with terracing of the landscape, was a substantial innovation from which agriculture benefited substantially (Sect. 4.2.2). With this, more land in arid areas, and even on steep slopes, could be put under cultivation. The first irrigated landscape was created approximately 7,000 years ago in Egypt in the fields along the Nile. The Sumerians began to construct irrigation and drainage canals 5,000 years ago in Mesopotamia (Euphrates and Tigris region), on which their entire agricultural production was dependent.

2.3.3 *Fertilizers and Pesticides*

Since the beginning of arable plant production, one problem is that the fertility of cultivated soil declines over time, because with each harvest the agroecosystem is deprived of nutrients. This situation requires restoration of these losses; this can be achieved by means of prolonged fallow periods or by fertilization. Among fertilizers, a basic differentiation can be made between

- **organic fertilizers**, or example manure, liquid manure, harvest residues, compost, green manure (Sect. 2.3.3), and human excrement, which are usually produced within an agricultural enterprise and thus called farmyard manure, and
- **inorganic fertilizers** or **mineral fertilizers**, which are either mined from natural reserves (primarily rock deposits), for example phosphorus (Sect. 3.7.5) and potassium, or produced synthetically, for example mineral nitrogen fertilizers (Sect. 3.7.4). Because mineral fertilizers are usually purchased by agricultural enterprises, they are termed commercial fertilizers.

Long past the medieval period, different types of rotation system, including fallow land (Sect. 2.4.1), were dominant in Europe. This method served to regenerate soil fertility, and fertilizers were available in very limited quantities only. In the nineteenth century, production and trade in fertilizers began, for example guano (excrement of sea birds harvested from large deposits, particularly on the islands off the coast of Peru), sodium nitrate (mainly from Chile), or bone meal. A further improvement in the situation was achieved as a result of the Haber–Bosch process for production of ammonia in 1910 (Sect. 3.7.4.2). Industrially produced nitrogen fertilizer could now, at least in the industrialized countries, completely provide for the needs of agriculture. This solution of the fertilization problem led to substantial increases in yields.

Since the mid-twentieth century, use of synthetic pesticides (Box 2.2) has contributed to the security of yields. In 1939, the insecticide DDT was developed and was used in agriculture after the end of the Second World War. Shortly thereafter the development of numerous other substances began (Sect. 5.2.1). The first synthetic herbicides (Sect. 5.1.1) also became commercially available at this time.

Box 2.2 Pesticides

Pesticides are natural or synthetic compounds that protect plants or plant products from damaging organisms (pests and diseases) and substances that kill undesired plants (weeds) or that affect the life processes of plants (e.g. growth regulators, germination inhibitors). "Plant protection" therefore only refers to protection of domestic plants and crops or the harvest products.

Pesticides are primarily classified with reference to the target organisms. Pesticides (Sect. 5.2) can be targeted against animal pests, primarily insects (insecticides), mites (acaricides), and nematodes (nematicides), but also against other pests, for example snails (molluscicides) or rodents (rodenticides). The substances that protect plants against phytopathogens (Sect. 5.3) primarily include fungicides and bactericides. Substances used to control weeds are called herbicides (Sect. 5.1.1).

2.3.4 Plant Breeding

Plant breeding has the objective of changing the genetic properties of plants and adapting them to the requirements of humans. The first domestic plants were developed by **selective breeding** practised by farmers over thousands of years (Sect. 2.2.1). The Austrian monk Gregor Mendel (1822–1884) recognized the regularity of inheritance in his breeding experiments with peas and thus formed the foundation of the modern science of genetics. In contrast with selective breeding, which primarily served for improvement of the properties of individual varieties, it now became possible to purposefully combine the genomes of two different varieties and thus to obtain varieties with new properties. In addition to selective breeding, this **cross breeding** is the basis of every plant-breeding program since the beginning of the twentieth century.

Another development of cross breeding is **hybrid breeding**. By repeated, artificial self-fertilization, inbred lines with particular properties are developed, which usually produce low yields themselves as a result of the in-breeding depression. Such inbred lines are then crossed with each other, whereby the so-called hybrids are created. In comparison with the parent lines, growth and yields of these hybrids are usually substantially higher. This phenomenon is called **heterosis** or **hybrid vigour** and is the opposite of the in-breeding depression. The increased yields are almost exclusively achieved by the first generation and decline in subsequent generations. When using hybrid varieties, the farmer must therefore purchase new seed for every growing season. Hybrid breeding is important in cross-pollinated crops, for example maize (Fig. 2.16), rye, sugar beet, sunflower, and many vegetable species.

The development of molecular genetics and cell biotechnology in the 1980s resulted in new perspectives in plant breeding as a result of genetic engineering. It became possible to transfer genes across species boundaries and thus bestow new traits on plants. The introduction of foreign DNA into the genome of an organism is called transformation, resulting in **genetically modified organisms (GMO)** or, for crops, **GM crops** (Box 2.3).

Box 2.3 Genetically Modified (GM) Crops

Depending on the purpose of a particular genetic modification, GM crops can be categorized as having either first, second or third-generation characteristics:

- First-generation transgenic crops were created to improve the agronomic properties of the plant, but not the quality of its product. First-generation GM crops include plants with greater resistance to pests or herbicides, or to environmental (abiotic) stressors, for example salinity, drought, or extremes of temperature.
- Second-generation plants were modified to obtain new food or feed properties, including crops with substances that are beneficial to human

(continued)

Box 2.3 (continued)

health (functional food). Characteristics include increased nutrition or enhanced quality, for example improved omega-3 fatty acid production in oil seeds, starch modification, or improved mineral or vitamin content. An example of this is "golden rice" which is engineered to produce beta-carotene (pro-vitamin A).

- Third-generation plants are expected to be used in the future for industrial production of non-plant products. This involves the creation and cultivation of transgenic plants that can produce biofuels, biodegradable plastics, enzymes, lubricant oils, or pharmaceutical substances, for example hormones and vaccines. Future approaches will also include the development of crops with reduced dependency on fertilizers and water, promoting the environmental sustainability of agricultural production systems (Sect. 8.1.2.3).

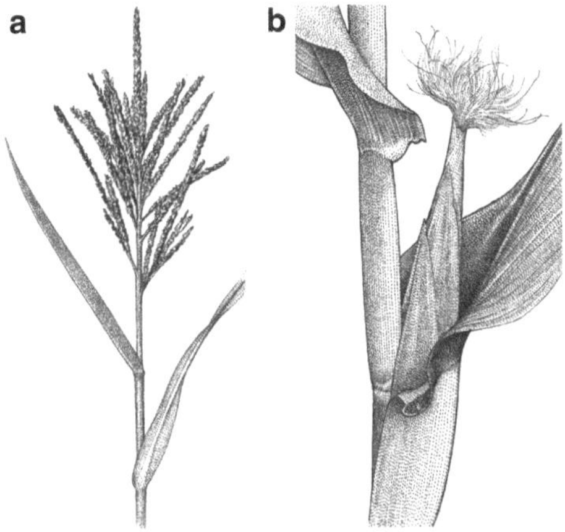

Fig. 2.16 In contrast with most other grass species, maize (*Zea mays*) has unisexual flowers that are located on separate inflorescences. The male flowers (**a**) are combined with the terminal panicle. The female flowers (**b**) are axillary and wrapped in bracts. Because maize is cross pollinated, fertilization does not occur within the same plant, but rather with pollen from another plant

The most economically important commercially produced transgenic crop varieties are currently soybean, maize, cotton, and rapeseed. Most of the transgenic varieties currently cultivated are resistant to herbicides (Sect. 5.1.2) and/or to specific phytophagous insects (Chap. 5.2.2). Since the beginning of the cultivation of transgenic crops in 1996, global production area has increased from 2 million ha to 148 million ha in 2010 (Fig. 2.17).

2.3.5 *Livestock Breeding*

The principle, methods and objectives in animal or livestock breeding are basically the same as in plant breeding, but differ in their use of terms. Domestic animals

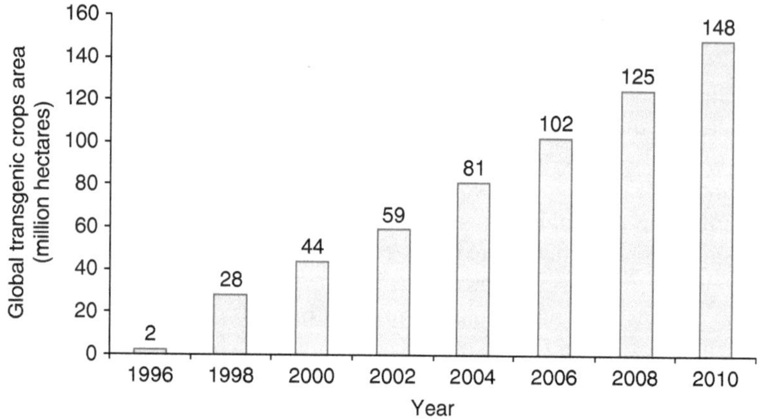

Fig. 2.17 Increase in the global area of transgenic crops between 1996 and 2010 (Based on James 2010)

originate from controlled selective breeding (artificial selection) for desirable traits from the human perspective, for example increased meat, milk, or egg production.

The concentration and maintenance of desired characteristics through several successive generations is called **line breeding**. This process increases genetic uniformity and is based on the mating of related animals. The relationship is normally less close than first degree, and therefore line breeding is a mild form of inbreeding. Consequently, a line represents a trait that is uniquely expressed by that population of a species, which is called **purebred** (or purebreed). The ancestors of a purebred animal are usually recorded over a number of generations, and the animal is then recorded as being pedigreed.

Cross breeding, the biological opposite of line breeding, is also a common method in livestock breeding applied to all important species. Similar to plants, crossbreeding of animals is conducted with purebred parents of two different breeds. Crossbreeding has two distinct advantages over purebreeding. The first is the possibility of creating offspring that shares or combines desirable traits of both parent lineages (**complementarity**). The second is the creation of offspring with hybrid vigour (heterosis), which are therefore superior to their purebred parents. They perform at a level above the average of their parents with regard to such characteristics as disease resistance, fertility, growth rate, etc. In general use, such animals are termed "crossbreeds", whereas the term "hybrid" is used in plant breeding. However, the term hybrid is also used to describe crosses between animals of different species, for example the mule (female horse × male donkey).

As a breeding practice, crossbreeding does not denote the indiscriminate mixing of breeds. Rather, it is the systematic and selective process of identifying breeding

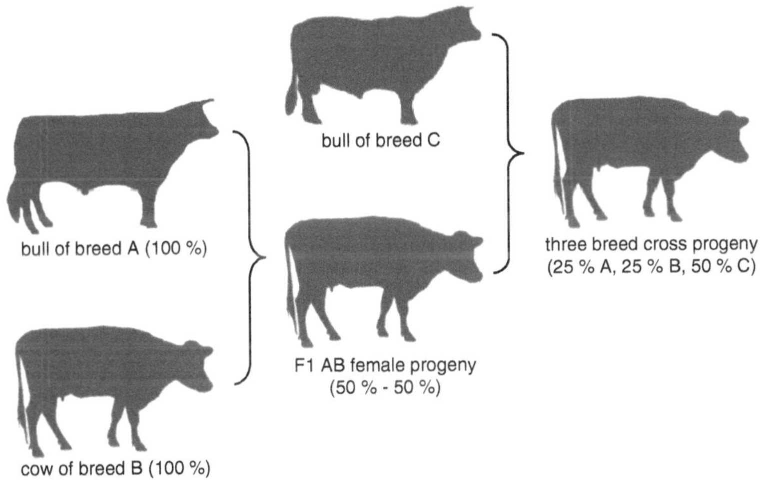

Fig. 2.18 Three-breed cross in cattle. It is obtained when all the females of a two breed cross are mated to a bull of a third, unrelated breed

animals with superior genetic merit for heritable, economically important traits. Depending on the breeding objective, there is a variety of crossbreeding systems, which include:

- **two-breed cross** (production of first cross offspring)
- **three-breed cross** (females from a two-breed cross are mated with a male of a third, unrelated breed; Fig. 2.18)
- **backcross** (females from a two-breed cross are mated with a purebred male of either of the original breeds)
- **rotational cross** (males of two or more breeds are mated with crossbred females)
- **composite breed** (matings among crossbred animals resulting from crosses of two or more breeds, used to form new or "composite" breeds designed to retain heterosis in future generations).

In addition to breeding, reproductive techniques for improvement of livestock traits have been developed in recent decades. **Reproductive cloning** became established in livestock production since this method was first successfully applied to a mammal (Dolly the sheep) in 1996. Reproductive cloning is used to produce identical genetic copies of whole animals. The most common cloning technique is somatic cell nuclear transfer (SCNT). For this, the nucleus of a somatic cell (a cell from the body) from the donor organism to be cloned is inserted into an egg cell from which the nucleus has been removed. As a result of chemical or electrical stimulation in an artificial environment, the egg cell develops into an early-stage embryo which then is implanted into the uterus of a female animal in which the clone grows before being born.

Reproductive cloning is a mechanism by which superior animals can be produced. It is usually used to improve the breeding stock with highly valuable sires, but not to produce animals for human consumption directly, because it is too expensive and too inefficient. Cloning success is very low and often results in a high incidence of death after birth or in severe physical deformities which violate animal welfare.

Although cloning itself is not a method of genetic engineering, it is a basis for the creation of **transgenic animals**, by introduction of alien genes into target animal embryos to alter the production characteristics of the animal (e.g. disease resistance, growth rate) or the quality of its products. With exception of some fish species, however, no genetically modified farm animals are commercially available at present.

2.3.6 Conventional Agriculture and Alternative Concepts

The objectives of the so-called **conventional** or **industrial agriculture** that developed in the second half of the twentieth century were intensification of production, essentially on the basis of products of the agrochemical industry (mineral fertilizers, synthetic pesticides), the use of uniform high-yield hybrid crops (including genetically modified crops), and the application of modern technology for cultivation of the land.

This has a variety of effects on the environment, which are not limited to agroecosystems but also affect the landscape and its structural and species diversity. Fertilizers, primarily nitrate and phosphate, pollute waters and soil (Sect. 3.7.4.2), and pesticides enter the food chain. The use of machinery in agriculture necessitates large and, preferably, homogeneous fields. As a consequence of mechanization, habitats such as hedgerows, ridges, and wetlands are increasingly lost from the landscape, and heavy machinery also leads to compaction of the soil.

Furthermore, most livestock species, for example chicken, turkeys, pigs, and cows in conventional agriculture are raised indoors at high densities and with little space to produce as much meat, eggs, or milk as possible at the lowest cost. These **factory farms** contribute substantially to the environmental impact of agriculture. Large amounts of manure are produced, stored, and released to the environment in slurry form (a liquid mixture of urine and faeces), causing nitrogen pollution of land, water, and air and greenhouse gas emissions. In addition, factory farming increases risk to human health as a result of excessive use of antibiotics to mitigate the spread of disease and to stimulate the growth of animals. This leads to the development of virulent, antibiotic-resistant pathogens which render antibiotics useless for treatment of human diseases. Overall, factory farming contributes substantially, and in different ways, to the most significant global environmental problems (Chap. 8).

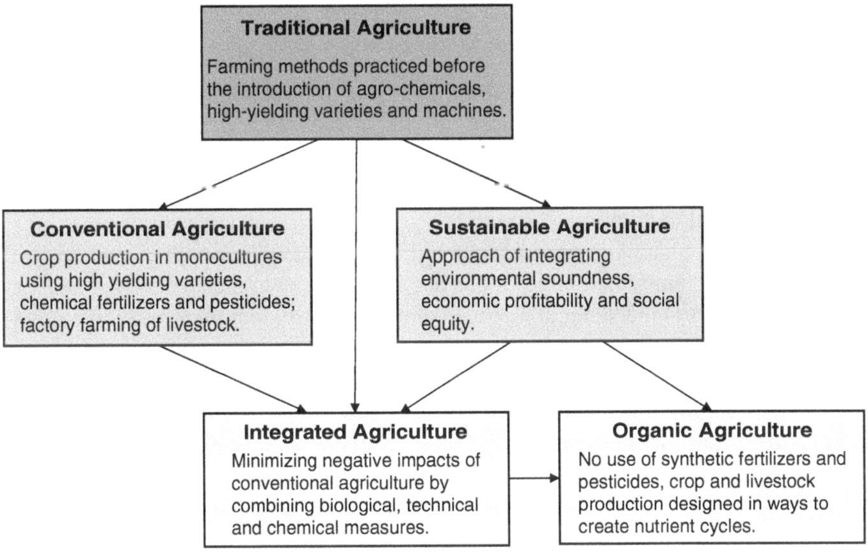

Fig. 2.19 Main concepts of agricultural production

To minimize the adverse effects of conventional agriculture, different alternative concepts of production have been developed (Fig. 2.19).

The overall idea of such alternatives is **sustainable agriculture**, which is based on the principle that agricultural production must meet the needs of the present without compromising the ability of future generations to meet their own needs. A variety of philosophies and perspectives from scientists to farmers and consumers have contributed to this vision, but there is no overall consensus on how to define and achieve sustainability in agriculture. However, most views agree that it integrates three main objectives: environmental soundness, economic profitability, and social equity. It does not, however, mean a return to traditional agriculture with the practices and low yields that characterized the nineteenth century. More closely, sustainable agriculture can be defined as an integrated system of plant and animal production practices that will, over the long term, satisfy human food and fibre needs, conserve environmental quality and natural resources, and enhance the quality of life for farmers and society as a whole by considering and integrating ecological cycles and controls.

In practice, there are two main approaches to achieving the objectives of sustainable agriculture:

1. **Integrated agriculture** principally makes use of the techniques of conventional agriculture, but tries to minimize its negative effects by combining biological, technical, and chemical measures that enable production of a high-quality harvest

while conserving natural resources. Integrated agriculture can therefore be regarded as a method between conventional and organic agriculture and includes:

- site-adapted design of the production area by use of suitable species and varieties of crops and livestock;
- crop rotation that mitigates weed, disease, and pest problems and provide alternatives sources of soil nitrogen;
- careful soil cultivation that avoids soil compaction and erosion;
- fertilization based on need, ideally with a combination of organic and mineral fertilizers, and which reduces the risk of water contamination;
- application of mechanical methods of weed control to reduce herbicide input; and
- use of pest-control strategies that reduce the need for pesticides by integration and promotion of natural enemies; synthetic pesticides should be used only when other methods of control cannot prevent a threshold of damage being crossed.

2. **Organic agriculture** is a way of farming that entirely avoids the application of synthetic fertilizers and pesticides. Its objective is to create nutrient cycles within the farm that are as closed as possible, which means that crop and animal production are designed such that fertilizers and feed are mostly produced and utilized on the farm. Of major importance is the conservation and promotion of soil fertility (Sect. 4.3.5.3). This is achieved by means of site-adapted crop rotation (Sect. 2.4.1.1) in which the use of legumes for fixation of nitrogen (Sect. 3.7.4.1) is important, as also is the utilization of organic fertilizers (compost, manure). For compensation of nutrient losses, purchased organic fertilizers, untreated rock meal, and bone meal, may be used. To avoid an excess of nutrients, the number of animals and the use of feedstuffs from outside the farm are limited. Overall, organic agriculture aims at using ecological principles to create synergies among the system components and to improve sustainability. Other principles of organic agriculture include:

- a ban on the use of genetically modified crops and other applications linked to genetic engineering;
- species-appropriate animal husbandry and a ban on the addition of hormones and antibiotics to feed; and
- the conservation and improvement of the structure and biodiversity of the landscape.

In contrast with integrated agriculture and integrated plant protection, organic agriculture is subject to government regulations which are defined in an EU regulation and in other national regulations around the world. These rules address, among other aspects, the application of fertilizers and pesticides. Plant or animal products of organic agriculture are identified by the legally protected terms "eco", "bio", or "organic". Other declarations regarding such concepts as integrated, inspected, or environmentally friendly production, do not guarantee that the product was produced without synthetic fertilizers and pesticides.

2.4 Classification of Agroecosystems

Depending of the objectives of the farmer, the availability of capital, energy, and technology, and the institutional and infrastructural context, a great diversity of crop and livestock production systems exist. In addition, there are also differences related to historical and geographic factors. The latter primarily result from the climate, the natural vegetation, and the soils of different zones of the world (Chap. 7).

2.4.1 *Cropping Systems*

Cropping systems can be classified on the basis of a combination of three key characteristics, each of which can be subdivided into two groups:

1. **Life-form of the crop**
   Crops, similar to other plants, can usually be classified into two groups according to their life-form:
   (a) **Annual crops**, for example cereals, require one vegetation period to complete their development and have, accordingly, a short life span. They must, therefore, be sown again every growing season. However, depending on the species and region, more than one harvest per year is possible. For example, in the humid tropics, rice, with a vegetation period of 120 days, can be cultivated three times per year.
   (b) **Perennial crops** have a lifespan of several years and deliver, depending on the species, a harvest in one or more of these years.
2. **Intensity of production**
   (a) In agroecosystems that are subject to **extensive** management, production is essentially based on the natural site conditions, which means that only a limited quantity of materials are imported into the system, and these usually consist of limited quantities of farmyard manure.
   (b) Agroecosystems that are subject to **intensive** management are characterized by the application of technology which includes the use of machines, mineral fertilizers, and pesticides. In such systems, plant production is not based solely on solar energy but also on fossil energy, which is required for operation of machines and for production and transport of fertilizers and pesticides.

   These two forms of land use are not absolute opposites, but are instead connected by a wide range of production intensity. On the one hand, the different transition forms imply the gradual improvement of production methods that has occurred during the development of agriculture over the course of thousands of years. On the other hand, they show the existing differences between production conditions and the potential of farming in the industrial and developing nations.

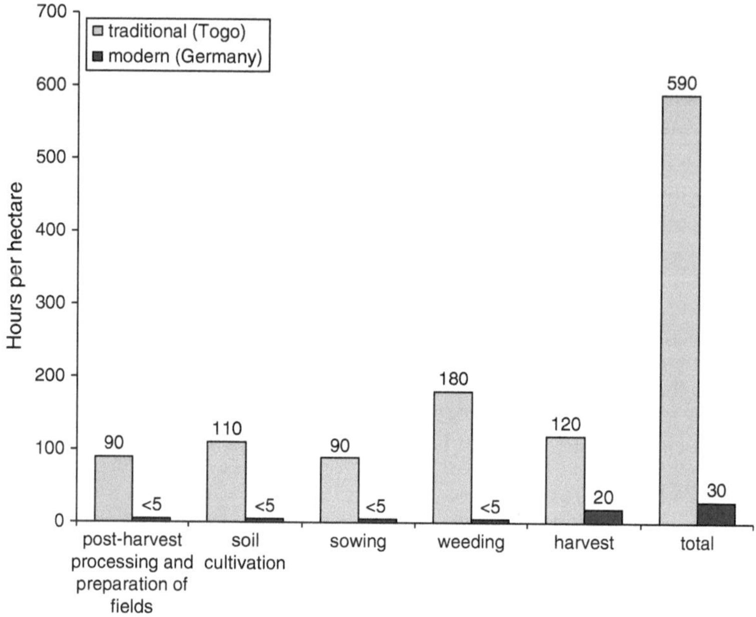

Fig. 2.20 Labour requirements for maize production in Togo and Germany (Based on Koch 1994)

With increasing intensification of agricultural production, the input of human labour usually decreases as the need for manual labour for activities such as soil cultivation, sowing, weeding, and harvesting, decreases. This results in significant differences between the labour required for extensively and intensively managed agroecosystems (Fig. 2.20).

3. **Cropping period**

   (a) In extensive land use, production of crops on the same area is possible for a limited period of time only (usually between 2 and 4 years). Then, the nutrient reserves in the soil no longer deliver satisfactory yields. For regeneration of the soil fertility, an extended fallow period (several years to decades) is therefore necessary, before crop production can start again. Systems in which crop production is interrupted by such a phase are called **rotation systems**.

   (b) By intensification of production, not only an increase in crop yield per field area but also a longer duration of use of the same field, up to permanent use, becomes possible. Thus, in such **permanent systems**, there are no multi-year fallow periods but, at most, seasonally dependent interruptions to production.

A more detailed description of rotation and permanent systems is presented in the following sections. An overview is given in Fig. 2.21.

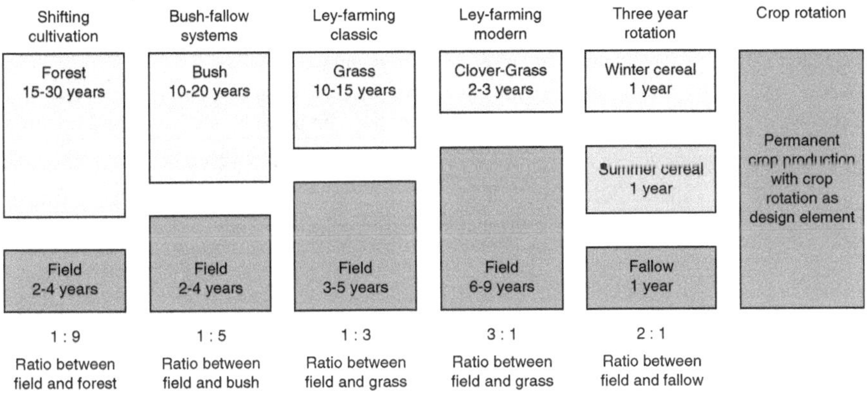

Fig. 2.21 Overview of agricultural rotation and permanent systems

2.4.1.1 Rotation Systems

Production systems with long fallow periods mark not only the beginning of agriculture but are still an important form of crop production in many regions.

Shifting Cultivation

Production systems in which short phases of field use alternate with long forest phases are called shifting cultivation or slash-and-burn farming. To establish a 1–4 year phase of crop production, the trees are felled and often burnt. Because the tree roots generally remain in the soil, cultivation of the fields between the remains of the forest vegetation is conducted in the simplest manner, for example with a digging stick. With declining soil fertility, the land is abandoned and left to natural succession, to be used again at a later date. In forested Europe, this form of production was the start of agriculture and was predominant into the early medieval period.

In the tropics, shifting cultivation is still performed by more than 250 million people (Metzger 2003). With increasing population density and the corresponding increasing demand for agricultural products, the forest phases between those of agricultural land use will become shorter. Finally, this leads to overexploitation or nutrient mining of the site, through which the growth of the trees is limited and the forest eventually disappears completely. Often, a cleared site can no longer be used for production because the fertile upper soil layers have been eroded by heavy tropical rainfall (Sect. 4.3.6).

Bush-Fallow Systems

Rotation systems in the form of bush-fallow systems primarily exist in the savanna regions. After a short period of agricultural land use, bush vegetation often becomes

Table 2.1 Types of ley-farming

Unregulated	Irregular alternation between crops and grassland
Regulated	Regular alternation between crops and grassland
Classic	Establishment of grassland after the crop phase by natural succession
Modern	Establishment of grassland after the crop phase by sowing of seed mixtures

established on fallow fields, which serves to regenerate the soil as in shifting cultivation, and is cleared again after a specific period of time.

Ley-Farming

Alternating use of an area as crop field and as grassland (meadow or pasture) is called ley-farming, whereby the grassland serves for soil regeneration. As in many other production systems, a broad range of historical and regional types exist (Table 2.1).

In the early medieval period, from approximately the sixth century onward, ley-farming became established in Central Europe and, in contrast with shifting cultivation, had the advantage that cultivation was not hindered by tree roots. Reforestation of the grassland was generally prevented by grazing.

Basically, one can distinguish between regulated and unregulated ley-farming. In the regulated type, the change in use occurs after a defined number of years. This type of system was usually found in the vicinity of the homestead; crop production was performed for 3–5 consecutive years and then alternated with at least 10 years of grassland use. In the unregulated type, the transition between crop field and grassland occurs irregularly and was typically practised on land distant from the homestead. In this case, grassland use extended from 10 to 40 years and was interrupted by a 1–2-year phase of cereal production.

In many regions of Europe, ley-farming was replaced in the ninth century by three-year rotation (see below), but was continued in marginal regions of production (e.g. the Black Forest, northern Germany, the Alps, and Scandinavia) until today. Whereas in classic ley-farming natural establishment of grasses occurred, the modern type includes sowing (usually of a clover–grass mixture) for production of animal feed. Different types of ley-farming are often among the predominant land-use systems today in the wet–dry tropics, but also in many regions of the temperate latitudes.

2.4.1.2 Permanent Cropping Systems

The population density of Europe increased substantially since approximately the eighth century. This process not only required expansion of agricultural land but also intensification of crop production on existing fields. This occurred through the introduction of a **3-year rotation**, which can be seen as intermediate between

Fig. 2.22 Spelt (*Triticum spelta*)

rotation and permanent systems. In its original form, this system consisted of 2 years of crop production (winter cereals in the 1st year, summer cereals in the 2nd year) and fallow in the 3rd year. The most important cereal crops were barley, rye, and oats. In southwestern Germany, spelt (*Triticum spelta*; Fig. 2.22) was also important. In the nineteenth century, an improved form of the 3-year rotation emerged in which the fallow year was used for the production of feed legumes (primarily clover) or root and tuber crops (primarily potato). The development of permanent cropping without fallow years was also promoted by the increasing availability of fertilizers (compare Sect. 2.3.3).

Because of the need to feed a growing population, permanent systems are also gaining increasing importance in the densely populated regions of the tropics. However, production there often occurs without the possibility of compensating for nutrient losses by sufficient fertilization. As a consequence of this, yields decline. For the savanna regions of West Africa, annual nutrient deficits are estimated to be between 15 and 25 kg nitrogen, up to 2 kg phosphorus, 15–20 kg potassium, and 5 kg magnesium per hectare (Buerkert and Hiernaux 1998).

Permanent Arable Cropping

Worldwide, permanent crop production is predominantly performed in the form of **one-crop systems**, which means the exclusive production of a single crop species in a field. In agricultural land use that is increasingly oriented toward intensification and yield maximization, specific crop species are also grown on the same field for several years in a row. Such **monocultures** are only possible with autotolerant plants. These are plants for which repeated, consecutive production in the same field does not lead to declines in yield. This is true for maize and many other important food plants, including rye and potatoes. In contrast, autointolerant species cannot be repeatedly grown on the same field without interruption, because otherwise yield losses result. This also often happens when an adequate supply of nutrients is not ensured. Species that are autointolerant include sugar beet, oats, wheat, pea, and rapeseed. The causes of this phenomenon, which is also termed soil sickness, are unknown in many cases. For some species (e.g. rapeseed) the reduction in yield is primarily the result of an increased incidence of plant diseases.

For other species, for example alfalfa and some rice varieties, allelopathic effects have been demonstrated (Sect. 4.4.2).

A proved and tested principle of permanent crop production is **crop rotation**, which means a temporal sequence of production of different crop species (mostly one-crop cultivation) in the same field. Crop rotation not only contributes to conservation or improvement of soil fertility (Sect. 4.3.5.3) but are also to a measure of weed management (Sect. 5.1.3) and pest management (Sect. 5.2.3.1), because the population development of weeds or pests can be affected by the change in crop species. In crop rotation, a distinction is made between cereals and leaf crops. The latter are all species other than cereals, for example rapeseed, root and tuber crops (potato, beet, field vegetables), and legumes, but also green and silage maize and fodder grasses. Production of cereals (C) and leaf crops (L) can be designed in different ways. Common systems include simple rotation (C – L – C – L) and double rotation (L – L – C – C). The seasonally dependent temporal gaps between the respective main crops can be filled with the production of **cover crops** which can serve as green manure or animal fodder. Green manure plants are grown to biologically bind nutrients (primarily nitrogen) left in the soil by previous crops, thus protecting them from leaching. Examples of cover crops include legumes, for example lupines and clover species, grasses, and different Brassicaceae (mustard family), for example white/yellow mustard (*Sinapis alba*), radish (*Raphanus sativus*), and field mustard or turnip (*Brassica rapa*). Another cover crop species is the lacy phacelia (*Phacelia tanacetifolia*), which is a member of the Hydrophyllaceae (waterleaf family). It was introduced to Europe from California and also serves as a feeding resource for bees. Summer and winter cover crops are distinguished, depending on the sowing date. In contrast with one-crop systems, **intercropping** involves at least two crop species grown simultaneously in the same field. They are grown in alternate rows or wide strips but often also without a planned order. A particular type of intercropping is the establishment of underseeds (undersown crops). This is done by sowing a plant species into the stand of a main crop. Underseeds used in cereals include clover and grass species which serve as animal fodder.

Whereas in the industrialized nations large one-crop systems predominate, intercropping is a traditional type of land use in many regions of the tropics and continues to be of importance today. Examples include the mixed cropping of maize, beans (*Phaseolus* species), and squash in Central America and the mixed cropping of cereals (sorghum, pearl millet) and legumes, for example cowpea, pigeon pea (*Cajanus cajan*), and/or peanuts in the savanna regions of Africa.

Intercropping enables a better utilization of nutrients and water from the soil, e.g., by growing shallow and deep-rooted crops together. In addition, intercropping can contribute to the suppression of weed growth and to reduced pest incidence (Sect. 5.2.4.5). Overall, intercropping serves to minimize the risk of losing the entire harvest. Even when one of the cultivated crops fails, the other may still produce a yield.

Because of improved resource utilization in intercropping, the yield per unit area is often higher than that from the same area with only one crop. Whether, and to what extent, intercropping furnishes better yields than one-crop cultivation can be

Table 2.2 Example of calculation of the land equivalent ratio (*LER*)

	Yield (t/ha)		
Crop	Intercropping (E_M)	Sole cropping (E_R)	LER partial
A	5.1	8.9	5.1/8.9 = 0.57
B	1.9	3.9	1.9/3.9 = 0.49
C	0.6	2.8	0.6/2.8 = 0.21
LER total	–	–	$\sum \frac{E_{Mi}}{E_{Ri}} = 1.27$

evaluated with the **Land Equivalent Ratio (LER)**. This ratio is calculated by comparing the yields of two or more crop species from intercropping with the yields produced with the same species in sole cultivation. LER is calculated by use of formula:

$$\text{LER} = \sum E_{Mi}/E_{Ri}$$

where E_M is the yield of one crop from intercropping and E_R is the yield of the same species from the one-crop system. For every crop, i, represented in the intercropping, the relationship between E_M and E_R is calculated. The resulting partial LER values are summed to produce the total LER value. A value >1.0 means there is a yield advantage from intercropping. In the example given in Table 2.2, the LER total value is 1.27, which means the area required in sole cropping to achieve the same yield of the three crops as is produced by intercropping would have to be 27% larger than in intercropping.

Perennial Crops

Permanent agricultural crops include all crop species with more than one year periods of use. These are perennial field crops, for example asparagus (*Asparagus officinalis*), hops (*Humulus lupulus*), and artichoke (*Cynara scolymus*), and the variety of shrub and tree crops used in agricultural production, for example grapes (*Vitis vinifera*) and tree fruit species.

Plantations (from Latin *plantare* = to plant) are typical permanent cropping systems. In the original sense this term referred to large scale agricultural enterprises in the tropics and subtropics in which plant products were usually grown in monocultures and produced for the world market (Table 2.3).

This type of operation was first established by the Europeans in their colonies and was inseparably connected with slavery from the beginning of the seventeenth century to the end of the nineteenth century.

Plantations are used in the production of:

- perennial field crops, for example sugar cane, sisal (*Agave sisalana*; Fig. 2.23), and pineapple;

Table 2.3 Important crop species of the tropics that are primarily produced in plantations for the world market

Species	Length of economic use period (years)	Main producers
Sugar cane	6–8	Brazil, India
Oil palm	50	Indonesia, Malaysia
Banana	20	India, Brazil
Coconut	80	Philippines, Indonesia
Pineapple	6–13	Thailand, Philippines
Rubber tree	30–40	Thailand, Indonesia
Coffee	30–100	Brazil, Columbia
Tea	50	India, China
Sisal	6–20	Brazil, Tanzania

Fig. 2.23 Sisal (*Agave sisalana*)

- shrub crops, for example coffee, tea (*Camellia sinensis*; Fig. 2.24), and cacao (*Theobroma cacao*; Fig. 2.25);
- herbaceous perennials, for example the banana (*Musa* × *paradisiaca*); and
- tree crops, for example oil palm, coconut palm (*Cocos nucifera*), and rubber tree (*Hevea brasiliensis*), and various fruit and spice trees, e.g. cloves (*Syzygium aromaticum*) and nutmeg (*Myristica fragans*).

Agroforestry systems are also permanent systems producing trees or other woody plants and agricultural crops together on the same plot. By combining such species, the diversity of products gained from the system can be increased and, in addition, various ecological functions are better fulfilled than in sole arable cropping systems. Woody plants provide better protection of the soil from wind and water erosion (Sect. 4.3.6) than annual plants and increase the biodiversity of the system. Their root systems reach greater depths than those of most arable crop species, which means that water and nutrients from deeper soil layers can be used. The organic material in tree litter reaches the soil's surface where it is decomposed by soil organisms, so the inorganic components of this "nutrient pump" become available to plants. Overall, the need for external fertilizers is thus reduced. The

Fig. 2.24 Tea (*Camellia sinensis*)

Fig. 2.25 Cacao (*Theobroma cacao*) is a cauliflorous tree species. Cauliflory refers to woody plants which flower and fruit from their main stems or branches as opposed to the ends of the twigs

root systems of woody plants also affect the soil structure and, thus, the rate of infiltration of precipitation and the water balance of the soil.

In addition to the positive effects mentioned above, woody plants can also have negative effects on the other plants of an agroforestry system, primarily as a result of competition for light, nutrients, and/or water. Therefore, for every system it is necessary to investigate which perennial and annual species are best suited for cultivation together, to achieve complementary effects. Factors such as sowing and planting dates of the species also affect the ecological interactions.

A particular form of agroforestry are **home gardens**, in which crops are integrated in natural tree stands. They are primarily found in Southeast Asia and serve to produce fruit trees, rattan palms, medicinal plants, and vegetables. Another type is the so-called agro-silvopastoral system (Latin *silva* = forest, and *pastor* = shepherd), in which forestry and agricultural uses are complemented by a livestock component.

2.5 Livestock Systems

In addition to, and sometimes in combination with cropping systems, a large variety of livestock production systems exist around the globe. Important factors involved in livestock production include the animal species and the type of product, the climatic conditions in the region of production, the type of land use, and the type and origin of feeding resources.

In consideration of these and with additional criteria, for example socio-economic conditions, several system classifications have been proposed. Among these, the classification scheme based on Seré and Steinfeld (1996) has become most widely accepted and is also used in the following discussion. Further information is provided by Steinfeld et al. (2006) and Robinson et al. (2011).

As shown in Fig. 2.26, this classification first distinguishes two main categories, i.e., solely livestock systems and mixed farming systems. These groups are then broken down into different categories considering animal type, land use, and climate.

2.5.1 Solely Livestock Production Systems

This group includes the two most extreme types of livestock production, ranging from old traditional systems using large areas of natural grassland to systems of intensive livestock production with high animal densities in factory farming systems with low space requirements.

2.5.1.1 Landless Livestock Systems

These systems are largely independent of climate and are defined as livestock systems in which more than 90% of the dry matter fed to animals is introduced from outside the farm, and in which annual average stocking is above ten livestock units (LU) per hectare of agricultural land. (In this context, a livestock unit is one individual for cattle or eight individuals for sheep or goats.)

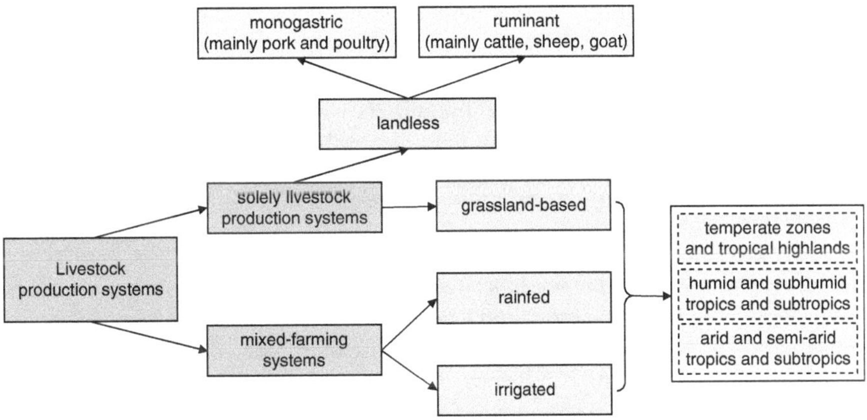

Fig. 2.26 Classification of world livestock production systems (Based on Seré and Steinfeld 1996)

Landless livestock systems can further be subdivided into:

- **Landless monogastric systems** (mainly pigs and poultry) are found predominantly in the industrialized countries of Europe and North America, with approximately half of global landless pork production and landless poultry production each. In pig production, Asia is second, with nearly one third of the world total. The predominant producer of pigs and poultry in Asia is China.
- **Landless ruminant systems** (mainly cattle, sheep, goats) exist in only a few regions of the world. For cattle, they are concentrated in eastern Europe; landless sheep production systems are found in western Asia and northern Africa only.

2.5.1.2 Grassland-Based Systems

Areas used for livestock grazing cover more than a quarter of the global land surface but are inhabited by approximately 4% of the world's human population only. Grassland-based livestock systems produce more than 10% of the dry matter fed to animals and have annual average stocking rates less than ten livestock units per hectare of agricultural land.

Overall, livestock grazing covers a wide range of systems which are defined as follows:

- **Total nomadism**: herders have no permanent place of residence; they move with their animals in a way that normally avoids depleting pastures without regular cultivation.
- **Semi-nomadism**: a permanent place of residence exists; supplementary cultivation is practised, but for long periods of time animal owners travel to distant grazing areas.

- **Transhumance**: herders have a permanent home, and their herds are sent to distant grazing areas, usually on seasonal cycles, e.g. in montane regions the movement between higher pastures in summer and valleys in winter.
- **Partial nomadism**: characterized by farmers who live continuously in permanent settlements and have herds at their disposal that graze in the vicinity.
- **Stationary animal husbandry**: animals remain on the farm or in the village throughout the year.

Solely livestock production on grasslands is of different importance in different climatic regions. Most of the world's large natural grasslands of the temperate zones have conditions suitable for crop production and, therefore, have been developed for arable farming, especially in the North American Prairie, the South American Pampas, and the East European Steppe. The largest areas of natural grassland used for grazing in the temperate zone are found in Central Asia including Mongolia, where 80% of the country is used for extensive grazing.

In the humid and sub-humid tropics, the grassland-based livestock production system is found mostly in the tropical and subtropical lowlands of South America, including the llanos of Colombia and Venezuela and the cattle-ranching lands of the Amazon basin which have developed from cleared and burned rainforest.

In many regions of the arid and semi-arid tropics, extensive grazing systems, usually known as **rangelands**, are the dominant form of land use. Extensive grazing of rangelands for livestock production is called **pastoralism**. In northern Africa and in western Asia, pastoralism is a traditional way of subsistence for an important part of the population. In dry regions, especially, fuel wood is often scarce, leading to an increased role for animals as providers of manure for fuel, in addition to their provision of meat and milk and as a means of transport. In Australia, parts of western United States, and southern Africa, commercial rangeland enterprises are the modern form of land use in dry regions.

2.5.2 Mixed-Farming Systems

Mixed-farming systems are agricultural production systems in which the waste products of one enterprise (residues from crop production) are used by the other enterprise (livestock production) which returns its own waste products (manure) to the first enterprise. In these types of system, more than 10% of the dry matter fed to animals comes from crop by-products or stubble, or more than 10% of the total value of production comes from non-livestock farming activities.

Globally, mixed-farming systems contribute more than half of total meat production, compared with approximately one-third by landless systems and less than 10% by grazing systems.

Mixed systems can combine crops and livestock to different extents, which include on-farm systems with more or less closed ecological cycles, spatially separated systems of between-farm mixing (exchanging resources between different farms within a region), or temporally changing activities and crop–livestock interactions related to seasonal differences (winter/summer or rainy/dry season). In smallholder systems, livestock, in addition to crop production, may provide food and income, draught power for crop production, manure to improve soil fertility, and financial insurance in times of scarcity.

Depending on the water resources available for crop production, mixed systems can be further subdivided into rain-fed and irrigated systems. Most mixed farming systems are rain-fed, and are widespread in semi-arid and sub-humid areas of the tropics and in temperate zones.

Overall, if the land area used for grazing and that used for feed crop production for landless livestock systems are added together, livestock production is the world's largest agricultural sector with the highest proportion of land use.

Driven by human population growth, urbanization, and increased income, the demand for animal-source food products is increasing, especially in the developing countries. In the future, production will increasingly be affected by competition for natural resources, particularly land and water, and competition between human food and animal feed production. Environmental problems of livestock production and potential solutions are raised in Sect. 6.2.2.

References

Bayerische Landesanstalt für Landwirtschaft (2000) Integrierter Pflanzenschutz. Rapsschädlinge. www.Lfl.Bayern.de/Publikationen

Buerkert A, Hiernaux P (1998) Nutrients in the West African Sudano-Sahelian zone: losses, transfers and role of external inputs. J Plant Nutr Soil Sci 161:365–383

Denham TP, Haberle SG, Lentfer C, Fullagar R, Field J, Therin M, Porch N, Winsborough B (2003) Origins of agriculture at Kuk swamp in the highlands of New Guinea. Science 301:189–193

Diamond J (1998) Guns, germs and steel: a short history of everybody for the last 13,000 years. Vintage, London

Diamond J (2002) Evolution, consequences and future of plant and animal domestication. Nature 418:700–707

James C (2010) Global status of commercialized biotech/GM crops. ISAAA (International Service for the Acquisition of Agri-biotech Applications) Report 2010, www.isaaa.org

Koch W (1994) Grundzüge des Pflanzenschutzes mit dem Versuch einer Bewertung. PLITS 12(4), W. & S. Koch Verlag, Stuttgart

Metzger JP (2003) Effects of slash-and-burn fallow periods on landscape structure. Environ Conserv 30:325–333

Piperno DR, Stothert KE (2003) Phytolith evidence for early Holocene *Cucurbita* domestication in Southwest Ecuador. Science 299:1054–1057

Price EO (1984) Behavioral aspects of animal domestication. Q Rev Biol 59:1–32

Robinson TP, Thornton PK, Franceschini G, Kruska RL, Chiozza F, Notenbaert A, Cecchi G, Herrero M, Epprecht M, Fritz S, You L, Conchedda G, See L (2011) Global livestock production systems. FAO/ILRI, Rome/Nairobi

Seré C, Steinfeld H (1996) World livestock production systems: current status, issues and trends. Animal production and health paper 127. FAO, Rome

Smith BD (1997) The initial domestication of *Cucurbita pepo* in the Americas 10,000 years ago. Science 276:932–934

Steinfeld H, Wassenaar T, Jutzi S (2006) Livestock production systems in developing countries: status, drivers, trends. Revue Scientifique et Technique de l'Office International des Epizooties 25:505–516

Zeder MA (2008) Domestication and early agriculture in the Mediterranean Basin: origins, diffusion, and impact. Proc Natl Acad Sci 105:11597–11604

Chapter 3
Patterns and Processes in Ecosystems

3.1 Biotic Interactions

The different individuals, populations, and species occurring together in a habitat do not exist independently of each other, but instead affect each other unilaterally or reciprocally in numerous relationships. The effects of such interactions have a substantial effect on the different structural and functional characteristics of an ecosystem, for example the number of species present, the sizes of their populations, the temporal development of the communities, and the flows of energy and nutrients.

3.1.1 Food Webs

The fundamental importance of food webs to all species is not only because of the trophic dependency of particular resources, but also because of the possible presence of natural enemies. The latter are also a biotic factor that affects the living conditions and population density of a species. With regard to sources of food, organisms can be subdivided into autotrophs and heterotrophs.

3.1.1.1 Autotrophic Organisms

Species that do not require organic substances as their foodstuffs, but instead have the ability to synthesize organic compounds from carbon dioxide and water, are called **autotrophs**. In ecosystems, they fulfil the function of **producers**. When the organisms obtain the required energy from sunlight, they are called **photoautotrophs** and perform photosynthesis (Sect. 4.1.1). In addition to some bacteria (e.g. cyanobacteria, formerly called blue–green algae), primarily green plants are capable of photosynthesis. **Chemoautotrophic** organisms derive the

K. Martin and J. Sauerborn, *Agroecology*, DOI 10.1007/978-94-007-5917-6_3,

energy needed for the production of organic compounds not from light, but instead by oxidation of inorganic compounds and thus are said to perform chemosynthesis. These organisms include nitrate, sulfur, methane, and iron-bacteria.

3.1.1.2 Heterotrophic Organisms

Animals, fungi, and most bacteria depend on organic compounds produced by other organisms for their nutrition and are called **heterotrophs**. This group includes consumers and detritivores.

Consumers

All species that feed on living organisms are consumers. They include two groups:

First order consumers obtain their food or nutrients from autotrophic organisms. The consumers of plants are:

- **Phytophages** (plant-eating animals), which can be further subdivided on the basis of the plant parts they use (Sect. 4.5.1).
- **Phytopathogens** (agents of plant disease), which primarily include fungi, bacteria, and viruses (Sect. 4.6).
- **Parasitic plants**, which obtain organic compounds and/or nutrients, and water from their host plants (Sect. 4.4.3).

First order consumers cause damage to plants that can lead to losses of fitness and yield (Box 3.1).

Second order consumers feed on heterotrophic organisms and can be divided into three groups:

- **Predators** are defined as organisms that kill other organisms to acquire food. Predation is found primarily among animals, but carnivorous (meat-eating) plants should also be regarded as predators.
- **Parasitoids** are insects (primarily ichneumon wasps; Sect. 5.2.4.1), that lay their eggs in the bodies of arthropods at particular stages of development (egg, larva, pupa, or adult) The parasitoid larvae live in the host's body (endoparasitoids) or attack the host's body from the outside (ectoparasitoids) and feed on its tissues. When the parasitoid has completed its development—after pupation or hatching of the adult (imago)—the host dies. The predatory life strategy of parasitoids is limited to the larval stage. In contrast, the adults are usually flower visitors, feeding on nectar or pollen.
- **Parasites** include groups of animals that live temporarily or permanently on or in the organisms of another animal species to obtain food from them (e.g. mosquitoes, fleas, tapeworms, leeches). In contrast with parasitoids, they do not usually kill their host and are, therefore, not predators.

Box 3.1 Fitness and Yield

For the host plant, attack by phytophages, phytopathogens, and/or parasitic plants is associated with loss of tissue and/or substances (primarily carbohydrates, nitrogen compounds, and water). As a consequence, physiological processes in the plant may be impaired, which can affect the plant's photosynthetic performance, its uptake of water and nutrients, and its use and distribution of photosynthetic products (e.g. as reserves). In reaction to an attack, the plant activates different energy-intensive defence mechanisms (Sect. 4.5.5). Overall, the resulting damage and energy losses can impair growth, development, and, finally, the reproductive success of the plant. **Fitness** serves as a measure of the reproductive success for all organisms. It is defined as the relative contribution of an individual to the offspring of a population. In the spermatophytes (seed-producing plants), this contribution is usually measured as the number of seeds the plant produces over the course of its lifetime. The effect on fitness is thus an objective criterion with which to measure the damage caused by a consumer to a wild plant.

In relation to agricultural crops, the term fitness usually has no meaning because stands of such species are not populations whose individuals reproduce themselves according to natural rules. Rather, their continued existence is determined by humans, for whom **yield** is the crucial factor. "Yield" refers primarily to the quantity and quality of products harvested, which for many species are not the seeds but rather other plant parts (e.g. for leafy vegetables and tubers).

Decomposers

Decomposers are species that, in contrast with consumers, obtain their foodstuffs from the dead remains of organisms. The dead organic matter which results from all organisms at the end of their existence is termed detritus. Two functional groups can be distinguished among the decomposers:

- **Detritivores** utilize organic substances in the same way as consumers and are involved in the decomposition of dead organic material in ecosystems.
- **Reducers** or **mineralizers** transform organic substances into inorganic end products, from which they obtain energy. Mineralization is mostly performed by microorganisms, primarily fungi and bacteria (Sect. 4.3.5.1).

3.1.1.3 Trophic Levels and Food Webs

Producers, first and second-order consumers, and detritivores each represent one distinct **trophic level**. At the same time, they are the links of the **food chain**. Food

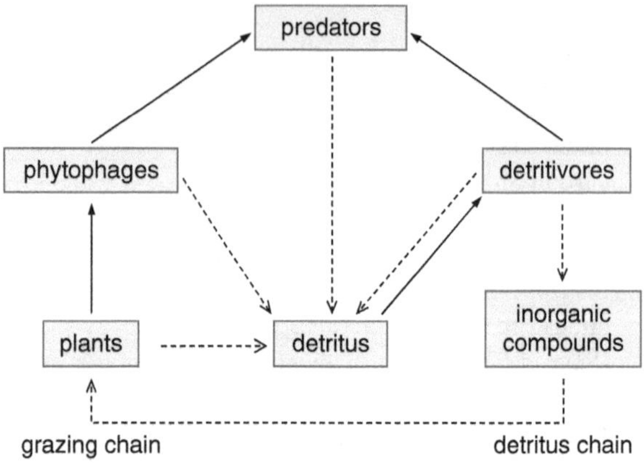

Fig. 3.1 Simplified representation of the food chains in ecosystems. The *dashed arrows* show the pathways of non-living materials

chains can be divided into two main types, which are related to each another and are usually found together in terrestrial ecosystems (Fig. 3.1):

1. The **grazing chain** begins with plants and leads, via the phytophages, to the predators.
2. The **detritus chain** is based on the dead organic matter that is produced at all trophic levels. It is consumed by the decomposers, on whom specific predators feed.

The trophic levels consist of numerous species. Most consumers use not only one, but several, species as their food source. Therefore, the trophic relationships between individual species do not form a simple chain, but rather a complex **food web**. As an example, Fig. 3.2 shows a detail of a food web based on rice.

Some species cannot be definitively assigned to a particular trophic level, because their range of foods includes several trophic levels. These species are called **omnivores** and may be, for example, phytophages and predators at the same time (e.g. some birds and small mammals) or may have different feeding strategies at different stages in their development (e.g. many insects with different feeding strategies as larvae and as adults).

3.1.2 *Competition*

The contest between two (or more) individuals or populations for food, space, or other resources of limited availability, which leads to unilateral or reciprocal negative effects on the participating organisms, is called **competition**. This definition applies to the competition between different species (**interspecific**), and to the competition between individuals of the same species (**intraspecific**).

Fig. 3.2 Detail of the food web of a rice field community in the Philippines. **1** = rice plant (*Oryza sativa*); **2** = yellow rice stemborer (*Scirpophaga incertulas*, Pyralidae), (**a**) adult, (**b**) larva, which lives in the rice stalks; **3** = parasitoid ichneumonid wasp (*Temelucha philippinensis*, Ichneumonidae) which attacks *Sciropophaga*; **4** = ladybird beetle (*Micraspis crocea*, Coccinellidae); **5** = brown planthopper (*Nilaparvata lugens*, Delphacidae), (**a**) adult, (**b**) nymph; **6** = damselflies (Coenagrionidae); **7** = plant bug (*Cytorhinus lividipennis*, Miridae); **8** = big-headed fly (*Pipunculus* sp., Pipunculidae), parasitoid of planthoppers/leafhoppers; **9** = grasshopper (*Acrida* sp., Acrididae); **10** wasp spider (*Argiope* sp., Araneidae); **11** = rice bug (*Leptocorisa* sp., Alydidae) sucks on milk-ripe kernels; **12** = rice farmer

Interspecific competition occurs when three conditions are fulfilled:

1. the species occur in the same habitat,
2. they utilize one or more of the same resources,
3. the availability of these resources is limited.

The intensity of interspecific competition is determined by the quantitative availability of resources and the size of the populations competing with each other.

The effects of interspecific competition are usually asymmetric. In other words, the disadvantages resulting from competition are greater for one of the populations involved than they are for the other. The negative effects are usually evident in the more or less obvious impairment of the development of individuals, which often leads to a reduction in reproduction and, eventually, to lower population growth.

In the extreme case, interspecific competition can become so intense that, over time, access of one species to a resource is completely prevented, or a species becomes locally extinct, because of the other, more competitive species. Such a **competitive exclusion** occurs when populations of two species live in the same habitat and occupy the same ecological niche, with one species reproducing more rapidly than the other. It is hardly possible to determine whether and how often this occurs, or has occurred, in nature. However, competitive exclusion can be observed in some instances when species that have the same pattern of resource use and have evolved in geographic isolation from each other, encounter each other. Such situations are often the result of human activity, for example transport of plant and animals species from one continent to another where they then displace native species from their habitats (Sect. 3.3.3). In many cases, the greater competitive ability of the introduced species can be explained on the basis of their finding more suitable development conditions in their new environments, and because they often encounter fewer natural enemies and diseases than in their original environments.

3.1.3 Mutualism

The term **mutualism** refers to interactions between species that are advantageous to one or both and are not associated with disadvantages for either side. Such "positive" relationships between individuals or populations take many forms and may range from associations promoted by one side that are seldom specific, to associations that are close, or even essential, and very specific. The latter are often referred to as **symbiosis**, whereas the term mutualism is an overall term that includes the full continuum of positive interactions.

Mutualistic relationships are found between very different species. For example, they occur between flowering plants and specific animals that serve as pollinators and seed dispersers and are therefore rewarded by the plant with nectar, pollen, or fruit flesh. Mutualism between bacteria or fungi on the one hand and animals or plants on the other, are also widespread. These include, for example, microorganisms, the mycorrhizae (Box 3.2), which break down cellulose in the digestive tract of some phytophages, and the relationships between nitrogen-fixing bacteria and legumes (Sect. 3.7.4.1). In addition, numerous other more or less mutualistic interactions can occur between two different species of organisms; these are most often related to the acquisition of food, to reproduction, or to protection from enemies.

Box 3.2 Mycorrhiza

The connections between fungi and the roots of higher plants are referred to as mycorrhiza (from the gr. *mykes* = fungi and *rhiza* = root). Such interactions are mutualistic relationships undertaken by more than 80% of plant species and approximately 6,000 species of fungi. The fungi facilitate better supply of nutrients (especially phosphorus) and water to the plant than would be possible with the plant's roots alone. This can partly be explained by the increase in soil space, via the fungi hyphae, available to the plants, and because uptake of nutrients through the hyphae is more effective than the uptake of nutrients by plant roots. The fungi receive all of their organic compound needs from the plants, which provide between 4 and 26% of their photosynthetic products for this purpose (Miller et al. 2002). Two major forms of mycorrhizae can be identified:

The fungal mycelium (mass of branching hyphae) of the **ectomycorrhiza** completely enclose the roots and colonize the root cortex but do not penetrate the root cells. This form is typical of many tree species in the temperate latitudes. The fungi involved are primarily mushrooms of the subdivision Agaricomycotina, which, e.g., include boletus, chanterelle, and fly agaric.

The fungal hyphae of the **endomycorrhiza** penetrate the cells of the root cortex. Such associations are found in many herbaceous species, grasses, and shrubs, but also in tropical tree species. The most common type of endomycorrhiza is the so-called **vesicular-arbuscular mycorrhiza (VAM)**. The name is derived from the structures of the fungal hyphae in the root cells, which may be either tree-like and highly branched (arbuscules) or bubble-like (vesicles). These fungi are representatives of the zygote fungi (Zygomycetae).

Evidence of mycorrhiza, primarily VAM, is also observed for many crop plants (e.g. barley, potato, maize, wheat). The fungi are not only important in the promotion of nutrient uptake but often also reduce the susceptibility of the plant to phytopathogenic fungi and nematodes. Representatives of the Brassicaceae (e.g. cabbage, canola), Chenopodiaceae (e.g. spinach, beet), and many other plant families do not usually have relationships with mycorrhizae fungi.

Species that have mutualistic relationships with one another are usually subject to many effects and factors. It cannot, therefore, be assumed that the partners, irrespective of their relative abundances, always benefit from each other in the same manner. Most mutualistic relationships are not very specific. Thus, for example, in animal–plant interactions such as flower pollination or seed dispersal, strong dependencies rarely emerge between individual species. The closest relationships are formed when the species are fully dependent on each other for their existence. In many cases, such symbioses are based on exchange of chemical compounds between the partners.

In interspecific competition and in most mutualistic associations, the relationships between the species involved are asymmetric, which means that they are not of equal importance to all the species involved.

3.2 Communities

An ecological **community** (also called **biocoenosis**) is an assemblage of species whose populations are in a relationship with each other and thus effect each other unilaterally or reciprocally through feeding relationships, competition, mutualism, or combinations of these interactions. The term "biocoenosis" was first used by the German marine zoologist Karl Möbius in 1877. He introduced the term in his description of oyster banks in the North Sea in his book "*Die Auster und die Austernwirtschaft*" (Möbius 1877).

Most species that coexist in a habitat have a great diversity of relationships with each other. Plants are usually attacked by more or less large numbers of phytophagous species (Sect. 4.5.1), may have mutualistic relationships with pollinators and microorganisms (e.g. mycorrhizae), and may also have relationships with the natural enemies of the phytophages (Sect. 5.2.4.3). They are, in addition, often subject to competition with other plants (Sect. 4.4.1). Predators feed on a particular selection of prey organisms (Sect. 5.2.4.2); most are themselves eaten, and often also have competitors and mutualists. Taken together, these relationships are the biotic factors which, with the abiotic factors, determine the development of populations. So-called indirect effects may also emerge as a result of this. Such effects result when the impact of one species on another further affects the interactions between other species. Examples of this are found in Sect. 5.2.4.

As functionally defined units, communities usually have no definite spatial boundaries. They cannot be defined by a specific site through abiotic characteristics and also cannot be clearly separated from other communities. Habitats in a landscape which can obviously be distinguished on the basis of their vegetation (e.g. forests, meadows, fields), still have relationships among each other through species and their interactions. Animals, through their living requirements, may be closely tied to particular plant stands or to other habitats, but most animal species, in contrast with the plants, are mobile and can change their location. This applies not only to the insects and birds capable of flight, but also to many soil-inhabiting species, for example ground beetles. Even spiders are capable of dispersing and of colonizing different habitats with the help of their silk and the wind. Such circumstances clearly show that **agricultural communities** are not closed units limited to specific cultivated areas, even though these human-established sites are clearly separated from the surrounding vegetation.

Most of the species occurring in agroecosystems, which include a variety of wild plants, phytophages and predators, also exist, at least temporarily, in other habitats of the landscape, from which they colonize the crop stands.

Sharp boundaries, which can be drawn between different systems on the basis of abiotic characteristics, do not often exist for communities. Thus, flowing or standing waters and their neighbouring terrestrial habitats are in contact with each other through diverse relationships between the organisms found in the respective habitats. Creeks and rivers are supplied with organic material (primarily leaves and other plant residues) produced by terrestrial species. These organic materials are an important resource for the development of aquatic consumer food webs, which primarily include crustaceans, insect larvae, molluscs (snails and mussels), and fish. Many of the insect species have aquatic larval stages but terrestrial adult stages. The latter may serve as prey of terrestrial animals (e.g. of birds and bats), but may also be predators themselves (e.g. dragonflies) and feed on terrestrial prey. For predatory fish species (e.g. rainbow trout) not only aquatic but also terrestrial arthropods, which land on the water's surface, serve as important food sources. The fish are themselves prey of terrestrial predators, for example herons, otters, and anglers.

Overall, it becomes clear that only parts of a community, rather than an entire community, can be the subject of study. These parts can be defined on the basis of different aspects:

1. **Spatial fractions** of a community can, for example, focus on a deciduous forest, a stream, a hedgerow, or a field, whereby the respective boundaries are defined in accordance with biotic or abiotic characteristics. Within these areas, various types of community can be identified on the basis of the following criteria:

   (a) **Systematic groups** (Taxa; Box 3.3) form a **taxocoenosis**. For example, they can include all representatives of the plant kingdom, in which case they are a plant community or phytocoenosis. The totality of all animals forms the animal community or zoocoenosis. Both can be divided into further systematic groups. Examples for taxocoenoses of the animal kingdom are the ground beetles (Carabidae), spiders (Araneae), and aphids (Aphidina).

   (b) **Functional groups or guilds** include species that—irrespective of their systematic classification—can be defined because they are similar in relation to their resource use, or in their interactions with other species or with the abiotic environment. Broad categories are represented by the individual trophic levels, which can be further subdivided on the basis of their specific characteristics (e.g. different functional groups of the phytophages; Sect. 4.5.1). In agroecosystems, the direct and indirect relationships that different species have with crops are of great importance. On the basis of their functions, they can be divided into pests (Sect. 4.5.4), pollinators (Sect. 4.5.3), natural enemies (antagonists; Sect. 5.2.4.1), weeds (Sect. 4.4), and phytopathogens (Sect. 4.6).

   (c) **Structural groups** include species that are found in specific spatial subunits of the ecosystem. For example, these include the soil organisms (soil biological community; Sect. 4.3.5) and the vegetation layers (strata) for example the herb, shrub, and tree layers. In agroecosystems, for example, strata are formed by the crop stand and the undersown crops (Sect. 2.4.1).

2. **Functional fractions** of communities are defined not by spatial boundaries but rather by feeding relationships. Thus, they form a food web and represent a **trophocoenosis** or trophic community. These can, for example, include all species that are in relationship with a specific plant species through different trophic levels, as shown in the example of rice in Fig. 3.2. The food web is then not limited to species that are found in the cropping area but, instead, also includes feeding relationships with species outside that area. For example, such species can include alternative hosts for phytophages or alternative prey for predators, which are used outside the cropping season (see Sect. 5.2.4.2).

Box 3.3 Biological Systematics (Taxonomy)

The objective of biological systematics or taxonomy is to identify and classify the various organisms. In other words, it tries to assign all organisms according to their presumed evolution and kinships in hierarchically organized categories. The first comprehensive classification of organisms was presented in 1735 in the work "*Systema naturae*" by the Swedish natural scientist Carl von Linné. Today, the classification of species into the following major categories (using the example of the seven-spotted ladybird) is still based on his work:

Kingdom	Animalia (animals)
Phylum/division	Arthropoda (arthropods)
Class	Insecta (insects)
Order	Coleoptera (beetles)
Family	Coccinellidae (ladybirds or lady beetles)
Genus	*Coccinella*
Species	*Coccinella septempunctata* Linnaeus 1758 (seven-spotted ladybird, Fig. 3.3)

Denominated categories, for example Coccinellidae or Coleoptera, represent a taxon (plural: taxa). The naming (nomenclature) of species follows international rules. The species name consists of two elements (binomial nomenclature) of which the first identifies the genus. Sometimes a third name follows which refers to a subspecies. The scientifically complete name also includes the name of the author that first described and named the species, and the year of publication. In scientific literature, genus and species names are written in *italics*.

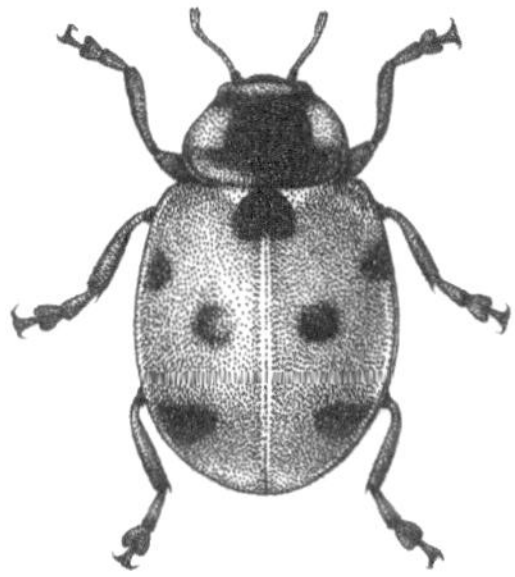

Fig. 3.3 *Coccinella septempunctata* Linnaeus 1758 (seven-spotted ladybird beetle)

3.3 Biodiversity

The term **biodiversity** means "the diversity of life" and includes:

- the diversity within a species or population (**genetic diversity**; see also Box 2.1),
- the number of species within a defined area or within a spatial or functional subunit of a community (**species diversity**), and
- the different components or habitats of a landscape (**landscape diversity**).

3.3.1 Species Diversity in Natural Systems

The number of species in an area or in a community is determined not only by the ecological conditions of the site but also by geographic factors. Viewed from a global perspective, the natural species diversity is very unevenly distributed. In principle, an increase in species diversity from the poles to the equator is evident and true for various groups of organisms, for example trees, mammals, birds, reptiles, and different orders or families of insects, for example ants (Table 3.1).

Tropical rain forests can have far more than 100 tree species per hectare, whereas in forests of the temperate latitudes a maximum of 30 species (North America) or 15 species (Central Europe) can usually be found per hectare. In a boreal forest, a maximum of five tree species is found in the same area. An additional gradient in biodiversity exists in mountainous regions, where the number of species decreases with increasing altitude.

Worldwide, the most diverse ecosystems, which primarily include rain forests and coral reefs, are found in the tropics.

The causal relationships between species diversity and latitude or altitude are unclear. However, it may be assumed they are primarily based on climate factors.

Table 3.1 Numbers of ant species found at different geographic latitudes of the Americas and Europe

Region	Latitude	Number of ant species
America		
Alaska (arctic part)	65–70° N	3
Alaska (total)	58–70° N	7
Iowa, USA	41–43° N	73
Utah, USA	37–42° N	63
Cuba	20–23° N	101
Trinidad	10–11° N	134
São Paulo, Brazil	20–25° S	222
Missiones, Argentina	26–28° S	191
Tucuman, Argentina	26–28° S	139
Buenos Aires, Argentina	33–39° S	103
Patagonia (total)	39–52° S	59
Patagonia (western part)	40–52° S	19
Tierra del Fuego	53–55° S	2
Europe		
Norway (northern part)	>69° N	10
Norway (total)	58–70° N	34
Germany	47–55° N	62
Italy	34–47° N	104

Based on Kusnezov (1957)

These are essentially related to incoming solar radiation (light and heat) and water supply (quantity and distribution of precipitation). These factors are also involved in the distribution of species at small scales (e.g. within a landscape) and are affected by, for example, the topography and exposure of a site.

Another factor that determines the species diversity of a habitat is the availability of food resources (for animals) and the availability of nutrient resources (for plants). However, the number of plant species is not highest on sites with the highest nutrient availability, but generally declines with increasing soil nutrient content. In cultivated landscapes also, it can be observed that highly nutrient-rich sites usually have low species diversity. For example, the edges of moving waters are usually well-supplied with nutrients, because of flooding, yet plant stands are usually species-poor and, in Central Europe, often dominated by one species only, for example the stinging nettle (*Urtica dioica*) or the common butterbur (*Petasites hybridus*). In contrast, nutrient-poor habitats, for example marginal grasslands, are often particularly rich in plant species. These situations occur primarily because only a few plant species can actually efficiently utilize high nutrient quantities. Such species have a growth and competitive advantage over most species, which are adapted to relatively nutrient-poor conditions.

There are usually close connections between the number of animal and plant species in a habitat. The more plants species found at a site, the more phytophagous species can exist there, because most of these are feeding specialists (Sect. 4.5.2). This, in turn, increases the number of potential prey species for predators with diverse food requirements. Species-rich plant stands also have greater structural diversity than species-poor plant stands, which also increases the range of potential ecological niches for various organisms.

3.3.2 *Agricultural Biodiversity*

The main aspects of biodiversity of agroecosystems include the diversity of the varieties and breeds of domesticated plant and animal species, and the natural diversity of species related to crop production (wild plants, pollinators, pests, and their natural enemies). This is in addition to the relationships between agroecosystems and their surrounding species, habitats, and ecosystems at the landscape scale.

3.3.2.1 Genetic Diversity of Crop and Livestock Species

Approximately 7,000 species of plants have been cultivated since humans started arable farming. Currently, only approximately 30 crops provide 95% of human food energy needs; four of these (rice, wheat, maize and potato) are responsible for more than 60% of our energy intake (FAO 2012).

In addition, the genetic diversity of these crop species has dramatically decreased. Of 8,000 traditional rice varieties grown in China in 1949, only 50 remained in 1970. In 1950, India had 30,000 wild varieties of rice, but in the near future, only 50 are expected to remain. Mexico has lost an estimated 80% of its maize varieties (Fowler and Mooney 1990). Overall, it is estimated that 75% of the genetic diversity of crop plants was lost in the twentieth century (FAO 1997). The main reasons for this development were the introduction of high-yielding varieties after the Green Revolution (Sect. 8.1.2.2) and the overall intensification and homogenization of crop production. This resulted in a change of traditional farming practices in which a large number of different, often locally-adapted, landraces of different genotypes were planted in the same field or on the same farm.

Of the estimated 15,000 species of mammals and birds, only between 30 and 40 have been domesticated for food production and fewer than 14 species, including cattle, goat, sheep, buffalo, and chicken, account for 90% of global livestock production. Of the total of about 7,600 livestock breeds, approximately 20% are classified as at risk of extinction. The rate of extinction amounts to almost one breed per month. Similar to the losses in crop diversity, the reason for this development is the intensification of production, utilizing a narrow range of breeds. Global production of meat, milk, and eggs is increasingly based on a limited number of high-output breeds—those that are most profitably utilized in industrial production systems (FAO 2007).

Global reductions in crop and livestock genetic diversity have a variety of implications for food production and security in the future:

- Reducing the genetic diversity of crops and livestock renders them more vulnerable to pests and diseases, which are more likely to spread when varieties or breeds have a similar genetic base. Although genetic uniformity itself does not necessarily mean higher vulnerability, it increases the likelihood that a pest or disease could affect a great proportion of the cropping area or livestock herd compared with the simultaneous use of a greater number of genotypes, which reduces the probability of large-scale epidemics. For example, genetic uniformity contributed to the spread of the Southern corn leaf blight (*Bipolaris maydis*), a fungus disease which led to a 15% reduction in the US corn yield in 1970.
- Similarly, genetic uniformity can increase vulnerability to abiotic stresses. This becomes increasingly important in the face of global climate change with expected changes in temperature, rainfall patterns, and an overall increase in the likelihood of extreme weather, for example droughts, heavy precipitation, and heat waves (Sect. 8.1.3). Genetic variability within a species confers at least the potential to resist stress, in both the short and long term. Therefore, the variance in yields from a more diverse crop system will be less than for a system with only one variety. In addition, crop genetic diversity is of particular importance to marginal sites and regions, which are less suitable for growing high-yielding varieties.
- Plant genetic resources present in traditional landraces or their wild relatives constitute an important basis of plant breeding. Diversity in plant genetic resources provides plant breeders with options to develop new and improved varieties. These are adapted to specific abiotic conditions or have such desired characteristics as resistance to diseases and pests.

For such reasons, genetic resources are required as inputs into the continuing process of enhancement through selective breeding (including genetic engineering). In the face of rapid global changes in climate and land use, conservation of crop and livestock genetic resources becomes increasingly urgent and requires the collection, evaluation, and deployment of such genetic material.

In general, genetic resources can be conserved either **in situ** (in their natural setting or in the field) or **ex situ** (outside their natural setting in gene banks). Gene banks are based on collections of genetic material (especially seeds) mainly from centres of crop origin, that are stored under controlled conditions and periodically regenerated (planted and grown) to maintain seed viability. However, not all kinds of plant genetic resource can be easily conserved ex situ and must, therefore, be kept as living plants in situ, which is more costly and requires additional land and labour.

There are approximately 1,400 gene banks around the world that contain approximately six million accessions (samples) including landraces and wild relatives of domesticated crops. Most of the genetic material has been collected from the world's major crops (rice, wheat, and maize). It is estimated that more than

90% of the genetic diversity of these crops is preserved ex situ. However, most of the minor crops, especially those which are of local importance only, are poorly represented or are in no genetic collections.

3.3.2.2 Species Diversity in Cultivated Landscapes

Before the start of the first agricultural activity approximately 7,000 years ago, Central Europe was almost completely covered with forest. Clearing of the forest led to the creation of open sites (mostly arable fields and meadows) that differed from the forest primarily in the solar radiation that reached the ground and, therefore, in light and temperature conditions. Thus, habitats developed for plant and animal species that previously only occurred in warmer and woodless regions, especially the Mediterranean region and the grasslands of southeastern Europe and western Asia. Such species could spread from their original ranges and colonize the new open habitats. Animal and plant species that profit from human modifications of the environment, especially the creation of cultivated systems, are termed **hemerophiles** (Greek *hemeros* = cultivated and *philos* = friend). Meanwhile, forests continued to exist alongside the cultivated areas. Therefore, the landscape of Central Europe today has significantly greater species diversity than it did before interference by humans.

In addition to crop species and varieties, of which usually only one or a few are cultivated in a single field, other factors determine the biodiversity of agroecosystems and cultivated landscapes. These include

- geographic, climatic, and soil conditions;
- type and intensity of cultivation; and
- diversity, structure, size, and distribution of habitats in the landscape.

Since the introduction of agriculture, the number of wild plant species in agroecosystems in Central Europe has steadily increased (Fig. 3.4). In a similar way, the number of animal species living there has probably also increased.

However, in the second half of the twentieth century, a substantial decline in the species diversity of agroecosystems occurred, primarily among wild plant species on many sites. Their decline has amounted to 50% on average. The loss of biodiversity not only affects agroecosystems, but also the diversity of cultural landscapes and their resident species, which include birds and such mammals as the European hare (*Lepus europaeus*). The cause is the increasing intensification of agricultural production, in which the following practices are of importance:

- Structural changes in the agricultural landscape. These are primarily the result of enlargement of field size, which is often accompanied by removal of bordering landscape elements (e.g. field breaks, hedgerows, stone walls) which serve as habitats and refuges for many animal and plant species. In addition, an increase in planting density (especially for cereals) and a decline in the diversity of

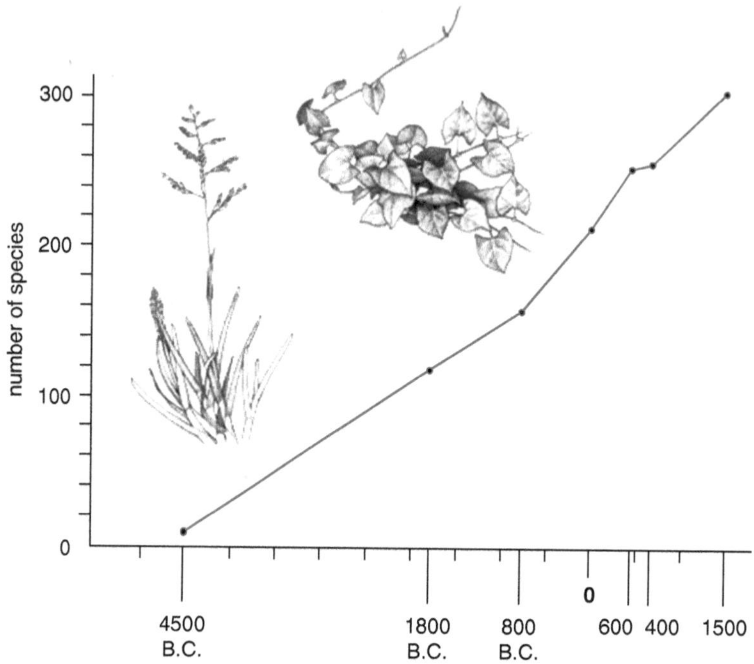

Fig. 3.4 Number of weed species recorded in Central Europe between the Mesolithic and Middle Ages (Based on Willerding 1986). As examples, the illustrations show annual meadow grass (*Poa annua*; *left*) and wild buckwheat (*Fallopia convolvulus*; *right*).

cultivated crop species has occurred, which further impairs chances of survival for many plant and animal species in the fields.

- Increased application of fertilizers. As in natural sites, the number of wild plant species in fields and grasslands is lower on sites with high nutrient supply (especially nitrogen) than on unfertilized control sites (Fig. 3.5a). In contrast, biomass production by the plants is higher on fertilized sites than on unfertilized sites (Fig. 3.5b).
- Weed and pest management. Not only herbicides (Sect. 5.1.1) but also other methods of weed management led to a decline in the diversity of wild plant species. Some of these species are highly dependent on dispersal via crop seeds, in which their seeds are stored with the harvest of the crop species and distributed back on to the fields with the next sowing. This pathway has been strongly limited by modern methods of seed cleaning. Therefore, species such as the common corn cockle (*Agrostemma githago*), have become very rare in the agricultural landscapes of Central Europe. Indirectly, many animal species are also affected by weed management because a decline in wild plant species results in a decline in resources. Insecticides (Sect. 5.2.1) affect not only pests but also other species of agricultural habitats, and therefore reduce species diversity overall.

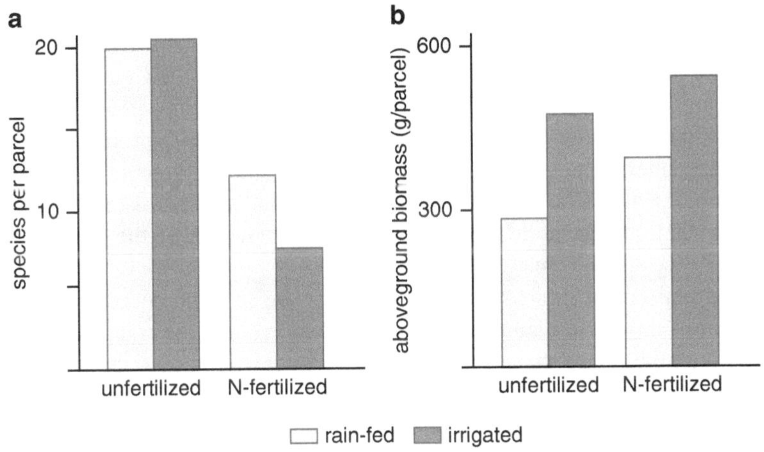

Fig. 3.5 Effects of nitrogen fertilization on (**a**) the number of plant species and (**b**) the aboveground biomass of irrigated and rain-fed parcels of 1-year fallow land in Michigan, USA (Based on Goldberg and Miller 1990)

- Grazing and mowing of grasslands. In addition to fertilization and weed management, plant species diversity of meadows and pastures is affected by mowing or grazing. Livestock species, for example cattle and sheep, can modify the vegetation of grasslands in different ways. At low stock density, they often maintain or increase the number of plant species, but high livestock densities can reduce diversity. These effects result from modification of the competitive relationships between plants—selective grazing of dominant grass species improves conditions for formerly suppressed species but high grazing pressure can eliminate sensitive species. In addition, diversity is affected on small scales by urine and faeces deposition and by trampling, affecting plants directly or indirectly via soil compaction. Furthermore, large-scale overgrazing can lead to different forms of land degradation (details are given in Sect. 6.2.1).

Paradoxically, modern agriculture is partially responsible for the decline in species whose presence in Central Europe was made possible only by the establishment of agroecosystems.

With reductions in plant and animal diversity, intensification of agroecosystems also affects the **functional biodiversity** of a system. Generally, functional biodiversity refers to the different types of species interactions and food web structures which are related to different processes including productivity, yield and other properties of the agroecosystem. Important aspects are further discussed in other chapters and include interactions between pests and their antagonists in the context of biological pest control (Sect. 5.2.4), pollinator services (Sect. 4.5.3), and invasive species (Sect. 3.3.3).

Anthropogenic modifications of ecosystems and agroecosystems can change the distribution ranges of species. This not only includes expansion of species' range, as occurred for many species after the introduction of agriculture to Central Europe, but can also limit the range, which can lead to the complete disappearance of species. Globally, changes in species diversity are determined by two processes only: the evolutionary emergence of new species, which can take millions of years, and the extinction of species. The latter is currently occurring, with estimates ranging from 17,000 to 100,000 species lost per year. A substantial part of the reason for this decline in species is the creation of agricultural land.

The cultural landscapes of the humid tropics are in extreme contrast with the Central European cultural landscape, which developed over the course of thousands of years. With the exception of some regional examples, tropical cultural landscapes are young, most of which were created in the twentieth century by removal of forest. Because tropical forests are among the most biodiverse ecosystems on Earth, their destruction is accompanied by substantially higher losses of biodiversity than is the case with the forests of Central Europe. On agricultural sites in the humid tropics, which are no longer suitable for agricultural use because of nutrient depletion, forests generally do not re-establish. Instead, extremely species-poor vegetation develops, which is dominated by species such as the grasses *Imperata cylindrica* and *Saccharum spontaneum* or bracken (*Pteridium aquilinum*). These plants are literally "cosmopolitan" and are found on all continents. Nature protection, as conducted in Central Europe to maintain the species of the cultural landscape, is superfluous in such regions. Appropriate measures in the tropics must concentrate on the conservation and regeneration of forests.

3.3.3 Invasive Species

As pointed out above, the global expansion of agriculture has led to shifts in the distribution and range expansion of many plant and animal species. This process started with many of the cultivated crops which have been actively introduced to new regions of the world. Further human activity, especially the expansion and intensification of global trade, resulted in the (largely accidental) exchange of different species between regions and continents. Worldwide, numbers of introduced wild terrestrial, freshwater, and coastal marine animal and plant species are continuously increasing (Fig. 3.6).

Overall, species that have reached regions where they did not previously occur as native species, are termed **alien species** or **exotics** in their new environment. The occurrence of most of these species is limited to small plots or areas where they have become established. A special category of exotics, however, are species which are able to expand their populations by invading new habitats. By this means they have negative effects on native species, communities, or ecological processes in natural or agricultural systems. Species with such characteristics are called **invasive species**. Besides this ecological definition, an invasive species can be termed "*an*

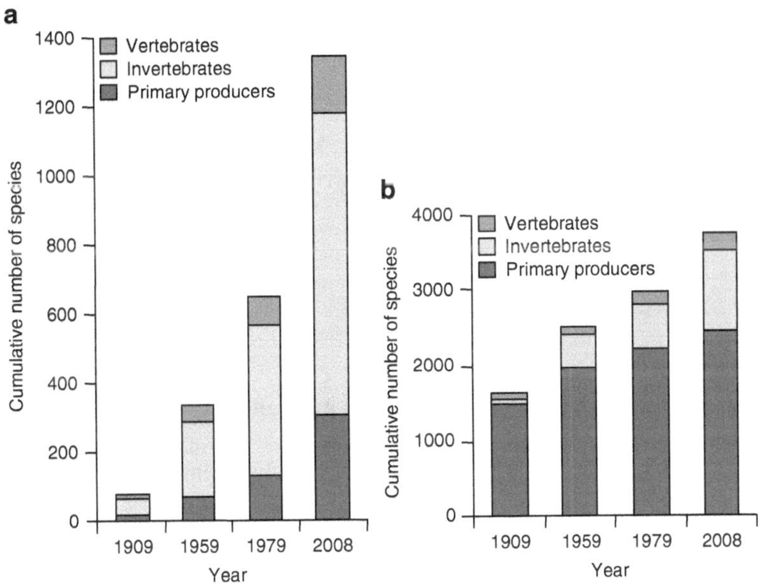

Fig. 3.6 Cumulative number of alien species established in (**a**) European marine/estuarine waters and (**b**) terrestrial environments in 11 European countries (Based on EEA 2010)

alien species whose introduction does or is likely to cause economic or environmental harm or harm to human health" (Beck et al. 2008).

The list of the world's worst invasive species (ISSG 2012) includes animal and plant diseases (e.g. rinderpest virus and phytophthora root rot, *Phytophthora cinnamomi*) insects (including five ant species), other invertebrates (e.g. giant African snail, *Achatina fulica*), feral animals (e.g. rabbit, *Oryctolagus cuniculus*), and numerous species of plants, including algae, herbs, and trees.

3.3.3.1 Impact of Invasive Species

The spread of invasive plant and animal species is believed to be the largest threat to global species diversity, besides direct human effects on natural ecosystems. Biological invasions have contributed to the decline of 42% of the endangered and threatened species of the United States (Wilcove et al. 1998). In other parts of the world, 80% of the endangered species are threatened as a result of invasion by alien species (Pimentel et al. 2005). Oceanic islands suffer especially high rates of species decline and losses. Invasive rats (*Rattus* species) have been implicated in the extinction or decline of hundreds of island endemic vertebrates. Predation by the brown tree snake, (*Boiga irregularis*) introduced to the island of Guam, caused nearly complete extermination of the island's native forest birds. In many cases, the invaded species cause subsequent changes in the native communities by distorting

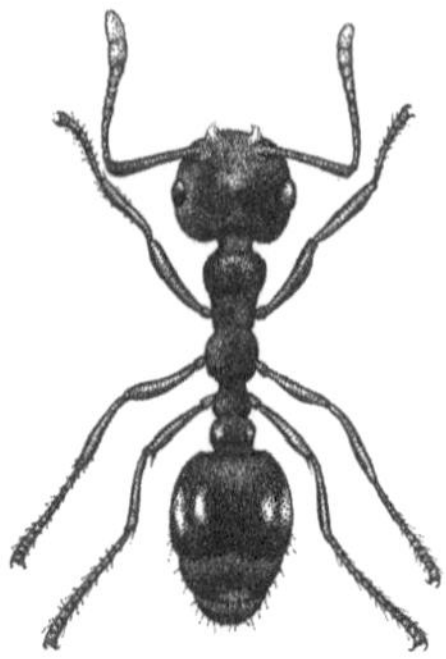

Fig. 3.7 Red fire ant (*Solenopsis invicta*)

food web interactions with cascade effects on many other species, or by changes in ecosystem structures and functions (e.g. of nutrient cycles, productivity, or habitat structure). Such modifications can facilitate the establishment of additional invasive species.

Further effects of invasive species related to land use and agricultural production are shown by the following examples:

- The red fire ant (*Solenopsis invicta*; Fig. 3.7) originates from South America and was inadvertently introduced to North America in the 1930s, where it rapidly spread throughout the southern United States.

More recently, the red fire ant was also introduced to Australia, China, and other regions. In the United States, the ants colonize open disturbed habitats including agricultural fields, rangeland, and gardens. Their colonies can reach 250,000 individuals. The species is omnivorous and acts as general predator and consumer of plant materials. As a predator, it affects all levels of species within the food chain, with effects on numerous invertebrates and ground-inhabiting vertebrates including birds, reptiles, and small mammals. They can reduce the species diversity of arthropod communities to 40% (Porter and Savignano 1990). Red fire ants are also reported to directly damage a variety of crops, for example tomato, corn, soybean, and sunflowers, by feeding on the seeds, seedlings, and developing fruits. Furthermore, red fire ants are very aggressive when disturbed and have a painful burning sting, which can cause allergic reactions. In addition to humans, fire ants also attack and harm large mammals, including wildlife, cattle, sheep, and horses on their grazing lands. Overall, the economic effect of fire ant infestations is enormous, with costs of control, medical treatment, and the various forms of damage running annually to several billion dollars in the United States.

- The aquatic golden apple snail (*Pomacea canaliculata*, Fig. 3.8) is native to tropical and subtropical South America where it lives in floodplains and seasonal wetlands. The snail is a polyphagous herbivore and feeds on floating and submersed higher plants. In the 1980s, it was introduced to Southeast Asia (first to Taiwan) with the purpose of establishing an additional source of food and income for farmers. The programme failed and, instead, the snails escaped

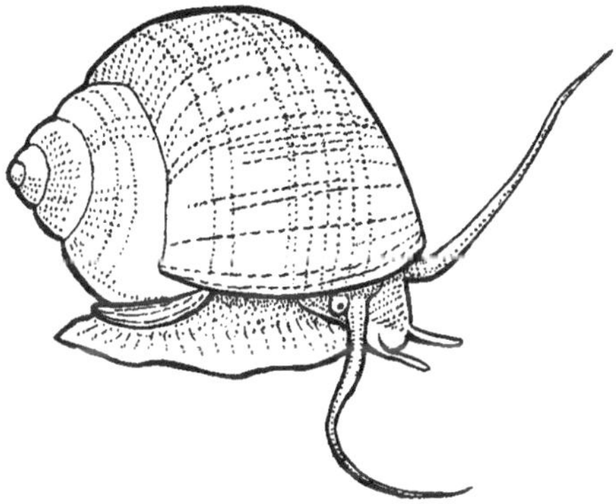

Fig. 3.8 Golden apple snail (*Pomacea canaliculata*)

and became dispersed to other countries, where they colonized irrigated rice fields. These provide a suitable habitat where the snails can also survive dry fallow seasons. Because wild aquatic plants are controlled by weed management in intensive rice production systems, *P. canaliculata* attacks young rice seedlings because of the lack of alternative food sources. Today, this invasive species occurs in nearly all rice-production areas of tropical and subtropical Asia and has developed into a major pest in rice production. The yield loss caused by *P. canaliculata* in rice has been estimated to vary from 5 to 100% depending on locality and population density.

- Kudzu (*Pueraria montana*) is a perennial leguminous vine (Fabaceae) native to East Asia (China, Japan, Korea). The plant was introduced to the United States in the late nineteenth century for a variety of purposes: it was used as an ornamental plant, as fodder plant for cattle, and as a cover plant to prevent soil erosion. In the southeastern United States, the plants escaped control and became distributed on different types of open land. Preferred habitats are forest edges, abandoned fields, and shrubland. Kudzu grows rapidly and is able to cover and smother orchards and plantation crops, including young forest plantations, and leads to the dieback of the overgrown vegetation. This reduces diversity and causes economic loss, especially in forestry.
- Invasive weeds are major threats to the biodiversity and livestock production of rangelands across large parts of the world. In the western United States, rangelands previously dominated by perennial bunchgrasses have been converted, primarily as a result of overgrazing, to annual grasslands that are susceptible to invasion by introduced dicots. According to DiTomaso (2000), there are more than 300 rangeland weeds in the United States. Some of the most problematic species include at least 15 species of the genus *Centaurea* (Asteraceae, knapweeds), for example *Centaurea diffusa* (Fig. 3.9). These and other weed species are low in palatability and avoided by livestock, and some are actually poisonous to livestock. Overall, they reduce the quality of forage, increase the cost of managing and producing livestock, slow animal weight gain, and reduce the quality of livestock products. Furthermore, many rangeland weeds have deep taproot systems which modify the water balance and nutrient availability of the system.

Fig. 3.9 Diffuse knapweed (*Centaurea diffusa*)

In Australia, up to 2,000 alien wild plant species became naturalized by the end of the twentieth century, with an estimated continuing rate of increase averaging 10 species per year. Many are weeds of rangelands, including such invasive shrubs as *Parkinsonia aculeata*, *Mimosa pigra*, *Acacia nilotica*, and *Prosopis* species. There is also concern that alien plant species include "sleeper weeds" that may become invasive and expand their range as a consequence of climate change.

The chances of controlling invasive species are usually limited. There are only a few examples of successful suppression of an invasive species, for example biological control of St. John's wort (*Hypericum perforatum*), an invasive rangeland weed of North America in the first half of the twentieth century (Sect. 5.2.4.7). However, complete eradication of an alien species can be achieved only during the very early stages of an invasion. Therefore, most attempts to control invasive species focus on measures to prevent their spread and impact. This requires detailed knowledge on the biology and ecology of the target species to develop an overall strategy which may cover a combination of management techniques including mechanical, chemical, and biological control and the control of invasive species transportation pathways.

3.3.3.2 What Makes Invasive Species So Successful?

Only a relatively small proportion of exotic species develop into invasive species in the region of introduction. For example, 21% of the North American flora consists of exotic species, but only 2% of these have developed to become pests (Doorduin and Vrieling 2011). Williamson (1996) proposed the "10:10 rule" to explain the probability of a species becoming a successful invader, meaning 10% of introduced species become established and 10% of those will become invasive.

As many as 29 individual mechanisms or processes have been proposed to explain why some alien species can achieve higher densities, higher rates of spread, and greater establishment success than others (Catford et al. 2009). Basically, these refer to (1) specific characteristics and traits of the invasive species and/or (2) characteristics of the invaded environment:

1. Biological traits of plants and animals that provide competitive advantage over native species and which are of advantage for successful invasion may include one or more of the following: (a) high rates and specific strategies of reproduction, (b) broad tolerance of temporarily unfavourable environmental conditions and disturbances, (c) high phenotypic plasticity, (d) the ability of adaptive evolution, (e) high resource use efficiency, and (f) effective (chemical) defence mechanisms against natural enemies.
2. Characteristics of the new environment consist of both abiotic and biotic factors that affect invasion success. These include (a) the overall climatic conditions and the available resources, (b) the presence and abundance of natural enemies, including diseases, (c) the ability of resident species in a community to reduce growth and performance of invading species populations (i.e. biotic resistance, which may depend on ecological interactions, food web structure, and species diversity of a native community), and (d) the amount of anthropogenic disturbance of vegetation, soil and other ecosystem characteristics and functions.

However, empirical evidence suggests that most of these species-specific and environmental factors can explain the success of some invaders only to some extent in some circumstances. Rather, invasion success is likely to be context-dependent and a result of a combination of different factors and mechanisms which can hardly be predicted for individual species. In addition, an initial determinant of invasion success can be the number of individuals of the species introduced and the frequency of introductions. The different factors and potential pathways that can result in a successful species invasion are shown in Fig. 3.10.

3.4 Succession

The process of directed, continuous and season-independent changes in communities, which are accompanied by the colonization of new species and the disappearance of previously present species, is called **succession**. This process consists of the sequence of different communities which may occur over the course of short or long term periods (a few years to millennia) and is determined by biotic and abiotic factors. Examples include changes in the communities of abandoned fields, tree fall, clear-cuts and fires in forests, sand dunes, or the development of new sites created through landslides.

The endpoint of such a development, reached under the given climatic conditions of a site, is called the **climax community**. Examples of the climax stage of different climate zones include tundra, taiga, savanna, steppe, tropical rain

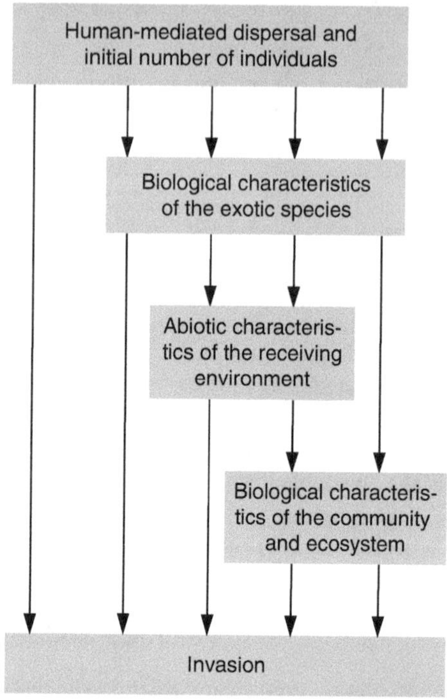

Fig. 3.10 Different factors and potential pathways that can result in a successful species invasion (Based on Catford et al. 2009)

forest, and deciduous forest in the temperate regions (Chap. 7). However, the climax community of individual sites within these climate zones can vary substantially in terms of species composition. This is determined primarily by abiotic site factors and interactions of the colonizing species.

Under nearly constant environmental conditions, the climax conditions are relatively stable, i.e., only short-term, undirected, and temporary fluctuations in the composition of the community occur. The causes of such **fluctuations** are, primarily, weather conditions, that can vary on a seasonal or yearly basis. Large changes in the species composition of climax communities occur only as a result of specific external effects or disturbances. For example, these can include changes in climate, the rise or fall of the water table, or inputs of nutrients. Depending on the effect of these factors, different developments may occur as a consequence.

Agroecosystems are stages of succession that do not undergo further development because of regular agricultural measures, for example sowing, harvest, and soil cultivation. This is observed for grasslands in regions of natural forest vegetation (for example the temperate regions of Europe) which only remain in this stage because of regular mowing or grazing. After abandonment of use, such systems undergo natural succession. On fallow fields in Central Europe, the following phases may be identified in such a succession:

1. In the beginning, annual wild plants dominate. These primarily include species that are found in cultivated landscapes as weeds (Sect. 4.4).

2. Next, perennial, herbaceous plants become established and gradually displace the annuals. Plant species diversity is highest in this phase.
3. In the third phase, woody plants begin to become established (shrub or bush phase). With increasing shade, the herbaceous species are displaced and the plant stand becomes more species poor.
4. In the final stage, trees are dominant and a closed forest develops.

The development of such a community can vary substantially in terms of species composition and the duration of different stages of development, depending on the site. Detailed predictions of the course of succession are therefore hardly possible. The development of agricultural fallow land is primarily determined by three groups of factors. These are:

- previous cultivation, the effect of which is largely determined by the last cultivated crops and agricultural practices, for example fertilization, soil cultivation, and use of herbicides. These factors significantly affect the growing conditions of the plants, and the seed bank found in the soil.
- natural site conditions, which are primarily determined by the climate, soil properties, the exposure, and slope. Generally, drier and cooler site conditions mean that succession proceeds more slowly.
- the vegetation in the direct and more distant vicinity of the site. The frequency of individual plant species, their distance from potential sites of colonization, and the dispersal ability of their seeds affect species composition, especially in early stages of succession. The presence of shrubs which spread through vegetative stolons can accelerate the encroachment of scrub on to fallow land.

Not only plant communities but also animal communities undergo succession. For example, flower-rich, herbaceous plant stands, which develop in the first years on agricultural fallow land are correlated with the presence of butterflies and wild bees. In addition, they provide resources for specific antagonists of pests of cultivated plants (Sect. 5.2.4.5).

The intentional establishment of agricultural fallow land, as a component of the cultural landscape, can contribute to conservation of the species diversity of typical field flora and fauna.

3.5 Flows of Energy and Material

Energy and organic and inorganic materials are continuously transported and transformed within the compartments of an ecosystem, and between different ecosystems. These processes are carried out by organisms and as a result of abiotic factors, for example water and wind.

Table 3.2 The ecological efficiencies which determine the portion of net primary productivity that is used by consumers for production of biomass within an ecosystem

Utilization efficiency in different ecosystems	
Forests	2–7%
Cropping systems[a]	9–21%
Grasslands	5–60%
Oceans	40–99%
Assimilation efficiency of different consumers	
Wood-eaters (xylophages)	15%
Grass and leaf-eaters (herbivores)	30–50%
Seed-eaters (granivores)	70–80%
Predators	70–90%
Detritivores	20–40%
Production efficiency of different consumers	
Mammals and birds	1–3%
Insects and other invertebrates	10–55%

Based on Whitaker (1975), Wiegert and Owen (1971), Humphreys (1979) and other sources
[a]Yield losses of the most important food crops worldwide attributed to phytophages (Based on Oerke et al. 1994)

3.5.1 Energy Flows Through Food Webs

The energy flows in ecosystems proceed via food chains and food webs. Of the total energy fixed by photosynthesis in a plant, representing **gross primary production (GPP)**, only a portion is converted into biomass, representing **net primary productivity (NPP)**. The difference between GPP and NPP is the portion of energy that is needed by the plant for the maintenance of metabolic processes; this is released as heat. Primary production is measured in grams of carbon (C) or grams of biomass (fresh or dry weight) per unit area per year.

The NPP of natural vegetation types on Earth varies within a broad range. In different terrestrial ecosystems, NPP can range from just a few g C m^{-2} $year^{-1}$ in deserts to approximately 2,000 g C m^{-2} $year^{-1}$ in tropical rain forests. The reasons for this are not only the differences in the yearly radiation received from the sun by the different latitudes on Earth (Sect. 4.1), but also such other factors as water and nutrient availability to plants. The proportion of NPP utilized in the production of biomass by the various organisms within the consumer food web is determined by different ecological efficiencies:

- The **utilization efficiency** is the proportion of NPP of a plant stand that is taken up by phytophagous consumers as food. This efficiency varies significantly among different ecosystems (Table 3.2). In a forest, this value is almost 7% (the differences between temperate and tropical latitudes is slight) and is primarily consumed by insects. In grassland ecosystems, utilization efficiencies range from 5 to 60%. The highest values for terrestrial ecosystems are found in the African savannas, for which grazing mammals are mostly responsible. Of the

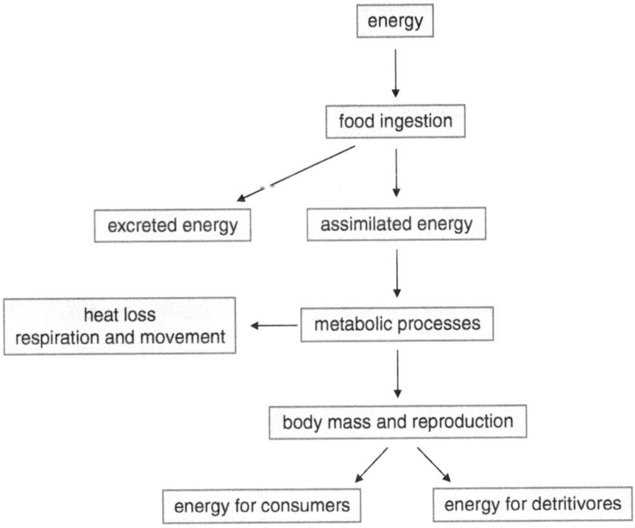

Fig. 3.11 Pathways of energy within a trophic level

globally most important crops in agroecosystems, between 9 and 21% are consumed by phytophagous pests.

- The **assimilation efficiency** represents the portion of the consumed biomass that is available to the consumer as energy source. Thus, it cannot be equated with the quantity of food that reaches the digestive system. Of this, a portion of variable size cannot be used by the organism and is therefore excreted. The assimilation efficiency depends on the quality of the food and the ability of the organism to make the food's energy content available during digestion (Table 3.2). In leaf and grass-eaters, the assimilation efficiency is approximately 30–50%, which means that at least half of the ingested food remains unused. Wood-eating species assimilate only approximately 15% of their food. The assimilation efficiency of seed and fruit-eaters and of secondary consumers (predators) is up to 80%. As in the producers, the assimilated energy is used primarily for two processes—respiration, in which it is released as heat, and production of biomass.
- The **production efficiency** is the proportion of assimilated energy that is available for production of biomass, i.e. for growth and reproduction. Groups of consumers can also be classified on the basis of this efficiency (Table 3.2). Mammals and birds, which must maintain a higher body temperature, can transform 1–2% of assimilated energy into biomass whereas insects and invertebrates can transform 10–55%.

The pathways of distribution of energy within a trophic level are shown in Fig. 3.11. The energy stored within the biomass of dead organisms provides the basis of the detritivore food web.

The energy that flows into the food web via the producers declines from one trophic level to the next. The related transfer efficiency varies with each ecosystem and the associated organisms.

3.5.2 Material Transport Through Water and Wind

In many landscapes, **water** is the most important transport medium for organic and inorganic material. Streams and rivers take up soil and detritus from their watershed. In flood plains, these loads are deposited and not only supply natural ecosystems, for example riparian forests, with nutrients, but also supply agroecosystems. For example, the sediments deposited during the flooding of the Nile had existential importance for the agriculture of Egypt for thousands of years. Removal of natural vegetation by humans increases soil erosion, which also increases the amount of material transported by rivers. The clearing of forests in Central Europe, which began with the spread of agriculture 7,000 years ago, led to filling of stream and river flood plains with fertile soil. In the tropics, because of greater precipitation, the clearing of forests often results in devastating landslides that can destroy entire villages in the valleys. In addition, the rivers transport the removed soil to the ocean, where coral reefs are buried by the sediment and subsequently die. As a result, populations of fish species directly or indirectly dependent on the reefs also decline, which in turn threatens the food supply and livelihoods of many coastal residents.

Material transport through **wind** is also of great importance for many ecosystems. During the Ice Ages, loess, i.e. silt produced by the movements of glaciers on rocks, was formed. On poorly vegetated areas, this material was subject to wind erosion and was, in some cases, deposited in great layers in other locations. Loess forms fertile soils that are among the best agricultural lands in many areas of the world (e.g. China, North America, Central Europe). Today, wind-related transport processes are still important. Over a distance of more than 5,000 km, millions of tons of sand and dust are transported from the Sahara to the Amazon region every year. This is a significant source of nutrients for the rain forest there.

Input of nitrogen from the atmosphere is an (additional) source of nutrients for several ecosystems (Sect. 3.7.4.2).

Ecosystems are not closed entities, but instead are connected with each other through the interactions of species, as well as through the flow of energy and materials.

3.5.3 Flows of Energy and Material in Agroecosystems

With regard to energy and material flows, intensively managed agroecosystems, in particular, have several characteristics distinct from those of natural ecosystems:

- Large proportions of the net primary productivity, i.e. the biomass of the crops, are removed from the system during harvest, which also results in a decline in the nutrient content of the soil.
- To compensate for these losses, fertilizers must be used (Sect. 2.3.3). Mineral fertilizers are imported into the agroecosystem from outside and must be produced and transported using energy, which in turn uses external resources, primarily fossil fuels.
- For the production of crops, further energy is required, specifically for soil cultivation, sowing, harvest, and possibly irrigation. This energy comes from manual labour, the use of animals or the application of machinery and is based on food, feed, or fuel, respectively. Usually, chemicals are also applied to control pests, weeds, and phytopathogens. These substances must also be produced, transported, and distributed. Their residues reach the food webs and ecosystems.
- Only a relatively small portion of the nutrients imported into an agroecosystem as fertilizers is taken up by the crop plants. For example, only approximately one third of the nitrogen applied to the soil in German agriculture is extracted via the plant and animal products. The rest is leached out or escapes as a gas into the atmosphere (Sect. 3.7.4.2). Sixty percent of the nitrogen input into the North and Baltic Seas originates from agriculture.
- Through global trade, agricultural products (e.g. cereals and soybean for animal production) reach other regions where they become a component of the material and energy flows of those ecosystems (e.g. in the form of animal excrement).

Intensively cultivated agroecosystems in particular are characterized by high material and energy flows and to only a limited extent by internal cycles.

3.6 Ecosystem Services

The present and future welfare of people, societies, and economies depend on the various goods and resources provided by nature. Beside the many kinds of plants and animals from agricultural and natural systems that sustain and benefit human life, this also includes a variety of services, often considered as granted and freely available. They are so-called public goods, which cannot be owned and have no markets and prices. Such services are based on natural ecosystem processes and functions, which result in the provision, regeneration, and long-term stability of such resources as water, soil, nutrients, organisms and biomass. Their value is associated with a gain or loss to human welfare, which often is not obvious until a service becomes critically disturbed and leads to negative effects, for example water scarcity, flooding, soil erosion, or land degradation.

Together, the goods and services generated by ecological functions and processes that directly or indirectly affect human wellbeing are called **ecosystem services**. According to the Millennium Ecosystem Assessment (2005), they can be subdivided into four main categories:

- **Supporting services** are those that are necessary for the production of all other ecosystem services. They include the basic functions and outputs of ecosystems, for example photosynthesis, biomass production, soil formation and retention, nutrient and water cycling, and habitat structures.
- **Provisioning services** are services that can be extracted from natural and agricultural systems for direct use by humans. They refer to all products derived from cultivated systems (for example food, forage, biofuel, fibre, and other renewable resources), in addition to those which are exploited from unmanaged systems, for example firewood, timber, or seafood. They also include services such as the provision of genetic resources from biodiversity and the delivery of services from species, for example pollination and the natural control of pests and diseases.
- **Regulating services** can be defined as those services which are necessary for the sustainable delivery of other services by regulation and the support of ecosystem processes. They contribute to the buffering of environmental fluctuations and risks, and to the resilience of ecosystems against different types of disturbance. Regulating services include erosion and flood control, climate regulation through carbon storage, water filtration, and waste decomposition.
- **Cultural services** are those services that contribute to the maintenance of human health and well-being by providing recreation, spiritual and aesthetic values and education opportunities, and include the creativity born from interactions with nature (music, art, architecture). These services can be different in different local and national human societies but are mainly associated with the presence of close-to-nature systems (e.g. nature reserves) or the diversity of a cultivated landscape.

Ecosystem services can be regarded as the "natural capital" of human welfare that yields a flow of valuable ecosystem goods or services into the future. In practice, however, it is very difficult to assign a "value" in economic terms to natural services. There have been a variety of attempts to measure the economic or monetary value of ecosystem services. Costanza et al. (1997) attempted to put monetary value on 17 different ecosystem services and calculated the total value as being between 18 and 54 trillion US dollars per year, compared with a total global gross domestic product (GDP) of 18 trillion US dollars at that time. Indeed, this figure gives an idea of the dimension of these values, but it remains questionable whether it is an appropriate and feasible approach to express the value of specific ecosystem services in pure monetary terms, because the types of value can be manifold. This aspect is considered in the concept of **Total Economic Value**

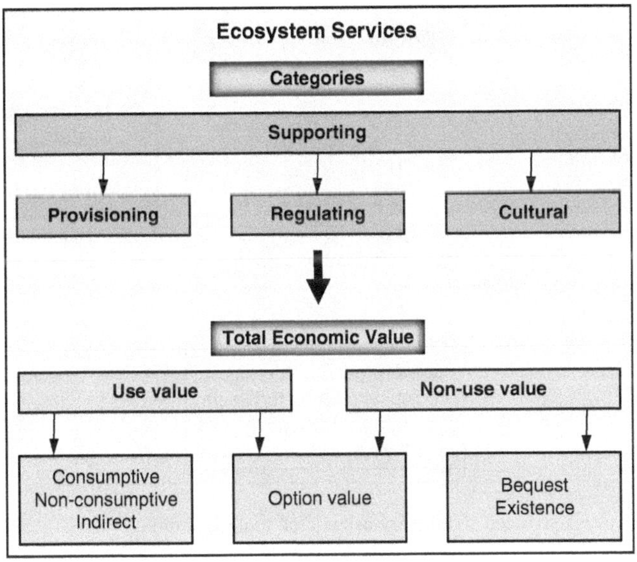

Fig. 3.12 Categories of ecosystem services and total economic values

(**TEV**), a widely used framework for classification of the different ecosystem goods and services. It distinguishes between two main categories—use and non-use values—which are subdivided into further groups (Pearce and Warford 1993; Fig. 3.12):

Use values refer to the use of natural resources by humans for consumption or production purposes. These include

- consumptive use values, given by the extraction of natural outputs (e.g. by harvesting, hunting, or fishing);
- non-consumptive use values, given by activities that do not require extraction of products, for example recreation, sports, wildlife watching, and other touristic activities; and
- indirect use values, which are used as intermediate inputs for production of final goods and services for humans, for example water and nutrient supply, and for such non-consumptive values as clean air, clean water for swimming, and low health risks.

Non-use values arise for ecosystem goods or services that people value for their existence, even though they may never make use of that resource or service (for example, the enjoyment of watching wildlife and landscape reports on television). Non-use values can be substantial for human wellbeing, but are difficult to quantify. This category includes:

- bequest values, which are values that people place on knowing that other people, including future generations, will have the option to enjoy the same goods and services, and

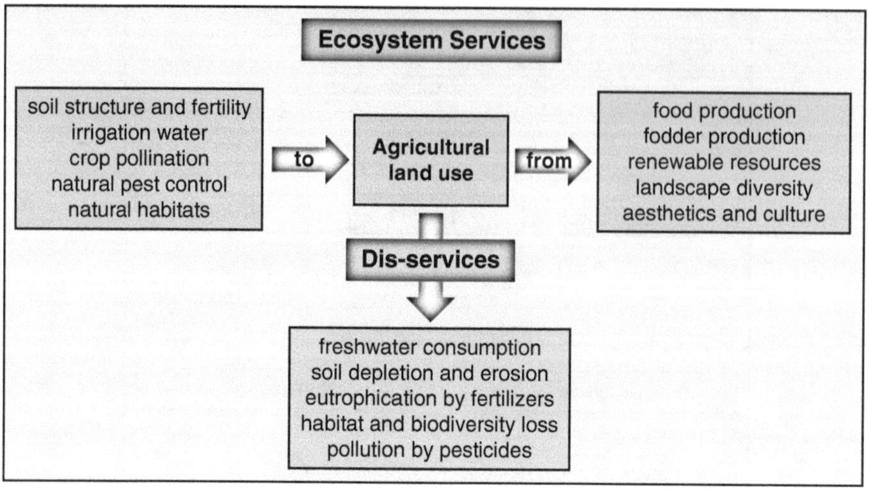

Fig. 3.13 Ecosystem services and disservices related to agriculture

- existence values, which relate to the satisfaction to humans from knowing that ecosystems or wildlife species, for example whales or elephants, continue to exist, based on the feeling that nature has an intrinsic value.

Option values cover both use and non-use values and refer to goods and services that are not used at present but may become important in the future. These especially include the value of biodiversity, for example the genetic diversity of varieties and breeds of domesticated species, and natural species diversity, which may provide important pharmaceutical resources.

3.6.1 Ecosystem Services Associated with Agriculture

Human activities increasingly modify natural ecosystem functions and processes and, consequently, ecosystem services. Ecosystem changes that have contributed substantial net gains to human well-being and economic development via agricultural production have been achieved at the growing cost of the degradation of other services. Approximately 60% of essential ecosystem services are being degraded or used unsustainably, increasing the likelihood of potentially abrupt changes that will seriously affect human wellbeing (Millennium Assessment 2005).

Agricultural production systems are involved in all categories of ecosystem services in three ways: they provide services, demand services, and affect other services (Fig. 3.13).

Ecosystem services from agriculture primarily include the provision of food crops, livestock products, fibres, biofuel and other plant resources, for example rubber. Provision of additional services is highly dependent on the type of land use and on management practices. Some agricultural landscapes provide cultural benefits because of their aesthetic values for recreation and tourism. These are often related to traditional land-use practices and/or the diversity of a cultivated landscape, but large monocultures in uniform landscapes have the opposite effect. Similarly, appropriate land use and management practices can improve soil properties or support biodiversity, but often affect such features negatively.

Ecosystem services to agriculture include several supporting and regulation services. Agriculture requires most of the global renewable water resources used for human activities (Sect. 3.7.1). Natural soil structure and fertility are of crucial importance in determining where different kinds of agriculture are conducted, and the productivity of the systems. Important provisioning services to agriculture include insect pollination of many crops (Sect. 4.5.3) and biological pest control provided by natural enemies, which depend on the availability of suitable habitats and other resources (Sect. 5.2.4).

Ecosystem disservices from agriculture are unwanted effects of agricultural production on other ecosystem services and the environment. These include a number of effects of global concern and are mainly related to agricultural intensification and land-use change. The main aspects are further discussed in other chapters and include

- irrigation of croplands, leading to depletion of freshwater resources;
- soil and land degradation by wind and water erosion;
- eutrophication of terrestrial and aquatic ecosystems as a result of excessive use of fertilizers;
- effects of pesticides on biodiversity and human health;
- expansion of production areas, leading to the loss of natural ecosystems; and
- emission of greenhouse gases from land-use change and livestock production.

Overall, ecosystem services associated with agriculture are closely interlinked and not independent of each other. This situation often requires ecological and economic tradeoffs in agricultural practice and management. Mitigation options should be considered on different spatial and temporal scales, including field, farm, landscape, or watershed, taking into account the lateral flows of ecosystem services.

Field measurements to reduce disservices may include selection of crops and land-use types adapted to the location (soil type, relief), use of cover crops and minimum tillage to reduce soil losses, and use of suitable intercrop and rotation practices to preserve soil fertility. At the farm scale, integration of crop and livestock production creates synergies by using crops and crop residues for livestock feed while capturing nutrients from livestock wastes for crop production.

The delivery and the provision of ecosystem services to and from agriculture is highly dependent on the structure of the landscape in which the agroecosystems are embedded. Production sites rely on services provided by nearby natural or semi-natural habitats, which may act as reservoirs for beneficial species and as buffers to reduce or regulate the escape of water and nutrients. This requires the maintenance of the integrity of multi-functional landscapes and is especially important in watersheds, where soil and water runoff brings the risk of landslides, floods, and sedimentation, which affect downstream ecosystems including coastal areas. Therefore, watershed management is necessary and must include the preservation of natural systems, for example hillside forests and riparian vegetation.

3.7 Global Material Cycles

For their metabolism, living organisms require not only organic carbon and oxygen, which are primarily made available through photosynthesis, but also a series of additional elements or compounds, for example nitrogen, phosphorus and water. These substances are part of biogeochemical cycles, which means they proceed through various living and non-living compartments in the bio, litho, pedo, hydro, and atmosphere. In contrast with energy, these materials are available to organisms again after passing through their respective cycles. The most important global material cycles are discussed in the following sections.

3.7.1 Water (H_2O)

The total quantity of water on earth can be regarded as constant and is estimated to be approximately 1.4 billion km^3. Approximately 71% of the Earth's surface is covered by the water of the oceans, which amount to more than 97% of the total water reserve. Less than 3% is fresh water. Most of the latter is not available to organisms because it is bound in the ice and snow of the polar regions and in glaciers, or is found in groundwater partly at inaccessible depths. Only approximately 0.6% of fresh water occurs as surface water, where it is also an important resource for humans and the landscapes dominated by humans (Fig. 3.14).

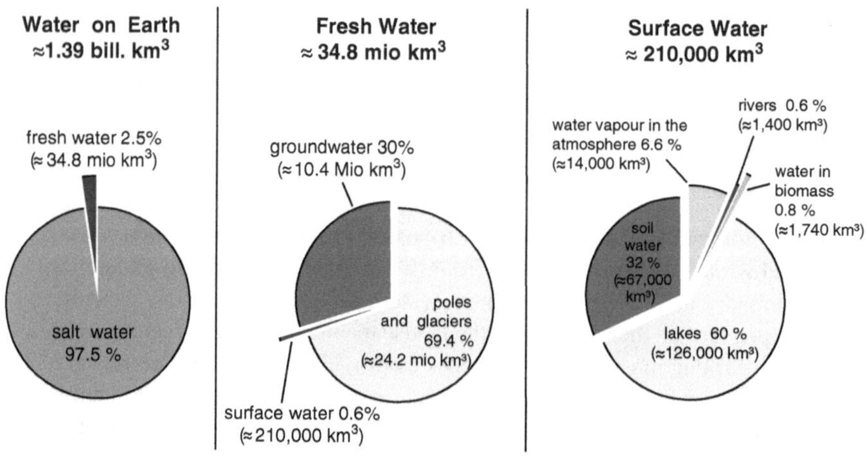

Fig. 3.14 Global water resources (Based on Wetzel 2001; Shiklomanov 1999)

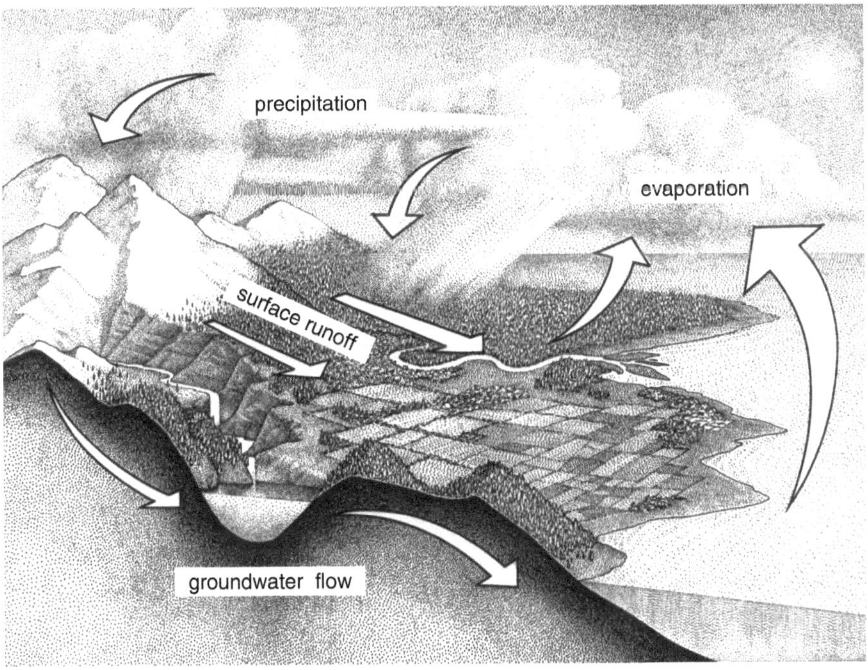

Fig. 3.15 The global water cycle

In most terrestrial ecosystems, water arrives primarily by precipitation. Because the average annual temperature decreases from the equator to the poles, this precipitation falls as snow for increasingly longer periods along this gradient. Precipitation is an essential factor in the definition of the different climate zones of the Earth (Chap. 6). In some deserts, rain only occurs at unpredictable intervals of several years. In contrast, the humid tropics usually receive more than 100 mm of rain every month. In the wet–dry tropics, more or less pronounced dry periods occur.

Not only the quantity of precipitation but also its temporal distribution have a substantial effect on the living conditions of organisms.

Water that reaches the soil surface can take a variety of pathways. Depending on the quantity of water, the topography of the site, and soil conditions, a larger or smaller portion of the water runs off over the surface. A further portion penetrates the soil and eventually reaches the groundwater by infiltration. Variable amounts of this water feed rivers and thereby reach the oceans. At the soil and water surface, evaporation also occurs. Solar energy is responsible for the transition from liquid to vapour, which also determines the quantity of evaporated water. The vaporized water reaches the atmosphere and condenses to form clouds, from which precipitation falls again (Fig. 3.15).

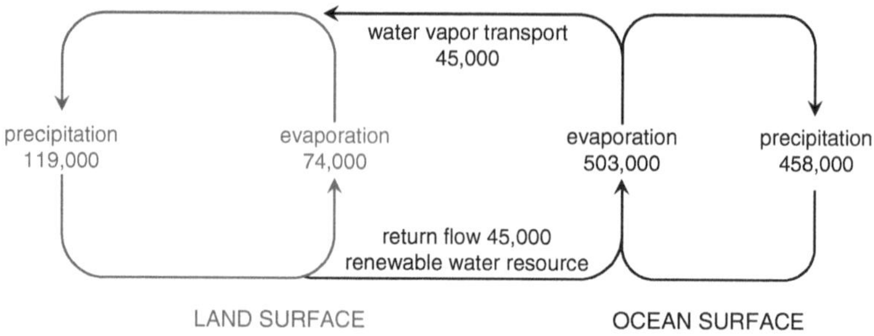

Fig. 3.16 Quantitative data for the global water cycle (km^3 $year^{-1}$; based on Shiklomanov 1999)

Approximately 503,000 km^3 of water evaporate from the ocean surfaces per year and an additional 74,000 km^3 from land areas; this is then returned to the Earth by precipitation (about 458,000 km^3 $year^{-1}$ over the oceans and 119,000 km^3 over land; Fig. 3.16). The difference between the precipitation and evaporation over land (119,000 – 74,000 = 45,000 km^3 $year^{-1}$) forms the **renewable water resources**. More precisely, this is the quantity of water that is transported by rivers and reaches lakes and groundwater.

Currently, approximately 8.5% of the global renewable water resource is used by humans. Approximately 66% of that is used in agriculture (primarily in irrigation) whereas 34% is used in industry and households. The total demand increased by approximately 2% annually during the twentieth century (Fig. 3.17).

3.7.2 Carbon (C)

At the global level, the carbon cycle consists of five large carbon stores, among which exchange of a variety of C compounds occurs. In decreasing order of carbon stored, these are the lithosphere (the rocky shell of the Earth), the oceans, the soil, the atmosphere, and the living biomass of terrestrial ecosystems. Biotic processes in the oceans and of terrestrial vegetation make up the largest portions of natural emission and absorption of **carbon dioxide** (CO_2).

The biological carbon cycle is closely linked to the energy flow of biological communities (Sect. 3.5.1). Figure 3.18 shows the pathways of carbon in natural terrestrial ecosystems.

The carbon source for producers is the atmosphere. It currently contains approximately 380 ppm (parts per million; 1 ppm = 1×10^{-6}), or 0.038% CO_2 gas, which plants use and reduce to other carbon compounds via photosynthesis. These compounds are used in the production of structural substances (primarily lignin and cellulose) and as components of additional metabolic products, for example fats, amino acids, and proteins, and specific plant products. On the other hand, producers require organic compounds for respiration, as do consumers and

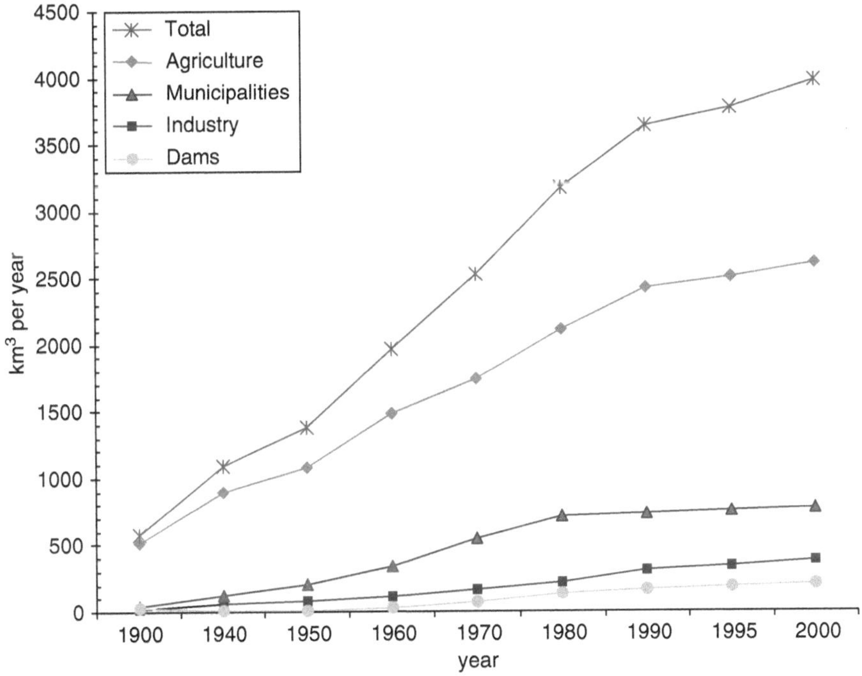

Fig. 3.17 Anthropogenic use of global renewable water resources in the course of the twentieth century (Based on Shiklomanov 1999)

detritivores, whereby not only the stored energy but also the bound CO_2 is released. For this reverse process of photosynthesis, most organisms use oxygen (O_2). Some bacteria can also utilize nitrate, methane, or hydrogen sulfide.

In plants, the ratio of photosynthetically fixed carbon to the carbon released through respiration is, on average, 2:1; but for younger individuals in the process of growing, the proportion of carbon fixed as biomass is substantially higher. The carbon compounds found in dead, predominantly plant, biomass in terrestrial ecosystems is decomposed in the soil by organisms, whereby CO_2 is released to the atmosphere (Sect. 4.3.5.1).

Theoretically, the CO_2 budget of undisturbed ecosystems is balanced, which means that the quantity of carbon fixed in net primary production is respired by heterotrophic organisms (e.g. microbial decomposition of organic substances and the respiration of animals) and released into the atmosphere. **Net ecosystem production (NEP)**, i.e. NPP minus the heterotrophic respiration, would thus be zero. However, in this balance, spatial and temporal variations may occur. During the succession of a system, the NEP is usually positive and approaches zero during the climax stage. As a result of removal of parts of the NPP during harvest, the NEP in agroecosystems becomes negative if no organic fertilizers are applied.

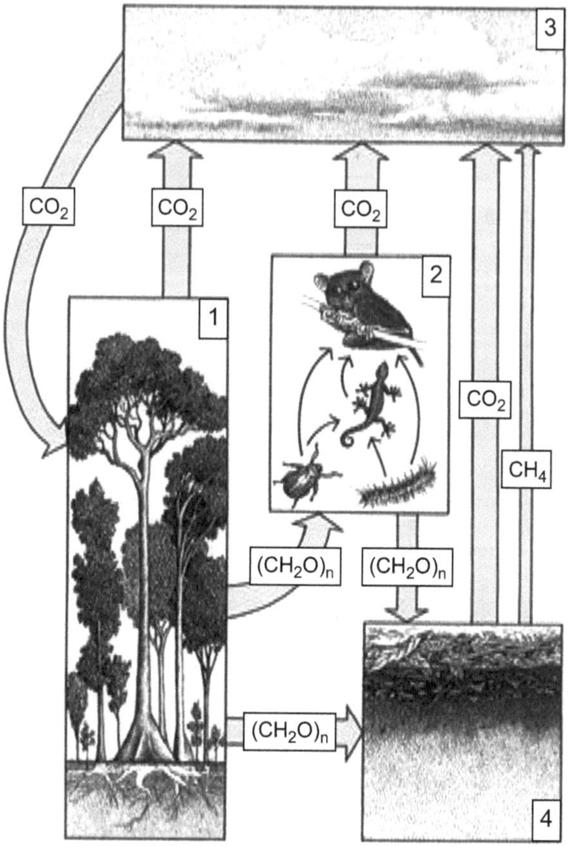

Fig. 3.18 The biological carbon cycle of natural terrestrial ecosystems (using the Philippine rain forest as an example). **1** = producers, **2** = first and second-order consumers, **3** = atmosphere, **4** = soil with decomposers

The activity of soil organisms, i.e. the rate of decomposition of carbon compounds in the soil and thus also the NEP, is highly affected by abiotic factors. These include temperature, water balance, availability of oxygen, and pH (Sect. 4.3.5.2). Microbial activity is primarily dependent on temperature and thus the rate of decomposition of carbon materials in the soil of the arctic tundra is approximately a factor of 20 lower than that in a tropical rain forest. The wet soils of swamps, rice fields, and other wetlands are generally anaerobic (i.e. free from oxygen). There, the carbon is predominantly released to the atmosphere in the form of **methane** (CH_4). However, the rate decomposition by anaerobic methane-producing bacteria is lower than that by aerobic microorganisms and can result in an accumulation of organic matter (peat). Anaerobic conditions were also the precondition for the development of **fossil fuels** (oil, gas, brown and hard coal) from dead marine organisms and land and/or wetland plants in a variety of geological eras. Soil cultivation is an anthropogenic factor affecting microbial decomposition of biomass. As a result, transformation of forests, grasslands, and wetlands into agricultural fields can lead to a substantial decline in the carbon content of the soil.

3.7.2.1 Anthropogenic Effects on the CO_2 Budget

The CO_2 content of the atmosphere has been subject to both small and large variations in the course of the Earth's history (Sect. 8.1.3.1). At the height of the last Ice Age approximately 21,000 years ago, the content was approximately 200 ppm and subsequently increased with increasing temperature. After the end of the Ice Age, approximately 10,000 years ago, 260 ppm had been reached and a slight increase led to approximately 280 ppm by 1800. Since the beginning of the Industrial Revolution, however, a substantial increase has been observed, and will probably continue beyond the current value of 380 ppm (Sect. 8.1.3.2).

Anthropogenic emissions, almost exclusively, can be regarded as the driver of the increase of more than 30% in CO_2 within the past 200 years. Specifically, the following human activities are known to have changed the natural carbon cycle:

- release of CO_2 by **burning of fossil fuels** (coal, oil, natural gas), which were previously not incorporated in the carbon cycle but were instead, so to speak, sealed in the lithosphere;
- **land use change**, which includes clearing of forests, burning of biomass, and release of CO_2 from organic soil stores as a result of agricultural practices; and
- **production of quicklime** or **burnt lime** (CaO), primarily used for production of concrete and fertilizers, by heating of limestone ($CaCO_3$), which results in release of CO_2.

Formation of quicklime at temperatures of approximately 1,000 °C:

$$CaCO_3 \rightarrow CaO + CO_2$$

Through these processes, CO_2 sources are created that did not previously exist or that did not exist to the current extent. Approximately 88% (the average for the years 2000–2009) of annual anthropogenic CO_2 emissions originate from fossil fuels and the production of cement. Approximately 12% originate from land use change. Of the total quantity, only 46% remains in the atmosphere. A further 27% is absorbed by terrestrial ecosystems and the same proportion (27%) by the oceans (Fig. 3.19). Table 3.3 shows the values of these data in comparison with the period 1980–1989.

Currently, the highest terrestrial absorption of CO_2 is probably occurring in the ecosystems of the temperate latitudes of North America, Asia, and Europe, which are, therefore, an important CO_2 sink. The main reasons for this could be changes in land use. These include reforestation, transformation of agricultural land into fallow and grasslands, and agricultural production processes that lead to an

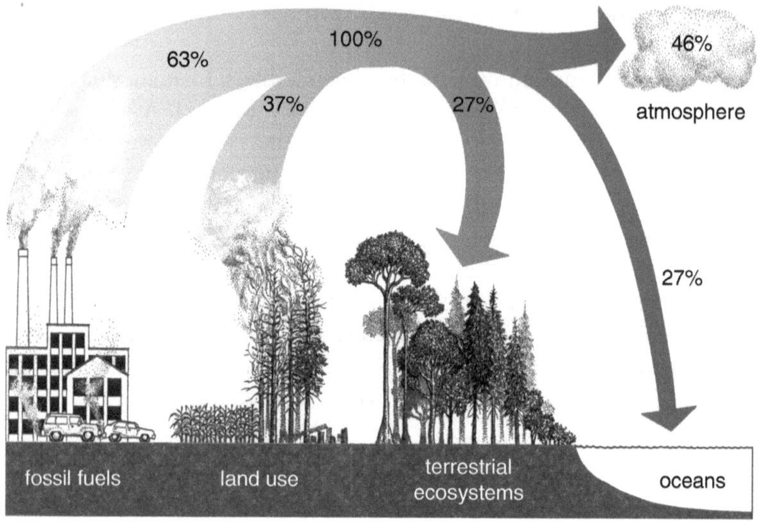

Fig. 3.19 Global proportions of anthropogenic CO_2 sources and their fate (annual averages for the years 2000–2009; data based on Global Carbon Project 2010)

Table 3.3 Global sources and sinks of CO_2 in the 1980s and 2000–2009

	Period	
	1980–1989	2000–2009
Sources		
Fossil fuels + cement	5.5	7.7
Land use	1.4	1.0
Sinks		
Atmosphere	3.4	4.0
Oceans	2.0	2.3
Terrestrial ecosystems	1.5	2.3

Values are gigatonnes (Gt) of carbon per year (1 Gt = 1 billion = 10^9 tonnes)
(Data from Global Carbon Project 2010)

increase in organic soil biomass. It is also possible there are already effects that can be traced back to the increased CO_2 content of the atmosphere and increased global temperatures, which lead to an increased biomass production by plants (Sect. 8.1.3).

3.7.3 *Oxygen (O)*

Oxygen is the most common element in the Earth's crust, and with a content of 21% in the form of O_2, is the second most common element in the atmosphere. Before the emergence of life there was almost no oxygen in the atmosphere. Only with the start of photosynthesis by cyanobacteria and, later, primarily by plants, did the O_2 concentration gradually increase to its current value (Sect. 8.1.3.1). Biological oxygen metabolism is closely related with carbon metabolism and proceeds in the quasi opposite direction: in photosynthesis, O_2 is formed that is required by plants, animals, and most microorganisms for aerobic respiration. Furthermore, oxygen is consumed in natural and anthropogenic burning of organic substances, in some land use changes by promotion of aerobic processes (e.g. draining of wetlands), and by non-biological oxidation. However, a decline in atmospheric O_2 concentrations as a consequence of human activity is currently not expected.

Another compound of oxygen is **ozone** (O_3), which is naturally concentrated 20–30 km above the Earth's surface in the stratosphere. Despite the relatively small concentration of, at most, 10 ppm, ozone fulfils a function of existential importance to life on Earth—up to 90% of solar UV-B rays are absorbed by the ozone layer. O_3 molecules are continuously created and destroyed by UV radiation, whereby the quantity of O_3 remains constant. The following reactions take place in this process:

1. splitting of O_2 by UV rays into free oxygen atoms, which bind with O_2 to create O_3; and
2. splitting of the O_3 molecules by UV rays into O_2 and free oxygen, from which O_3 is again created with release of heat (i.e. infrared radiation, IR). Overall, the UV-B radiation is thus transformed into harmless forms of energy.

Absorption of UV radiation in the stratosphere:

$$O_2 + UV \rightarrow O^- + O^-$$

$$2O_2 + 2O^- \rightarrow 2O_3$$

$$O_3 + UV \rightarrow O_2 + O^-$$

$$O_2 + O^- \rightarrow O_3 + IR$$

However, these stratospheric processes are affected by anthropogenic activity, and specifically by release of halogenated hydrocarbons, for example Freon (dichlorodifluoromethane; CF_2Cl_2). Such compounds were and are used in various ways in industry (as coolants in refrigerators and air conditioners, in synthetic materials, as cleaning products, in fire extinguishers, and as propellants in spray cans). In the stratosphere these compounds lose halogen atoms (e.g. chlorine), because of the action of UV radiation, which react with ozone with formation of O_2 and halogenated monoxides. The latter combine with free oxygen atoms, whereby O_2 is formed and the halogen atoms are again released. This is a catalytic process that repeatedly causes the destruction of O_3 molecules. This chain reaction leads to a decline in the concentration of ozone in the stratosphere and is also believed to be responsible for the annual emergence of the "ozone hole" over the Antarctic.

Splitting of halogenated hydrocarbons (e.g. Freon) and reaction of halogens in the stratosphere:

$$CF_2Cl_2 + UV \rightarrow CF_2Cl + Cl^-$$

$$Cl^- + O_3 \rightarrow ClO + O_2$$

$$ClO + O_3 \rightarrow ClO_2 + O_2$$

$$ClO_2 \rightarrow Cl^- + O_2$$

Ozone can also be formed close to the ground (low-level ozone). There, it does not occur here as a result of the splitting of O_2, because the radiation energy is too low. Instead, nitrogen dioxide (NO_2) for which the available light energy is sufficient to drive photolysis, serves as the supplier of oxygen atoms. By reaction of the free oxygen atoms with O_2, ozone is formed.

In nature, this reversible process is of little importance, because nitrogen dioxide is only available in small quantities. Under specific conditions, however, high ozone concentrations can form close to ground level, particularly in cities with heavy traffic in association with intense sunlight during the summer. Oxides of nitrogen, primarily nitric oxide (NO), are released as vehicle emissions. Another precondition for formation of low level ozone is the presence of light and volatile hydrocarbon compounds (e.g. exhaust, solvents), as a result of which, in a series of reactions, oxidation of NO to NO_2 occurs. This leads to a high concentration of NO_2, so ozone can be continuously formed in sunlight. In the dark, no ozone formation occurs, because the balance of the chain reaction shifts toward ozone degradation (oxidation of NO by O_3). The same effect can be achieved during the day if emission of nitrogen oxides or hydrocarbons is consistently reduced. As a result of wind drift, it is possible that increased ozone concentrations are also found outside the areas in which they are produced. In rural areas there are fewer sources of nitrogen oxides, which means that the nightly decomposition of ozone is significantly reduced. As a result, there is also the threat of damage to plants, primarily of the leaf surfaces. Reduced growth and increased susceptibility to pests and diseases can be related to this.

Formation of low-level ozone:

$$NO_2 + O_2 \leftarrow UV \rightarrow NO + O_3$$

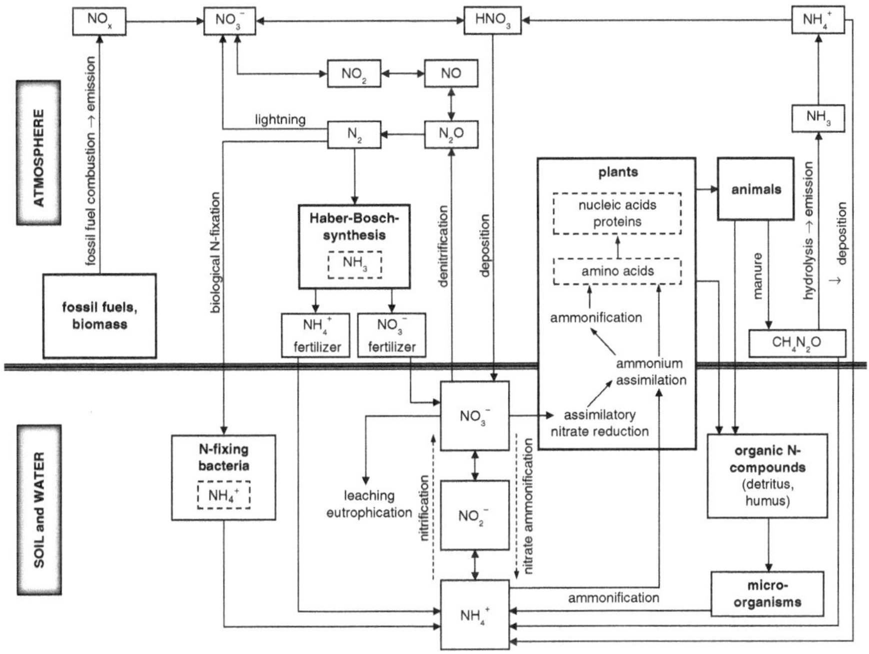

Fig. 3.20 The nitrogen cycle

3.7.4 Nitrogen (N)

Nitrogen, in the form of N_2, is the most common compound in the atmosphere, the proportion being 79%. This element is a component of a variety of organic compounds, for example amino acids, proteins, and nucleic acids, and is thus essential for all organisms.

3.7.4.1 Processes in the Nitrogen Cycle

As N_2, nitrogen is useful to neither plants nor animals. Thus, processes in which transformation of nitrogen compounds occurs are of central importance in the global nitrogen cycle (Fig. 3.20).

Nitrogen Fixation

The term nitrogen fixation includes those processes that lead to the formation of N compounds from atmospheric N_2. In nature, there are two ways in which this can occur:

1. **Lightning.** N_2 is a very inert gas. The energy that is released by lightning strikes leads to splitting of the nitrogen molecule. By way of various reactions with oxygen and water, nitrous acid (HNO_2) and nitric acid (HNO_3), or nitrite (NO_2^-) and nitrate (NO_3^-), are formed.
2. **Biological nitrogen fixation.** In contrast with plants and animals, some bacteria are capable of fixing atmospheric nitrogen, whereby N_2 is reduced to ammonia (NH_3) and ammonium (NH_4^+).

Among the nitrogen-fixing bacteria, free-living and symbiotic species can be distinguished. Representatives of different heterotrophic genera (e.g. *Aztobacter, Azospirillum, Klebsiella, Beijerinckia*) are found in the soil and can be concentrated in the vicinity of the rhizosphere of specific plants. This occurs, for example, with a variety of Poaceae, for example sugar cane, rice, maize, and other cereal species. The bacteria utilize organic compounds, which are released by the roots of these species as food sources. Cyanobacteria are another group that include nitrogen-fixers. Because they are autotrophs, they do not live in the soil but primarily in water and on moist surfaces and are of importance in agroecosystems, especially rice paddy fields.

Reduction of atmospheric nitrogen to ammonia:

$$N_2 + 6H^+ + 6e^- \rightarrow 2NH_3$$

Symbiotic nitrogen-fixers live in mutualistic relationships with other organisms (plants and some corals). They receive carbon compounds from these organisms and the partner profits from delivery of fixed nitrogen. From an agroecological perspective, the following symbiotic relationships between bacteria and plants are of the greatest importance:

- **Root-nodule-forming bacteria and legumes**. Bacteria of the genera *Rhizobium* and *Bradyrhizobium* (**rhizobia**) are found in soil; they carry out aerobic metabolism, utilizing a variety of organic carbon compounds, and nitrate and ammonium as nitrogen sources. In the presence of legumes (Fabaceae) these bacteria change their life strategy. They enter the roots of the plants, reproduce by cell division and form a micro-colony, typically visible as root nodules (Fig. 3.21). There, the bacteria differentiate morphologically into the so-called bacteriods

Fig. 3.21 Roots of a leguminous plant with typical root nodules

capable of fixing atmospheric nitrogen. The ammonium produced reaches the plant cells and is transported in the form of amino acids through a transport system that also supplies the nodule tissues with photosynthetic products. This N_2 fixation is performed with the aid of the oxygen-sensitive nitrogenase enzyme system. Leghaemoglobin is responsible for maintenance of low O_2 concentrations. This compound consists of a protein component (globin) that is delivered by the plant and a pigment component (protohaem) that originates from the bacteriod. Colonization of the plant also promotes reproduction of the rhizobia; i.e., after the death of the plant the number of bacteria in the soil increases. More than 90% of leguminous species are infected by nodule-forming bacteria; these include soybean, field beans, beans of the genus *Phaseolus*, pea, lentil, and peanut among the numerous agriculturally important crops. Important fodder plants include *Trifolium*, *Lotus*, and *Medicago* species.

- ***Azolla* and *Anabaena***. The genus *Azolla* are aquatic floating ferns of the family Salviniaceae, of which six species are known worldwide. They have small, distichously arranged leaves divided into two lobes. In the upper lobe, photosynthesis occurs; the lower lobe is in contact with water. The plant hosts the cyanobacteria *Anabaena azollae* in hollows on the underside of the upper lobe. The cyanobacteria can be found there at all stages of the fern's development. Because *A. azollae* has highly specialized cells, called heterocysts, it is capable of fixing atmospheric nitrogen. The resulting ammonium is taken up by the fern, which in turn supplies the cyanobacteria with photosynthetic products (primarily sucrose). Similar to the legumes in terrestrial systems, the *Azolla* species, via symbiosis with nitrogen-fixing bacteria, also have the potential to improve the supply of this nutrient to rice paddy fields and thus to increase rice yields. Use of such species (primarily *Azolla pinnata*) is of particular importance, with a long tradition, in southeast Asia.

Ammonium Assimilation

Uptake of ammonium (NH_4^+) and its transfer in compounds produced by an organism is called ammonium assimilation. Plants that live in mutualistic relationships with nitrogen fixers receive these compounds from the respective bacterial species. However, ammonium can also be taken up by the plant directly via the roots. In the soil, however, NH_4^+ is of quantitative importance only under acidic, moist, and oxygen-poor conditions. Generally, nitrate (NO_3^-) in the soil is the most important source of nitrogen for plants.

Ammonification

Separation of NH_4^+ or amino groups ($-NH_2$) from organic nitrogenous compounds is termed ammonification. This process occurs in plants and in ecosystems (in soils and water). In plants, NH_2 groups are enzymatically removed (deamination) and transferred to other compounds (transamination).

Ammonification as the first step in N-mineralization in ecosystems:

$$R\text{-}NH_2 + H_2O \rightarrow NH_4^+ + R\text{-}OH$$

In the soil, most of the nitrogen is present in the form of organic compounds (>95%), which primarily originate from dead organisms. Ammonification there is performed by a variety of bacteria and fungi and is the first step in nitrogen mineralization.

Nitrification

NH_4^+ released in the soil by ammonification is oxidized by some microorganisms in the presence of oxygen. This first produces nitrite (NO_2^-) and finally nitrate (NO_3^-). These processes are two steps in the mineralization of nitrogen whereby the two products are produced by different chemoautotrophic bacteria that gain energy in the process. Nitrite is primarily released by members of the genus *Nitrosomonas* is and subsequently oxidized to nitrate by *Nitrobacter* bacteria.

Formation of nitrite by Nitrosomonas:

$$NH_4^+ + 1.5\,O_2 \rightarrow NO_2^- + 2\,H + +H_2O$$

Formation of nitrate by Nitrobacter:

$$NO_2^- + 0.5\,O_2 \rightarrow NO_3^-$$

Under suitable conditions (neutral to slightly acidic soils, temperatures around 25 °C, moderately moist and sufficiently oxygenated), nitrification proceeds almost completely, which means that no significant amounts of ammonium or nitrite remain. Annual nitrification and mineralization of organic nitrogen compounds in soils range between approximately 1 and 5%.

Nitrate Reduction

Transformation of nitrate into an organism's own organic nitrogenous compounds, i.e. the reverse of mineralization, is called the **assimilatory nitrate reduction** pathway. This process is performed by plants and a variety of microorganisms, but not by animals. The reduction proceeds in two main steps, from nitrate to nitrite and from nitrite to ammonium. Subsequently, assimilation of ammonium occurs, i.e., incorporation of NH_4^+ into organic compounds, initially forming the amino acid glutamine.

In largely anaerobic, wet, and acid habitats, some bacteria can reduce nitrate to nitrite and further to ammonium. This pathway is called **nitrate ammonification** and might be regarded as a shortcut in the biological nitrogen cycle that formally combines denitrification and nitrogen fixation. In contrast with assimilatory nitrate reduction, this process does not include direct transformation of NH_4^+ into an organic substance, i.e. the ammonium is present in the ecosystem and available to plants.

Denitrification

Denitrification is another process in which nitrate is transformed into further compounds under anaerobic conditions. This is mostly by heterotrophic bacteria that can use the oxygen in nitrate and nitrite for respiration, instead of O_2. Reduction of nitrate occurs in several steps and the final product is molecular nitrogen (N_2).

Reduction of nitrogen in denitrification:

$$NO_3^- \rightarrow NO_2^- \rightarrow NO \rightarrow N_2O \rightarrow N_2$$

Denitrification is also promoted by high availability of organic carbon, a high soil water content, and high temperatures, among other factors, and occurs primarily in neutral soils.

3.7.4.2 Anthropogenic Effects on the Nitrogen Cycle

Similar to the carbon cycle, global turnover of nitrogen compounds has substantial anthropogenic components. At present, more than half of the fixed nitrogen cycling through the atmosphere and back to terrestrial surfaces results from human activity (Table 3.4).

The most important contributions to the transformation of the natural nitrogen cycle are:

- Production and use of fertilizers: industrial nitrogen fixation, largely for production of nitrogen fertilizers, was made possible by the Haber–Bosch process patented in 1910, which produces ammonia (NH_3) from atmospheric nitrogen (N_2) and hydrogen (H_2). The synthesis occurs under high pressures, at approximately 500 °C and with the help of a catalyst based on iron. From the resulting ammonia, urea (CH_4N_2O), and nitric acid and its salts (nitrates) can be produced and thus a variety of fertilizers can be manufactured. The quantity of industrial nitrogen fixation worldwide is estimated to be approximately 125 Teragrams (Tg) per year.

Reaction of the Haber–Bosch process:

$$N_2 + 3H_2 \rightarrow 2NH_3$$

- The production of some crop plants: by cultivation of legumes (N_2 fixation by rhizobia) and establishment of rice paddies (N_2 fixation by cyanobacteria), humans contribute to the annual amount of nitrogen fixed by approximately 20 Tg. In comparison, the quantity of atmospheric nitrogen fixed by natural, terrestrial ecosystems is estimated to be 120 Tg year^{-1} at most.
- Use of fossil fuels: mineral oil products and coal contain nitrogen compounds that are released in combustion, primarily in the form of nitrogen oxides (NO_x).

Table 3.4 Nitrogen budget for the global land surface

	Pre-industrial (natural)	Human-derived	Total
Biological nitrogen fixation	120	20	140
Industrial nitrogen fixation	0	125	125
Fossil fuel combustion	0	25	25
Lightning	5	0	5
Totals	*125*	*170*	*295*

Based on Schlesinger (2009). Values are Teragrams (Tg) nitrogen year^{-1} (1 Tg = 10^{12} g = 1 million = 10^6 metric tons)

At high temperatures, fixation of atmospheric nitrogen also occurs, as occurs naturally as a result of lightning. As a result of the use of fossil fuels, the natural nitrogen cycle receives an additional 25 Tg of fixed nitrogen per year.

These and other human activities, for example biomass burning and soil cultivation have far-reaching consequences in the atmosphere and biosphere. These are primarily related to the effects of nitrate, ammonia, dinitrogen oxide (laughing gas), and other oxides of nitrogen.

Nitrate (NO_3^-)

The only way to bind nitrate added to agroecosystems in the form of mineral fertilizers is uptake by organisms. Because soils have high adsorption capacity for cations only, excess nitrate (an anion) can easily end up in surface water and groundwater. The extent of such leaching depends largely on the quantity of fertilizer applied, the climate, the soil type, and the crop stand. At temperate latitudes, leaching is highest in the winter months, because of the reduced activity of plants and microorganisms. However, by cultivation of winter cover crops nitrate leaching can be reduced. Nitrate in groundwater and drinking water is problematic even though the compound itself is only toxic in high quantities. In the human body bacterial transformation of nitrate into nitrite occurs, which can lead to symptoms of poisoning, especially in infants. Nitrate reacts with amines and amides to form nitrosamine and nitrosamide, respectively, which are carcinogenic.

In surface waters, input of nitrogen and phosphorus can lead to **eutrophication**, i.e. increases in nutrient content. As a consequence of this, growth of algae increases, which can initially lead to an increase in oxygen content near the water surface. During decomposition of the dead biomass, this oxygen is consumed by detritivores. Further transformation of the organic matter then proceeds under anaerobic conditions, and such gases as hydrogen sulfide, methane, and ammonia are formed. The presence of organisms that require oxygen is, therefore, no longer possible. Lower inputs of nutrients, which are not sufficient to cause algal blooms in waters, can still result in a decline in biodiversity. The reasons for this are the same as for nutrient inputs in terrestrial ecosystems (Sect. 3.3.1).

Ammonia (NH_3)

Human activities are responsible for approximately 70% of the ammonia released worldwide. The largest part originates from agriculture, with more than 80% coming from animal excreta in livestock production. There, the most important process is the hydrolysis of urea found in liquid and solid manure.

Formation of gaseous ammonia occurs in alkaline environments and is promoted by heat. Because ammonia is highly water soluble, it reacts rapidly in the atmosphere to form ammonium. This is returned to terrestrial and aquatic ecosystems, where it is transformed into nitrate.

Hydrolysis of urea and formation of ammonium:

$$CH_4N_2O + 3H_2O \rightarrow 2\,NH_4^+ + HCO_3^- + OH$$
$$NH_4^+ + H_2O \leftrightarrow NH_3 + H_2O + H^+$$

Dinitrogen Oxide (N_2O; Laughing Gas)

Under specific conditions, nitrate reduction in the soil does not proceed to completion; as a result, not only is N_2 produced as a final product but also other products, primarily laughing gas (dinitrogen oxide, N_2O). Low temperatures, low availability of carbon compounds, high nitrate content (primarily as a result of fertilization) and acidic soils are some of the factors that promote this process. Laughing gas, similar to hydrocarbons, contributes to depletion of the ozone layer (Sect. 3.7.3) in the stratosphere and also acts as a greenhouse gas (Sect. 8.1.3.1).

Degradation of ozone by reaction with laughing gas:

$$N_2O + O_3 \rightarrow 2\,NO + O_2$$
$$NO + O_3 \rightarrow NO_2 + O_2$$

Nitrogen Oxides (NO_x)

Use of fossil fuels makes the largest contribution to global production of NO_x compounds (especially NO and NO_2). These compounds not only contribute to the ground level development of ozone (Sect. 3.7.3), but also to the development of acid rain. This happens via oxidation of NO and NO_2, whereby nitric acid and nitrate are formed. Via a variety of processes, ammonia also contributes to acidification (see above). Another, sometimes larger, contribution to acid rain is that of sulfur dioxide (SO_2) which, similar to nitrogen oxide, is primarily released during burning of oil and coal and forms sulfuric acid and sulfate (SO_4^-) via oxidation.

Atmospheric deposition of nitrogen compounds in Central European ecosystems amounts to approximately 10–30 kg ha^{-1} year^{-1} and makes a significant contribution to eutrophication and acidification of soils and waters.

3.7.5 *Phosphorus (P)*

Phosphorus is similar to nitrogen in its importance as an essential nutrient for all living organisms. It is a crucial component of nucleic acids, cell membranes, and of the energy carrier adenosine triphosphate (ATP). In the form of phytic acid, phosphorus is found as an energy source and reserve in plant seeds. Calcium phosphate is a component of the bones and teeth of vertebrates.

The largest phosphorus reserves on Earth are found in marine sediments and the resulting rocks, in rocks of volcanic origin (basalts), and in soils. The approximately 200 known inorganic compounds in which phosphate is found are almost all derivatives of phosphoric acid (H_3PO_4). The most common are calcium phosphate (apatite), which is found in the form of fluorapatite ($Ca_5(PO_4)_3$ F), hydroxyapatite ($Ca_5(PO_4)_3OH$), and chlorapatite ($Ca_5(PO_4)_3Cl$). Gaseous compounds of phosphorus are not of importance. Unfertilized soil contains, depending on a variety of factors (e.g. parent material, extent of weathering, clay and humus content) between approximately 100 mg kg^{-1} (sandy, strongly weathered soil) and 1,500 mg kg^{-1} (soil on basaltic rock) of total phosphorus. Total phosphorus is made up of a variety of organic and inorganic compounds. The latter comprise stable phosphates, which are found as salts (mineral phosphate), and labile phosphate ions (HPO_4^{2-}, $H_2PO_4^-$), which are attached to soil particles (primarily clay minerals). These forms make up the bound phosphate. Plants primarily take up phosphorus in the form of hydroxylphosphate ($H_2PO_4^-$), which is then incorporated into organic compounds. However, the quantities of phosphate present in solution in the soil and therefore available to plants are very limited and amount to only approximately 0.1% of the total phosphorus in the soil. To supply plants, phosphate must be delivered from the bound fraction, which only occurs when the concentration in the soil solution is low, i.e. when the soluble supplies have been taken up by the plants. In phosphorus-poor soils, mobilization of bound phosphate can proceed via organic acids, which are produced by free-living microorganisms, plant roots, or mycorrhizal fungi. Most phosphorus compounds introduced via litter and detritus,

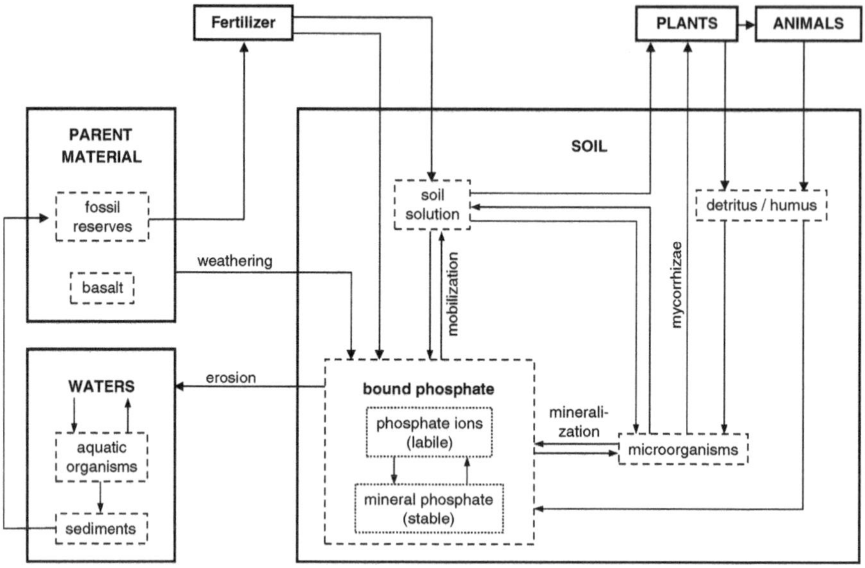

Fig. 3.22 The phosphorus cycle

which are rapidly decomposed by microorganisms, or are directly returned to the soil reserves via leaching (Fig. 3.22).

Phosphorus mineral fertilizers used to improve the phosphorus supply in agricultural soils primarily originate from rock reserves. However, these raw phosphates are usually poorly water soluble and must therefore first be modified. This is achieved by grinding or by treatment with sulfuric acid (in the production of superphosphate). Only a small proportion of the phosphate fertilizer applied to fields reaches the soil solution; most is transformed into bound phosphate. Over time, the phosphate content of many intensively used agricultural soils increases, because the bound phosphate is only subject to leaching to a limited extent, as a result of strong adsorption by mineral soils. For this reason, phosphate fertilization—in contrast with over-fertilization with nitrogen—has little effect on groundwater. As a result of erosion and runoff, however, phosphates reach rivers and lakes on soil particles, which, in association with nitrogen fertilization, leads to eutrophication of waters and to an increase in plant growth (Sect. 3.7.4.2). In Germany, approximately half of the phosphate that reaches flowing waters and, eventually, the ocean originates from agriculture.

In contrast with the situation in intensive agriculture at temperate latitudes, large areas of the tropics have a low soil phosphate content, which is a limiting factor in plant production. This is not only because of the insufficient availability and high cost of phosphate fertilizers but also the result of the high phosphate fixation of many soils (e.g. Ferralsols, Sect. 7.2.1) of these climate zones. This severely limits the continuous supply of phosphate to crops.

Modern agriculture is dependent on phosphorus derived from phosphate rock; this is a non-renewable resource and current global reserves may be depleted in 50–100 years. Although phosphorus demand is projected to increase, the expected global peak in phosphorus production is predicted to occur around 2030. Although the exact timing of peak phosphorus production might be disputed, it is widely acknowledged within the fertilizer industry that the quality of remaining phosphate rock is decreasing and production costs are increasing (Cordell et al. 2009).

References

Beck KG, Zimmerman K, Schardt JD, Stone J, Lukens RR, Reichard S, Randall J, Cangelosi AA, Cooper D, Thompson JP (2008) Invasive species defined in a policy context: recommendations from the Federal Invasive Species Advisory Committee. Invasive Plant Sci Manag 1:414–421

Catford JA, Jansson R, Nilsson C (2009) Reducing redundancy in invasion ecology by integrating hypotheses into a single theoretical framework. Divers Distrib 15:22–40

Cordell D, Drangert J-O, White S (2009) The story of phosphorus: global food security and food for thought. Glob Environ Chang 19:292–305

Costanza R, D'arge R, de Groot R, Farber S, Grasso M, Hannon B, Limburg K, Naeem S, O'Neill RV, Paruelo J, Raskin RG, Sutton P, van den Belt M (1997) The value of the world's ecosystem services and natural capital. Nature 387:253–260

DiTomaso JM (2000) Invasive weeds in rangelands: species, impacts, and management. Weed Sci 48:255–265

Doorduin L, Vrieling K (2011) A review of the phytochemical support for the shifting defence hypothesis. Phytochem Rev 10:99–106

EEA (2010) Invasive alien species in Europe. European Environment Agency, http://www.eea.europa.eu/data-and-maps/indicators/invasive-alien-species-in-europe/invasive-alien-species-in-europe

FAO (1997) The state of the world's plant genetic resources for food and agriculture. Food and Agriculture Organisation of the United Nations, Rome

FAO (2007) The state of the world's animal genetic resources for food and agriculture. Food and Agriculture Organization of the United Nations, Rome. ftp://ftp.fao.org/docrep/fao/010/a1250e/a1250e.pdf

FAO (2012) Biodiversity for a world without hunger. Food and Agriculture Organization of the United Nations, Rome. http://www.fao.org/biodiversity/components/plants/en/

Fowler C, Mooney P (1990) Shattering: food, politics, and the loss of genetic diversity. University of Arizona Press, Tucson

Global Carbon Project (2010) Carbon budget and trends 2009. www.globalcarbonproject.org/carbonbudget

Goldberg DE, Miller TE (1990) Effects of different resource additions on species diversity in an annual plant community. Ecology 71:213–225

Humphreys WF (1979) Production and respiration in animal populations. J Anim Ecol 48:427–453

ISSG (2012) 100 of the world's worst invasive alien species. Global invasive species database, http://www.issg.org/database/species/search.asp?st=100ss

Kusnezov N (1957) Numbers of species of ants in faunae of different latitudes. Evolution 11:298–299

Millennium Ecosystem Assessment (2005) Summary for decision makers. In: Ecosystems and human well-being. Island Press, Washington, DC

Miller RM, Miller SP, Jastrow JD, Rivetta CB (2002) Mycorrhizal medicated feedbacks influence net carbon gain and nutrient uptake in *Andropogon gerardii*. New Phytol 155:149–162

Möbius K (1877) Die Auster und die Austernwirtschaft. Verlag Wiegandt, Hempel und Parey, Berlin
Oerke EC, Dehne HW, Schönbeck F, Weber A (1994) Crop production and crop protection. Elsevier, Amsterdam
Pearce DW, Warford JW (1993) World without end: economics, environment, and sustainable development. Oxford University Press, Oxford
Pimentel D, Zuniga R, Morrison D (2005) Update on the environmental and economic costs associated with alien-invasive species in the United States. Ecol Econ 52:273–288
Porter SD, Savignano DA (1990) Invasion of polygyne fire ants decimates native ants and disrupts arthropod community. Ecology 71:2095–2106
Schlesinger WH (2009) On the fate of anthropogenic nitrogen. Proc Natl Acad Sci 106:203–208
Shiklomanov JA (1999) World water resources and their use – a joint SHI/UNESCO product. http://webworld.unesco.org/water/ihp/db/shiklomanov/index.shtml
Wetzel RG (2001) Freshwater ecosystems. In: Levin SA (ed) Encyclopedia of biodiversity 3. Academic, San Diego, pp 75–87
Whitaker RH (1975) Communities and ecosystems, 2nd edn. Macmillan, New York
Wiegert RG, Owen DF (1971) Trophic structure, available resources and population density in terrestrial vs. aquatic ecosystems. J Theor Biol 30:69–81
Wilcove DS, Rothstein D, Dubow J, Philips A, Losos E (1998) Quantifying threats to imperilled species in the United States. Bioscience 48:607–615
Willerding U (1986) Zur Geschichte der Unkräuter Mitteleuropas. K. Wachholz Verlag, Neumünster
Williamson M (1996) Biological invasions. Chapman & Hall, London

Chapter 4
Crops and Their Environment

4.1 Radiation and Energy

The primary energy source for almost all ecosystems is solar energy, which reaches the Earth continuously. At the boundaries of the Earth's atmosphere, the sun's radiation efficiency is approximately 1,360 W m^{-2}. This value is nearly constant and is, therefore, termed the **solar constant**. The portion of the radiation that reaches the Earth's surface, the so-called **insolation (incoming solar radiation)**, is lower, because approximately half of the solar energy is lost by absorption and reflection in the atmosphere. Another factor that determines the insolation of a particular site is the angle at which the sun's rays meet the Earth. Therefore, the insolation is significantly higher at the equator than at the poles. The worldwide highest values (>250 W m^{-2} annual average) are reached in tropical and subtropical region rarely covered by clouds (e.g. the Sahara desert). The mean annual insolation in the temperate latitudes is approximately half of that value and amounts to 127 W m^{-2}. In Stuttgart-Hohenheim in Germany, for example, the value is highest in July (230 W m^{-2}) and lowest in December (50 W m^{-2}).

4.1.1 Photosynthesis

The sun's electromagnetic radiation covers a range of wavelengths from 100 to 5,000 nm (nanometres) and can be divided into three parts (Fig. 4.1): ultraviolet radiation (100–380 nm), light visible to humans (380–740 nm), and near infrared radiation (740–5,000 nm). The visible region of the electromagnetic spectrum is nearly identical with the spectrum that plants can utilize for photosynthesis and is termed **photosynthetically active radiation (PAR)**.

K. Martin and J. Sauerborn, *Agroecology*, DOI 10.1007/978-94-007-5917-6_4,

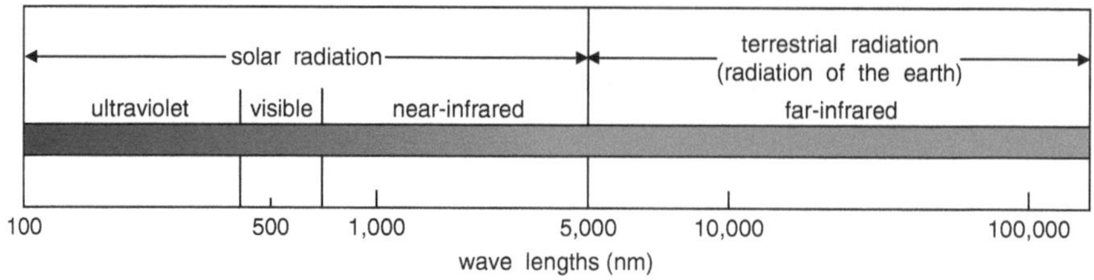

Fig. 4.1 The electromagnetic spectrum (solar and terrestrial radiation). 1 nanometre (nm) = 10^{-9} (one billionth) metre

Photosynthesis is a process that enables green plants and some bacteria to produce carbohydrates (i.e. biomass), from carbon dioxide and water by use of light energy (Box 4.1). This process releases oxygen (O_2):

$$6CO_2 + 12H_2O + \text{light energy} \rightarrow C_6H_{12}O_6 + 6H_2O + 6O_2$$

Box 4.1 Carbohydrates

The basic units of carbohydrates are the simple sugars (monosaccharides), for example glucose and fructose. By condensation of two monosaccharides, a disaccharide is formed. Sucrose is made up of glucose and fructose and is the most important disaccharide produced by plants. Carbohydrates that are made up of more than four monosaccharides are termed polysaccharides. The starches are included in this group. Starches are made up of 200–1,000 glucose molecules and serve as carbohydrate reservoirs for plants. Starch is also the most important carbohydrate in human nutrition and serves as an energy source. The daily requirement of carbohydrates is 5–6 g kg^{-1} body weight. In the body, starches are broken down by enzymes into glucose, which is taken up by the blood and cells.

Numerous crops are cultivated for the production of starch, which can be derived from different plant parts. Examples:

Plant part	Crop species (examples)
Seed	Cereals, buckwheat (*Fagopyrum esculentum*)
Root	Sweet potato (*Ipomoea batatas*), cassava (*Manihot esculenta*), yam (*Dioscorea* species)
Tuber	Potato (*Solanum tuberosum*)
Rhizome	Taro (*Colocasia esculenta*)
Stem	Sago palm (*Metroxylon sagu*), ensete (*Ensete ventricosum*)
Fruit	Plantain (*Musa* species), breadfruit (*Artocarpus altilis*; Fig. 4.2)

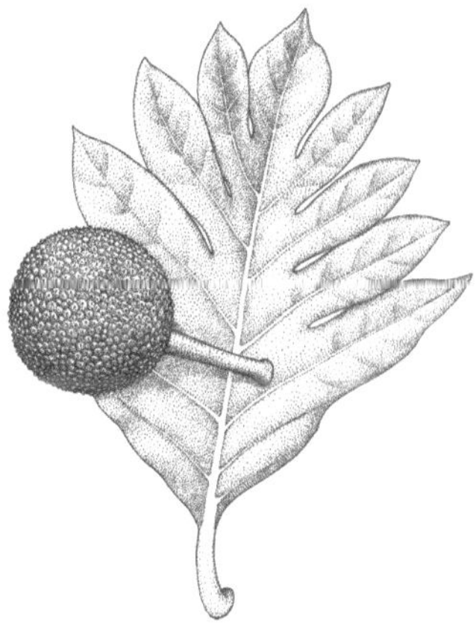

Fig. 4.2 Breadfruit tree (*Artocarpus altilis*), fruit and leaf

Plant photosynthesis takes place in the chloroplasts. These cell organelles contain pigment systems, the chlorophylls, which absorb light energy and transform it into chemical energy. The chlorophylls of plants have two absorption peaks within the visible light spectrum, one of which lies, approximately, between 400 and 500 nm and the other between 600 and 700 nm. Light in the intermediate region (i.e., 500–600 nm) can be used to a limited extent only. In contrast, cyanobacteria, which are also capable of photosynthesis, have other pigments with their absorption peak at wavelengths between 500 and 600 nm.

The CO_2 that is required for photosynthesis and taken from the air must be fixed by the plants, which means it must be incorporated into a molecule before it can be utilized for production of carbohydrates. Fixation of CO_2 does not occur in the same way in all plants and is associated with particular physiological and morphological characteristics of specific species. Overall, three photosynthetic types are identified; the C_3, C_4, and CAM plants.

4.1.1.1 C_3-Plants

For 95% of all plant species, the CO_2 taken from the air is fixed as phosphoglycerate. Because this molecule is made up of three C atoms, plants that perform synthesis in this way are termed C_3 plants. Binding of the CO_2 occurs via the enzyme ribulose-1,5-biphosphate-carboxylase/oxygenase (RuBisCo). However, this molecule has only low affinity for CO_2 and can also bind oxygen (O_2). Because of other reactions, some of the bound carbon (approximately 20–30%) is lost as of CO_2, which therefore reduces the effectiveness of photosynthesis. Because this process occurs in light, it is called **photorespiration**. With increasing light intensity and temperature, the associated carbon loss increases. The optimum temperature for CO_2 fixation by C_3 plants is between 15 and 25°C.

If the concentration of CO_2 is increased, the photosynthetic performance of C_3 plants can be increased by up to 30%. This "CO_2 fertilization effect" is used in greenhouse cultivation (e.g. of cucumbers and tomatoes) to achieve higher yields.

As a result of the increasing CO_2 content of the Earth's atmosphere, this effect will, in the future, probably also become relevant to the production of field crops (Sect. 8.1.3.2).

The gas exchange of plants occurs via the stomata (Greek *stoma* = mouth, opening) in the epidermis. For uptake of CO_2 and release of O_2, the stomata of C_3 plants are usually open during the day. At the same time, the stomata regulate plant transpiration, i.e. release of water or water vapour.

With increasing temperature, the rate of transpiration increases and, as a result, loss of water. When these losses cannot be compensated by water uptake through the roots, then the plant will suffer from water stress. Plants react to this by contracting or closing their stomata, which simultaneously results in a reduction in photosynthetic performance.

The photosynthetic performance of C_3 plants is not only affected by light and CO_2 availability, but also to a great extent by temperature and by water availability.

4.1.1.2 C_4-Plants

In approximately 2% of all plant species, fixation of atmospheric CO_2 does not occur via the enzyme RuBisCo, but instead via the enzyme phosphoenolpyruvate-carboxylase (PEPC). This process produces, depending on the plant species, either malate or aspartate. Because these acid compounds are each made up of four carbon atoms, plants with this mechanism are called C_4 plants. Fixation of CO_2 through PEPC occurs in the mesophyll cells of the leaves. The C_4 acids produced in the mesophyll cells are transferred into the bundle sheath cells. There, the respective acid (malate or aspartate) is chemically broken down into a smaller organic molecule, and CO_2 is released. This carbon dioxide then enters the chloroplast of the bundle sheath cell and is fixed a second time in the same way as in the C_3 plants (i.e. by RuBisCo).

Thus, in the C_4 plants initial CO_2 fixation occurs that is spatially separated from the other photosynthetic processes. The advantage of this mechanism is the significantly higher CO_2 affinity of PEPC in comparison with that of RuBisCo. Therefore, the carbon losses associated with photorespiration are substantially lower in C_4 plants than in C_3 plants. In other words, with higher light intensity, the net photosynthetic rate of the C_4 plants is significantly higher than in C_3 plants. Furthermore, high temperatures do not lead to significant reductions in the photosynthetic performance of C_4 plants. The optimum temperature of CO_2 fixation in C_4 plants is between approximately 30 and 40°C. As a result of the greater CO_2 affinity of PEPC, C_4 plants are capable of utilizing relatively low CO_2 concentrations. They can, therefore, perform effective photosynthesis with contracted stomata, thus limiting water losses through transpiration.

Because of these properties, C_4 plants are found primarily in warm and relatively dry areas of the tropics and subtropics. C_4 plants include, among others, many tropical grass species (including such important crop species as maize, sugar cane, sorghum, and pearl millet) and species of the Amaranthaceae (amaranth family).

In respect of photosynthetic performance, the C_4 plants have advantages over the C_3 plants at sites with high solar radiation, higher temperatures, and relatively lower water availability.

4.1.1.3 CAM-Plants

Another pathway of CO_2 utilization occurs in the CAM plants. CAM means *crassulacean acid metabolism*, because it was first discovered amongst species of the Crassulaceae (orpine family). However, this mechanism is also found in numerous other plant families—in approximately 3% of all plant species. These plants have the same two carbon-fixing steps as are present in C_4 plants, that is, initial CO_2 fixation occurs via PEPC and malate is produced, from which the CO_2 is later released and subsequently bound by RuBisCo. However, in the CAM plants there is no spatial separation of these processes, but rather a temporal separation between day and night. At night, the stomata are open to take up CO_2. As a result of the accumulation of malate, the pH inside the cells declines substantially (the term *acid* metabolism refers to this). During the day, the stomata are closed and photosynthesis is performed with the previously fixed carbon. Because photosynthesis and transpiration are largely decoupled, the CAM plants can prevent high water losses under high solar radiation. Although the advantages of the C_4 mechanism are primarily evident under conditions of high solar radiation, the CAM mechanism is, to a greater extent, an adaptation to reduce water loss. Many typical representatives of the CAM plants can therefore be found in the subtropical deserts and semi-deserts (Sect. 7.2.2.1), for example, the Cactaceae (spine plant family) and the Agavaceae. The best-known CAM crop plant is pineapple (*Ananas comosus*; Fig. 4.3), of the Bromeliaceae family.

Figure 4.4 shows an overview of the mechanisms of CO_2 fixation of the C_3, C_4, and CAM plants.

4.1.1.4 Solar Radiation in Agroecosystems

An important condition required for crop plants to achieve high biomass production and yields is optimum utilization of the available radiation energy. Because not all species have the same photosynthetic performance under the same conditions, the site of cultivation must meet the requirements of the species or variety. For production of C_4 plants, suitable sites include regions in which the optimum

Fig. 4.3 Pineapple (*Ananas comosus*)

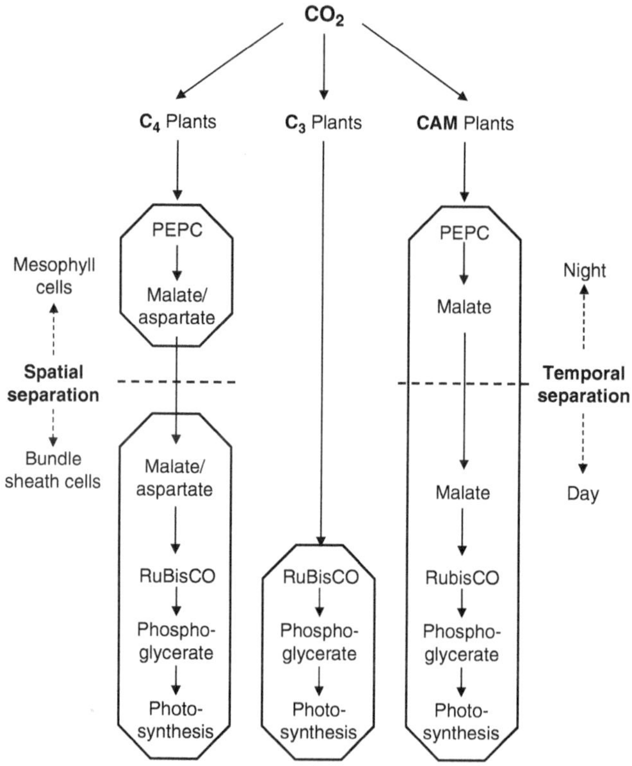

Fig. 4.4 CO_2 fixation in C_3, C_4, and CAM plants

temperatures for photosynthesis (>30°) and high radiation intensities are found. In contrast, C_3 plants have the best photosynthetic performance under relatively cool and moist conditions.

The distribution of solar radiation within a plant stand also affects photosynthetic performance. The light that reaches the plants is partially reflected by the leaves, to a large part absorbed, and partially passes through the plants. Overall, intensity is reduced as the radiation passes through a plant stand to the ground. The proportion of the radiation entering the stand that can actually be absorbed by the plants depends on the leaf area index and the architecture or type of plant. The **leaf area index (LAI)** describes the relationship between the total upper leaf area of a plant stand and the surface area of the ground of the stand (i.e., leaf area per unit land area). However, a high LAI alone does not guarantee optimum use of solar radiation, because the photosynthetic performance of plants is limited by hanging leaves and the shading of leaves within the same plant. For cereals, especially, varieties were therefore bred that better absorb solar radiation. The upper leaves of such plants stand upright so the solar radiation that enters the stand can also be utilized by leaves located further down the plant.

It is not only available solar radiation but also day length, or **photoperiod**, that affects the growing conditions of plants. The photoperiod is determined by latitude and season and is of crucial importance in inducing the formation of flowers by many plant species. The insolation ($W\ m^{-2}$) plays a much less important role, in other words, it makes very little difference to the plants whether there is sunshine or clouds during the photoperiod.

For the so-called **long-day plants**, a particular minimum of day length must be reached for the flowering of the plant. The required day length varies from species to species, but most often lies between 12 and 14 h. These conditions are met in Central Europe from approximately April to September. Important long-day crop plants include wheat, barley, potato, canola, and peas. Their sowing date must be chosen such that the flowering period falls within the long day phase. For other long-day crop plants, for example lettuce (*Lactuca sativa*), spinach (*Spinacia oleracea*), and Chinese cabbage (*Brassica pekinensis*), which are used as leafy vegetables, flowering is not desired. To prevent flowering, the crop's growing period must lie in a phase during which the days become shorter.

Flowering of the **short-day plants** is only, or primarily, induced when day length falls below a specific minimum, i.e. the days may not be longer than approximately 12–14 h. Species in this group include rice, soybean, sugar cane (Fig. 4.5), and coffee. However, there are many plant species that flower independent of day length (e.g. tomato and sunflower); these are called **day-neutral plants**. Variety-specific differences with regard to photoperiod are observed for some crops. Thus, there are short and long-day varieties of tobacco, maize, and potato.

The relationships between day length and flower formation are essentially determined by the climate of the plant's habitat. In the Mediterranean climate, where dry periods predominate in the summer, many plants flower and develop seeds in the long-day conditions of spring. Also most of the crop plants that originate from the Middle East are long-day plants. In areas of dry winters in the temperate latitudes, short-day plants are primarily found, including soybean, which originates from China. Plant species from the tropics are short-day plants and day-neutral because in these latitudes a day length of approximately 14 h is not exceeded at any point in the year.

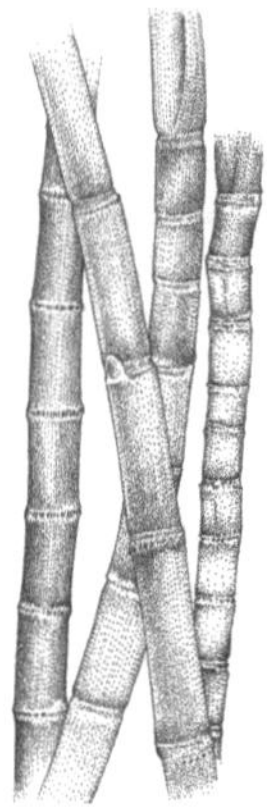

Fig. 4.5 Sugar cane (*Saccharum officinarum*)

4.1.2 Heat Energy

Only approximately 0.01% of the solar radiation that reaches the Earth's surface is used for photosynthesis. Most global radiation is absorbed by the Earth's surface and transformed into long-wave (infrared) radiation, which subsequently heats the air layers close to the ground. Some of the radiation emitted from the surface is absorbed by the atmosphere and reflected; another portion is released into space (Fig. 4.6). The reflected radiation is the so-called counter-radiation and is primarily determined by the water vapour content of the air and of clouds, and by the concentration of CO_2 and other trace gases. On clear nights with low humidity, the counter-radiation is low, which is why the ground's surface cools substantially. At the global level, the counter-radiation is responsible for the natural greenhouse effect (Sect. 8.1.3). Because of the relationship between latitude and global insolation, heating of the air by long-wave radiation is higher at the equator than at the poles. However, via air and ocean currents the energy is transported from the equator to the higher latitudes (Sect. 7.1).

4.1.2.1 Temperature and Location

The suitability of a location for the production of crops is not only affected by the average temperature of the site, but also to a great extent by the maximum and minimum temperatures. In particular, the occurrence of late frosts in spring limits production in regions that are characterized by an overall mild climate. The local temperature conditions are, therefore, not only affected by the latitude but also by additional factors. These include:

Topography. In a valley, the temperature can decline significantly more overnight than in higher locations, because of the sinking of cool, heavy air. The differences can, at temperate latitudes, amount to 10°C or more in summer.

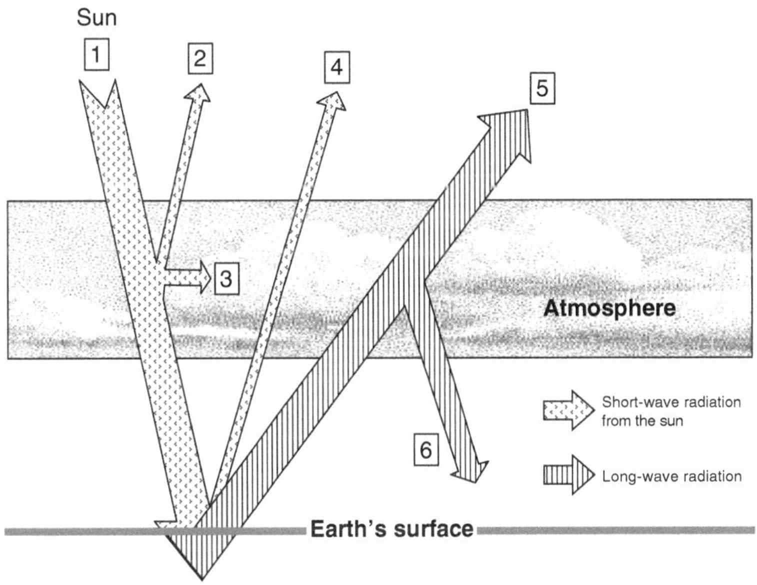

Fig. 4.6 Schematic presentation of the radiation budget of the Earth. **1**=solar radiation, **2**= reflection by the atmosphere, **3** = absorption by the atmosphere, **4** = reflection at the Earth's surface, **5** = long-wave radiation emitted from the Earth's surface, **6** = counter-radiation

Because of the threat of frost damage, valley bottoms of Central Europe are suitable for the production of specific crop plants only (e.g. fruit trees).

Altitude. There is a temperature gradient between the air layers close to the ground surface and those at higher elevations. That is why the average temperature decreases in mountainous areas with increasing altitude at a rate of 0.6–1.0°C per 100 m. In tropical and subtropical mountainous regions, it is, therefore, possible to cultivate crops that are otherwise only suitable for higher latitudes (Fig. 4.7).

Continent–ocean distribution. Oceans and continents differ in their heat budgets. Land surfaces are characterized by their strong warming in summer and strong cooling in winter. In contrast, oceans warm slowly and also cool slowly. Because of this, in winter the large land masses of the northern hemisphere are far colder than the oceans of the same latitude. In July, the opposite situation is found. The oceans have a balancing effect on the temperature of coastal regions and islands, i.e. temperature differences throughout the year are less extreme than in the continental regions at the same latitudes (Sect. 7.2.3).

4.1.2.2 Effect of Temperature on the Development of Crops

Plants are **ectothermic** (gr. e*ktos* = outside, *thermos* = warm) organisms, which means they must derive most of heat energy required for metabolism from their surroundings. Endothermic (gr. e*ndon* = inside) organisms, which include

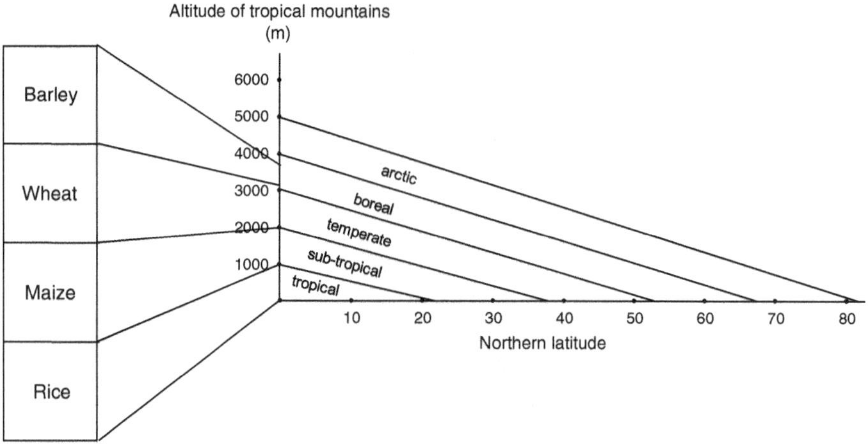

Fig. 4.7 Temperature-based suitability of different altitudes of tropical mountainous regions and different latitudes and/or climate zones for cultivation of particular crop species

mammals and birds, can maintain their body temperature by internal heat production, with a high energy turnover. The life processes of ectothermic organisms are subject to the **van't Hoff rule**. This rule states that the rate of biochemical reactions doubles with a 10°C increase in temperature (up to a point when proteins break down). As a result, a close relationship exists between the surrounding temperature and the time required by an ectothermic organism to develop (e.g. the time a plant requires to reach the flowering stage or to develop mature seeds, or that an insect requires for the completion of its larval stage). The time required for such a process is determined by the temperature sum that is reached over a period of time. For a development process to occur at all, a threshold temperature in the surrounding environment, the so-called base temperature, must be exceeded. When the temperature falls below a species-specific minimum, development in individuals of this species ceases.

The time required by an organism for its development can be expressed in degree-days (measured in °C). Degree-days (T_n) are calculated by use of the formula:

$$T_n = \sum T'$$

where n is the number of days and T' is the temperature sum reached in one day:

$$\begin{aligned} T' &= [(T_{\text{max}} + T_{\text{min}})/2] - T_{\text{b}} \\ T_{\text{max}} &= \text{daily maximum temperature} \\ T_{\text{min}} &= \text{daily minimum temperature} \\ T_{\text{b}} &= \text{base temperature} \end{aligned}$$

Table 4.1 Example of the calculation of degree-days

	Days						
	1	2	3	4	5	6	T_n
T_b (°C)	5	5	5	5	5	5	
T_{min} (°C)	8	9	6	8	7	9	
T_{max} (°C)	14	17	16	20	19	15	
T' (°C)	6	8	6	9	8	7	44

Fig. 4.8 Faba bean (*Vicia faba*)

An example of the calculation of degree-days is given in Table 4.1.

In agricultural practice, measurement of growing degree days (GDD) and temperature sums can be used to predict the period of time required between sowing and flowering or harvest for different growing periods.

Example: In the Mediterranean countries, faba bean (*Vicia faba*; Fig. 4.8) makes an important contribution to the protein supply for human and animal nutrition, and to improvement of soil fertility.

Sowing is usually performed in September or October, just before the start of the winter rainy season. On sites far from coastal areas with cold winters, for example northwest Syria, frost damage to flowering faba beans can result in significant yield losses. The threat is greatest in December and January. The risk of frost damage for particular sowing dates can be estimated by predicting the time of flowering of faba bean with the help of a simulation model that uses the concept of temperature sums, with temperature and day length data. This method shows that the risk decreases the further the sowing date is shifted into wintertime (Fig. 4.9). With sowing dates in January, the plants flower in April when there is no longer any risk of frost. As a result of the increasing temperature, the time span between sowing and flowering is significantly lower than for early sowing dates (Grenz 2004). However, with the end of the winter rains in May, the risk of yield losses increases because of water stress during the pod-filling stage. Therefore, to determine the optimum sowing date, more complex plant-development models are needed.

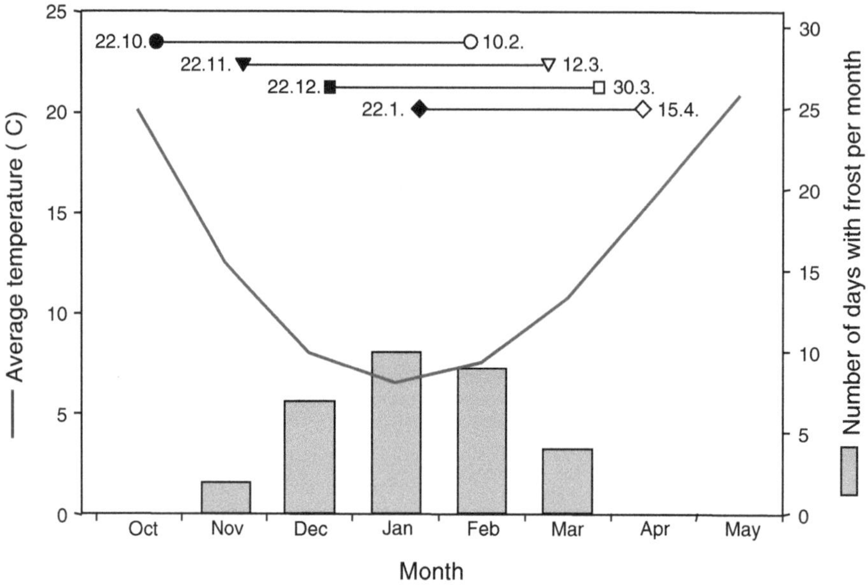

Fig. 4.9 Average monthly temperatures and number of days with frost (temperature below 0°C) per month, and the periods between sowing (*black symbols*) and flowering (*white symbols*) of faba bean in Tel Hadya, northwest Syria. The weather data are based on measurements from the years 1979 to 2000 and the flowering dates are based on computer simulation (Based on Grenz 2004)

4.2 Water

There is a close relationship between plant growth and water consumption. The **transpiration coefficient** serves as a measure of the water requirements of plants. The coefficient indicates how many litres of water are required to produce 1 kg of plant dry mass (DM). This value is species and variety-specific and is between 200 and 800 l kg^{-1} DM for C_3 plants and between 200 and 350 l kg^{-1} DM for C_4 plants.

An important criterion in the selection of a production site for a crop is, therefore, the water requirement of that species or variety. However, it is not only the precipitation conditions and the possibility for irrigation that are important for the water supply, but also particular properties of the soil which determine the soil water balance.

4.2.1 Soil Water Balance

Temporal changes in the water content of the soil, which depend on the absorption, storage, and release of water, are referred to as the soil water balance.

4.2.1.1 Infiltration

The percolation of surface water into soil is termed **infiltration**. Some of the infiltration water drains through the soil and reaches the groundwater as percolation water. Another portion, the **capillary water**, is held in the soil by capillarity and by binding to soil particles. The **field capacity** is the water content or soil moisture held by a soil after excess water has drained away and the downward movement of water has slowed. The field capacity is expressed in ml H_2O per 100 ml soil volume and depends primarily on the size of soil pores, in other words on the space that can be filled with water. Sandy soils have the lowest field capacity (because a large portion of the infiltrating water flows through the soil via the relatively large spaces). Clay soils, which have the smallest pores, have the largest field capacity (compare Sect. 4.3.1). The smaller a pore, the greater the potential with which the water is held, i.e., soil water tension. These conditions determine the ability of plants to utilize the water present in the soil. The portion of soil water that can be taken up by plants via their roots is termed the **plant-available water**. In those areas of the soil from which a root takes up water, water-free zones occur. As a result, a matrix potential gradient emerges between the connected soil pores. Consequently, water flows out of water-saturated areas of the soil into water-free zones and continuously supplies roots with water.

Specific quantities of water, which differ depending on the type of soil, are held by the solid surfaces of soil particles with greater force than can be overcome by the roots. This water, which is not available to the plants, is called **hygroscopic water**. When a plant has removed all the available water from a soil, it begins to wilt irreversibly, i.e., the **permanent wilting point** has been reached.

The movement of water through the soil of a field is shown schematically in Fig. 4.10.

4.2.1.2 Evaporation

Evaporation is the transformation of water from the fluid to the gaseous state. When this occurs from soil and water surfaces free from vegetation, the process is called **evaporation**. Evaporation from the surface of plants is called **transpiration**. In dense plant stands, transpiration is often significantly higher than evaporation. Together, these two forms of water release constitute the **evapotranspiration**. With regard to this, two aspects must be distinguished:

- The **potential evapotranspiration** indicates how high the annual evapotranspiration would be under the climate conditions of a given site, if an unlimited supply of water were available.
- The **actual evapotranspiration** is the actual annual quantity of water released to the atmosphere.

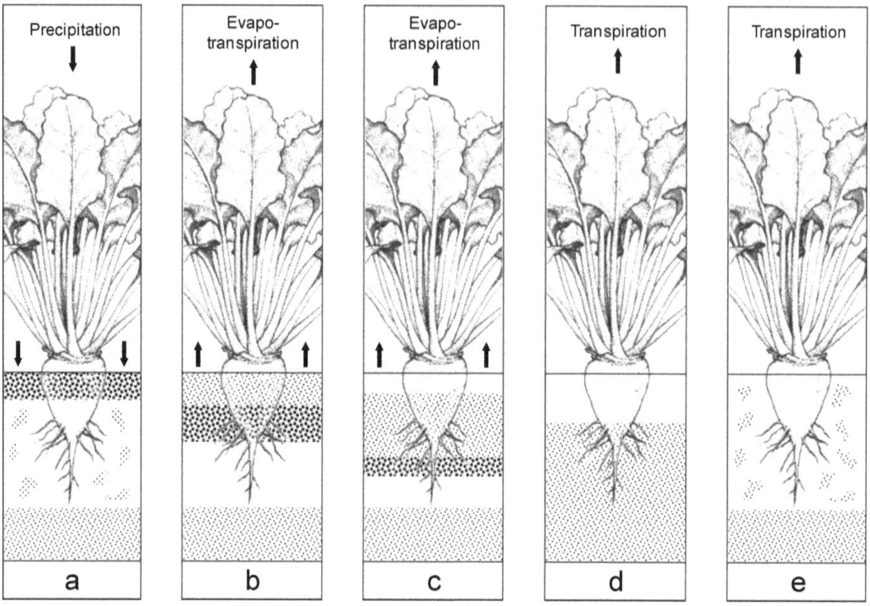

Fig. 4.10 Movements of water in an agricultural soil. (**a**) Precipitation water infiltrates a dry soil. (**b**) As percolation water, it reaches lower soil layers. Above this, capillary water is held back (field capacity). Simultaneously, water is removed from the soil via evapotranspiration. (**c**) As the percolation water moves deeper into the soil, the soil surface begins to dry out. (**d**) The percolation water reaches the groundwater, and capillary water is still present in the soil. (**e**) The plants remove the available water from the soil until the permanent wilting point is reached (Based on Gliessman 2000 and Daubenmire 1974)

Regions in which the potential evapotranspiration is higher than the average precipitation are defined as **arid**. Regions in which the quantity of water released to the atmosphere is less than that deposited by precipitation are called **humid**. In such areas, accumulation of groundwater occurs.

4.2.2 Irrigation

The climatic water balance, defined as the difference between precipitation and potential evapotranspiration, also determines the conditions of crop production in the different climate zones.

In the humid tropics, precipitation is usually sufficient for year-round production of most crop species. In regions with strongly seasonal rainfall, rain-fed crop production is limited to the phases of high precipitation, which is primarily the situation in the wet-dry tropics (Sect. 7.2.1.2) and the winter rainfall regions of the Mediterranean climate (Sect. 7.2.2.2). In regions with long dry periods or periods of limited or irregular precipitation, it is, at best, possible to utilize the water stored in the soil after rainfall for the production of crops. To increase yields, however, or even to make agricultural production possible at all in such regions, it is usually necessary to irrigate the land. In such cases, the water supply is secured via reservoirs in which rain water is captured, or from rivers, or from aquifers. The last of these may contain either renewable groundwater or fossil groundwater which accumulated during periods of high precipitation in the past and is not renewable.

Globally, the area that is irrigated has increased sixfold since the beginning of the twentieth century. Irrigation is applied to approximately 20% of the total available cropland, but is responsible for delivering more than one third of all food produced. Nearly two-thirds of irrigated crop land is found in India, China, the United States, Central Asia, and Pakistan. The most important crops grown on irrigated land are rice and wheat. Without irrigation, global cereal production would decrease by approximately 20%, so more land would be required to produce the same amount of food (Siebert and Döll 2010).

Approximately 90% of global rice production takes place on irrigated land, which is primarily found in south and Southeast Asia. Although rice can be grown under rain-fed conditions, the best production conditions are found on flooded fields. This explains the high water requirement of rice production which cannot usually be met solely by precipitation even in the humid tropics. In the wet–dry monsoon regions of south Asia (Sect. 7.1), rice production is primarily found in valley bottoms that are flooded once per year (e.g. Ganges, Mekong). In some areas of the humid tropics of Southeast Asia, terraces are used for rice production (Fig. 4.11).

On the one hand, irrigated agriculture contributes substantially to the global food supply and to the development of rural areas. On the other hand, it also contributes to major environmental problems. In many regions, more water is used for irrigation than is delivered by precipitation. As a result, in some regions of India and China the groundwater level is sinking by more than 1 m year^{-1}. This not only affects agricultural production, but also threatens natural ecosystems.

Another problem is salinization which primarily affects arid regions. In such areas irrigation results in higher rates of evaporation which leads to a gradual increase in the salt content of the soil, even if the water used has a low salt content. This process can only be prevented by sufficient drainage, by constructing drainage canals, or by installing drainage pipes, which ensures that excess water and the dissolved salts therein are drawn off.

"No irrigation without drainage" is the principle for irrigated agriculture in arid areas.

In Iraq, 20% of irrigated areas have had already to be abandoned because of soil salinization. In Pakistan, salinized soils have led to a 30% decline in yields. The salt content of soils impairs crops primarily as a result of osmotic effects (reduced water uptake), toxic effects of particular ions (e.g. chlorides and sulfates of sodium and

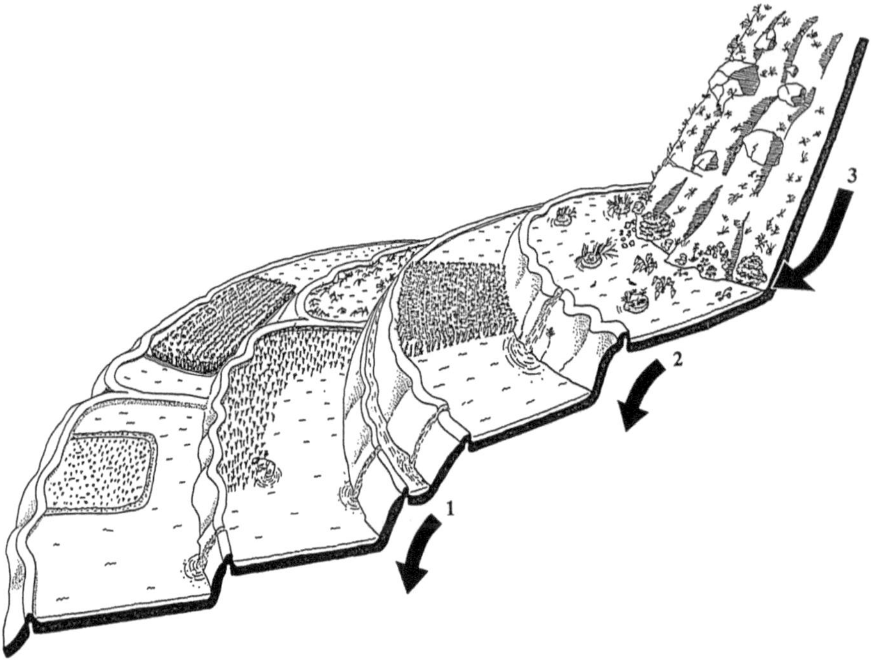

Fig. 4.11 Water supply in rice terraces of the traditional production system of Ifugao (Northern Luzon, Philippines). **1** = from diverted brooks, **2** = from field to field, **3** = with percolating spring water from the mountain side (From Martin 1994; drawn by Ch. Allgaier)

magnesium), and changes in soil properties (e.g. soil aeration and rooting ability). Individual crop species and varieties have different sensitivity to the salt concentration in soil solutions. Relatively tolerant species include barley and cotton whereas potatoes, wheat, and many legumes have low tolerance to salt. By breeding measures, including genetic engineering, attempts are being made to increase the salt tolerance of crop species.

The efficiency of irrigation methods is relatively small and amounts to, on average, less than 50%, which means that more than half of the applied water is not used by the plants, but rather percolates into the groundwater, runs off the surface, or evaporates. With special irrigation techniques, however, the quantity of water lost and, hence, the quantity of water used may be reduced. One way is to repeatedly use the water that flows out of an irrigated area for further irrigation; however, this leads to accumulation of salt in the water and thus to a reduction in its usability. Another method of water conservation is drip irrigation. In drip irrigation, pipes are laid either above or below ground, which then deliver water drop-wise to the plant roots through small openings in the pipes (sometimes nutrients are also delivered to the plants via the water, called fertigation). With such methods, water use efficiency of 95% can be achieved. The danger of soil salinization is only slight with this method because of the targeted delivery of the water. Drip irrigation is used primarily in fruit and vegetable production.

4.3 Soil

The upper layer of the Earth's crust, which is formed above the bedrock, is called **soil**. This layer is the product of the transformation of mineral materials and organic substances and is created by the effects of abiotic and biotic factors. Litho, bio, hydro, and atmosphere interpenetrate in the soil and form a unique component of the terrestrial ecosystem, the **pedosphere** (Greek *pedos* = soil; Fig. 4.12).

Soils fulfil numerous ecological functions:

- Soils are habitats for organisms. In this function, they directly serve the soil biotic community and the plants, which depend on the soil for rooting space and for uptake of nutrients and water. Indirectly, soils thus serve as the nutrient supplier for heterotrophic animal species and humans; the latter require soils for crop production.
- Soils are important in the regulation of different ecological cycles. By transformation and decomposition of organic and inorganic substances, soils significantly contribute to the cycling of carbon, nitrogen, and other elements.
- Soils serve as reservoirs, filters, buffers, and transformers of materials. An important function of soils is absorption of precipitation and its release to the atmosphere, to surface waters, and to groundwater. Related to this is the soil's ability to hold pollutants (e.g. pesticides, heavy metals) either mechanically or via adsorption by soil particles and thus to prevent or reduce the input of pollutants to water bodies. This function also includes the transformation of pollutants into non-hazardous compounds. The buffer action of soils is based primarily on the neutralization of introduced acids, which, e.g., originate from the atmosphere or fertilizers.

4.3.1 Soil Development and Soil Properties

The development and properties of soils are primarily determined and shaped by the parent material, topography or relief (e.g. slope), climate (temperature and precipitation), and organisms (Fig. 4.13). Soils are subject to constant change via transformation and translocation processes.

Rocks on the Earth's surface are subject to weathering, which means they are physically broken down into smaller pieces (e.g. by frost bursting) and chemically transformed (e.g. by oxidation and the action of acids). Through these processes nutrients are released, which enable the colonization of the rocks by organisms and the succession of communities. Subsequently, detritus, primarily from plant residues (litter), is increasingly deposited. From this litter, the activities of organisms produce the organic material of the soil, which is referred to as **humus**. Part of the humus is found as a more or less decomposed layer on top of the soil, and the remainder is

gaseous	air		
liquid	water		
solid	organic	living	underground plant parts
			soil organisms
		dead	detritus, humus
	inorganic	ions	
		minerals	

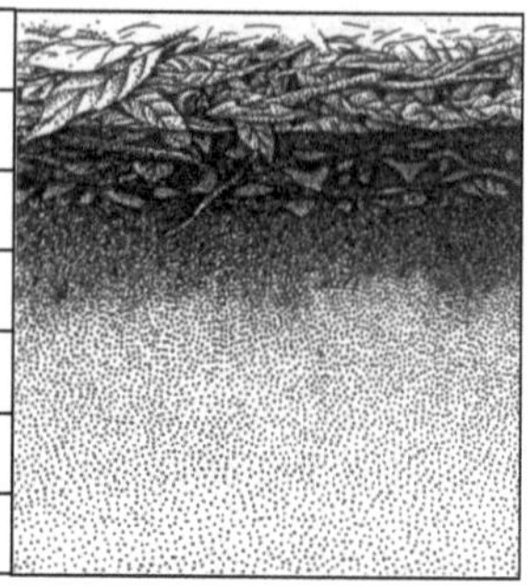

Fig. 4.12 Composition of the pedosphere

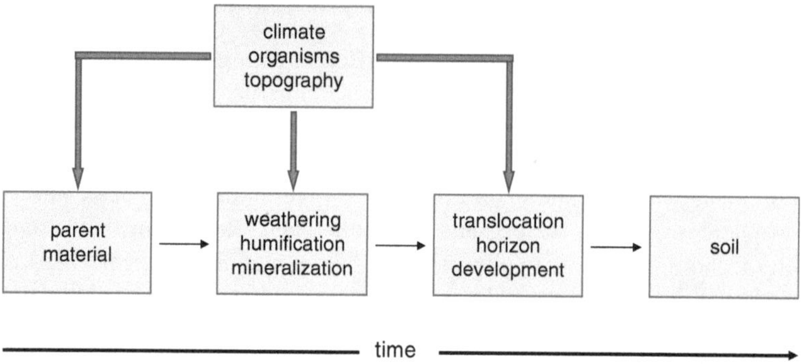

Fig. 4.13 Soil development (pedogenesis)

incorporated into the mineral soil. Depending on biotic and abiotic conditions, different types of humus are formed:

- **Mor** originates primarily from the litter of coniferous trees at relatively low temperatures and high precipitation.
- **Moder** forms primarily from the litter of deciduous forests under cool and moist conditions.
- **Mull** forms primarily from the easily decomposable litter of herbaceous plants of meadows and fields, and steppe vegetation under continental climate conditions.

Organic material that accumulates under wet conditions is called **peat** and is the raw material for peat soils.

Depending on the site conditions and the stage of development, soils depths may range from just a few centimetres to several metres. The development of deep and differentiated soils which are suitable for agriculture requires several thousand years. A fertile agricultural soil in Central Europe has a humus-rich upper layer that extends approximately 20 cm.

Table 4.2 Properties of different soil types

	Water infiltration and aeration	Water-holding and absorption capacity	Nutrient content	Root permeability	Soil condition for cultivation
Soils with a high sand content	High	Low	Low	Good	Good
Soils with a high clay content	Low	High	High	Moderate–poor	Poor
Soils with a high silt content	Medium	Medium	Medium–high	Moderate	Moderate–good

Soil development in Central Europe began at the end of the last ice age. Under moderate climate conditions, 100–300 years are required to build up a 1-cm layer of humus.

In addition to organic components, soils contain mineral materials that occur with a variety of particle sizes (grain sizes). The grains are differentiated into coarse (>2 mm) and fine soil fragments; the latter are further divided into sand (2–0.063 mm), silt (0.063–0.002 mm), and clay (<0.002 mm). The proportions and mixture of the fine soil fragments determine the **soil type**. In most soils, fragments of the different classes are found together. Depending on the proportion of each fragment class and on the soil texture, soil types are classified more specifically (e.g. loamy sand or silty clay). Loam is a mixture of the three classes sand, silt, and clay.

Depending on the mixture of fragments and their texture, soils have specific properties (Table 4.2) of different suitability for cultivation of crops. For root and tuber crops, for example potato, yam, and manioc, sandy, well-drained soils are usually favourable because the underground plant parts are less susceptible to diseases and are easier to harvest. Rice and taro (*Colocasia esculenta*) are adapted to clay-rich, waterlogged, or flooded soils. These plants have aerenchyma in their tissue (intercellular spaces or air channels which enable exchange of gases between the shoot and the root); this guarantees the supply of oxygen to the roots.

The physical properties of different soil types and the respective dominant climate conditions affect the growth conditions of plants in different ways. Sandy soils with large pores have a low water-storage capacity, which can lead to a drying of the soil in periods of low precipitation. Clay soils with small pores have a high field capacity which enables plants to avoid the danger of drought damage during longer periods without precipitation. Such conditions are found primarily in humid regions. In arid regions, the low precipitation, which occurs at great temporal intervals, often only supplies the upper layers of the soil with water. Because of different field capacities, the same quantity of water percolates deeper into sandy soils than into clay soils. By evaporation, the water at the surface of clay soils is rapidly lost whereas the moisture found deeper in sandy soils remains available longer and can be better utilized by plant roots (Walter and Breckle 1999).

In humid regions, favourable water conditions for plants are found on clay soils whereas in arid regions favourable water conditions are found on sandy soils.

4.3.2 Identification and Classification of Soils

In the course of their development, soils produce zones with different structures, which are termed **soil horizons** and are visible in a vertical section of the soil (soil profile).

The most important main soil horizons of naturally developed soils, by their position from top to bottom within the soil profile, are:

- **L** (litter) and **O** (organic) horizons above the mineral soil (organic layers of more or less decomposed plant residues)
- **A** horizon (mineral topsoil containing organic material, i.e. humus)
- **B** horizon (mineral subsoil of weathered parent material and accumulations from upper layers)
- **C** horizon (parent material, little affected by the soil-forming processes)

All of these categories can be further subdivided for more detailed description of the layer characteristics. The type, number and arrangement of horizons are criteria used for identification and classification of soils. However, there is no generally accepted system. Rather, various national and international systems based on different traditions exist for classification of soils. Important examples are:

- For the purpose of a global soil map, the FAO-UNESCO developed a soil classification system that differentiates between 30 major groups of soil (Table 4.3). The system is based on diagnostic horizons in consideration of specific characteristics and materials. All soil types and units described in this book are based on the FAO system.
- In North America, a classification system for soils (soil taxonomy) has become established that is defined by the presence or absence of defined diagnostic horizons and divides soils into 12 orders.
- In the German soil classification system, soils are primarily classified on the basis of the effect of different factors (e.g. parent material, vegetation, moisture, anthropogenic effects) on the development of the soil (Scheffer and Schachtschabel 2002).

Table 4.3 FAO-UNESCO (1988) soil classification

Major group	Characteristics	Etymology
Acrisols	Soils with subsurface accumulation of low-activity clays and low base saturation	Lat. *acris*, acidic, sour
Albeluvisols	Acid soils with a bleached horizon penetrating into a clay-rich subsurface horizon	Lat. *albus*, white Lat. *luvi*, eluviate
Alisols	Soils with subsurface accumulation of high activity clays, rich in exchangeable aluminium	Lat. *alumen*, Alum
Andosols	Young soils in volcanic deposits	Jap. *an do*, black earth
Anthrosol	Soils that have been formed or heavily modified by long-term human activity	Gr. *anthropos*, human
Arenosols	Sandy soils featuring very weak or no soil development	Lat. *arena*, sand
Calcisols	Soils with accumulation of secondary calcium carbonates	Lat. *calx*, lime
Cambisols	Weakly to moderately developed soils	Lat. *cambiere*, change
Chernozems	Soils with thick, blackish topsoil, rich in organic matter with calcareous subsoil	Russ. *cherny*, black; *zemlja*, earth
Cryosols	Soils with permafrost within 1 m depth	Gr. *kryos*, cold, ice
Durisols	Soils with accumulation of secondary gypsum	Lat. *durus*, firm, hard
Ferralsols	Deep, strongly weathered soils with a chemically poor, but physically stable subsoil	Lat. *ferrum*, iron; al from aluminum
Fluvisols	Young soils in alluvial deposits	Lat. *fluvius*, river
Gleysols	Soils with permanent or temporary wetness near the surface	Rus. *glej*, moist, heavy soil
Gypsisols	Soils with accumulation of secondary gypsum	Gr. *gypsos*, gypsum
Histosols	Soils composed of organic materials	Gr. *histos*, tissue
Kastanozems	Soils with thick, dark brown topsoil, rich in organic matter and calcareous or gypsum-rich subsoil	Lat. *castanea*, chestnut
Leptosols	Very shallow soils over hard rock or in unconsolidated very gravelly material	Gr. *leptos*, thin/ shallow
Lixisols	Soils with subsurface accumulation of low-activity clays and high base saturation	
Luvisols	Soils with subsurface accumulation of high-activity clays	Lat. *luvi*, eluviate
Nitisols	Deep, dark red, brown or yellow clayey soils having a pronounced shiny, nut-shaped structure	Lat. *nitidus*, shiny
Phaeozems	Soils with a thick, dark topsoil rich in organic matter and evidence of removal of carbonates	Gr. *phaios*, dark grey
Planosols	Soils with bleached, temporarily water-saturated topsoil on slowly permeable subsoil	Lat. *planus*, flat
Plinthosols	Wet soils with an irreversibly hardening mixture of iron, clay, and quartz in the subsoil	Gr. *plinthos*, brick
Podzols	Acid soils with blackish/brownish/reddish subsoil with alluvial iron–aluminium–organic compounds	Russ. *pod zola*, under ash
Regosols	Soils with very limited soil development	Gr. *rhegos*, cover
Solonchaks	Strongly saline soils	Rus. *sol*, salt
Solonetz	Soils with subsurface clay accumulation, rich in sodium	Rus. *sol*, salt
Umbrisols	Acid soils with a thick, dark topsoil rich in organic matter	Lat. *umbra*, shadow/ dark
Vertisols	Dark-coloured cracking and swelling clays	Lat. *vertere*, turn

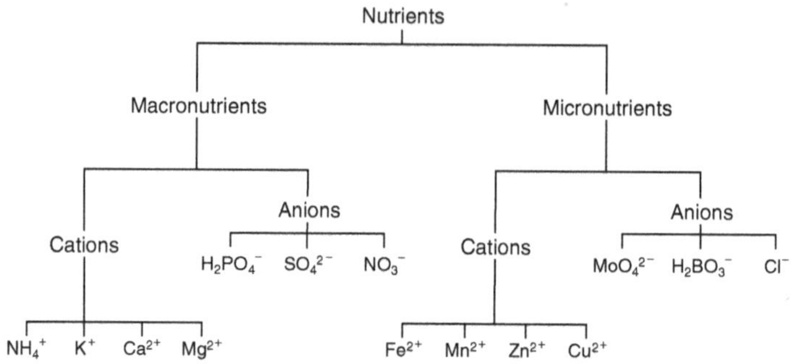

Fig. 4.14 Macro- and micronutrients of plants in the forms in which they are taken up by the plants

4.3.3 Plant Nutrients

In addition to sun's energy, CO_2, and water, mineral nutrients are also necessary for the growth and development of plants. These must be taken up by the plants as ions, and are usually drawn from the soil solution via the plant roots.

Plants have a relatively high requirement for the elements nitrogen (N), phosphorus (P), potassium (K), sulfur (S), calcium (Ca), and magnesium (Mg). These are the **macronutrients**. The elements iron (Fe), manganese (Mg), boron (B), zinc (Zn), copper (Cu), molybdenum (Mo), and chlorine (Cl) are required in smaller amounts and belong to the **micronutrients** (Fig. 4.14). Together, these nutrients are the essential (necessary for life) nutrients. They act either as components of organic substances or are required for particular physiological processes. Other elements, for example sodium (Na), selenium (Se), cobalt (Co), and nickel (Ni), are important for at least some plants as **trace elements**.

In agroecosystems, nutrients are rarely present in balanced proportions and in the necessary quantities. Not only a deficit but also excess of an essential nutrient can negatively affect the development of a plant. In addition to the essential nutrients, other elements are also found in the soil, in particular such heavy metals as cadmium (Cd), lead (Pb), mercury (Hg), and chromium (Cr). At high concentrations these elements are toxic to organisms because they inhibit physiological processes and the activity of enzymes.

When all essential nutrients are available in the quantities necessary for plants, development of the plant proceeds regularly. As soon as the plant is lacking one of these nutrients, signs of deficiencies will emerge and growth will be depressed. On the basis of this observation, Justus von Liebig (1803–1873) formulated the **Law of the Minimum**. It states that plant growth is controlled not by the total amount of nutrients available, but by the scarcest nutrient. In other words, yield is proportional to the amount of the most limiting nutrient, whichever it may be.

4.3.3.1 Availability of Nutrients

In the soil, the nutrients required by the plants are predominantly present in diverse bound forms, from which they are released by the following processes:

- bound in minerals of the parent material and released by weathering of the parent material,
- organically bound in detritus and humus and released during decomposition by soil organisms,
- adsorbed on soil colloids and released by ion exchange (see below).

Not only the quantities of nutrients in the soil, but also spatial and temporal availability are crucial in the growth and development of plants.

Because nutrients must be taken up by plants from the soil solution, they are only usable when the root zone is sufficiently moist. Lack of water can therefore limit uptake of nutrients by plants. The clay and humus content of the soil also has a significant effect on nutrient availability. Of greatest importance are particles with sizes less than 0.001 mm (**soil colloids**). In interaction with calcium, these also form the so-called clay–humus complexes. The surfaces of clay colloids have electrostatic negative charges on which cations (positively charged particles) from the soil solution become attached (adsorption). On humus colloids and oxidized particles, positive charges are found on which anions (negatively charged particles) are adsorbed. Conversely, release of adsorbed ions to the soil solution may occur (desorption) by displacement or exchange of ions. Ion exchange can occur only between ions of equivalent charge, for example 6 H^+ for 3 Mg^{2+} or 3 Ca^{2+}.

The ions that may be exchanged on colloids, which are, in part, of importance as plant nutrients, include the cations Ca^{2+}, Mg^{2+}, K^+, Na^+, H^+, and Al^{3+}, and the anions PO_4^{3-} and SO_4^{2-}. There are almost no specific binding sites for NO_3^- and Cl^-, which is why these ions are most often found in the soluble form.

Ion exchange in the soil, which is also of great importance to the nutrient status of plants, involves exchange processes between colloids, the soil solution, and the roots (Fig. 4.15). Roots release ions, primarily H^+ and HCO_3^-, which are formed during cell respiration, into the soil solution. This enables uptake of an equivalent quantity of nutrient ions. As a consequence, further ion-exchange processes occur between the soil solution and the soil colloids.

The total quantity of exchangeable cations and anions, in other words the maximum amount of ions that may be adsorbed in the soil, is termed the exchange capacity. In clay and humus-rich soils, largely cation exchange sites are present, and anion exchange sites are of minor importance only. Because many important plant nutrients are cations, usually only the **cation-exchange capacity (CEC)** is used as a measure of nutrient supply for plants. The CEC is expressed in charge

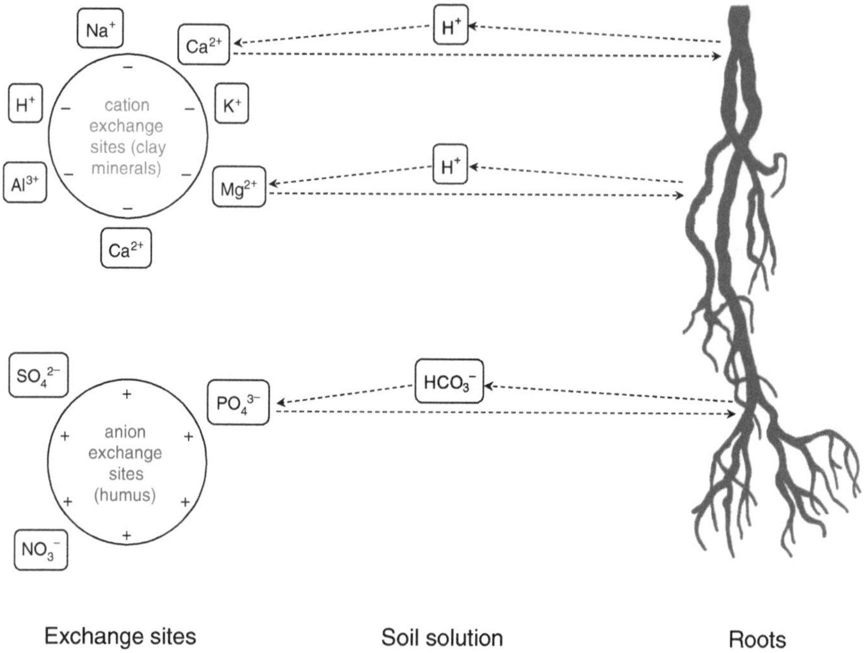

Fig. 4.15 Ion-exchange processes in the soil

equivalents or milligram equivalents (mol_c or $centimol_c = cmol_c$) per kilogram dry soil. The higher the CEC, the better is the ability of the soil to hold and exchange cations. The CEC is primarily dependent on the soil's content of mineral and organic exchange sites and, in most soils with approximately 2–3% humus, this lies between 5 $cmol_c$ (sandy soils) and 40 $cmol_c$ (clay soils).

The amount of Ca^{2+}, Mg^{2+}, K^+, and Na^+ ions with an alkaline effect on the exchange sites is termed the **base saturation**. The remaining amount comprises H^+ and Al^{3+} ions, which determine soil acidity (see below). A base saturation of, for example, 60% means that 60% of the ions adsorbed by the exchange sites are Ca^{2+}, Mg^{2+}, K^+, and Na^+ ions, and 40% are H^+ and Al^{3+} ions.

4.3.4 Soil Reaction (pH)

The acidic and alkaline effects of the soil solution are termed the **soil reaction**. This depends on the quantity of acidic H^+ ions and alkaline OH^- ions and is measured, on the basis of hydrogen ion concentration, as the **pH value** [Latin p = *pondus* (weight), H = *Hydrogenium* (hydrogen)]. pH is defined as the negative logarithm of the H^+ ion concentration in grams (g) per litre (l) solution (pH $x = 10^{-x}$ g H^+ l^{-1}). pH 7 means that 10^{-7} g H^+ ions are found in 1 l of soil solution. Because the quantity of OH^- ions is the same in this case, the soil solution is said to be neutral at pH 7. The pH scale ranges from 0 to 14. The higher the H^+ ion concentration, the lower the pH. When the pH is below 7, the soil solution is acidic; when the pH is more than 7 the soil solution is alkaline (basic). The range of pH values in Central Europe ranges from approximately 3 to 8. Most soils are slightly acidic (around pH 5–6.5).

pH significantly affects the amounts of nutrients adsorbed, and thus their availability. The availability of most nutrients is highest under slightly acidic to neutral conditions (pH 5.5–7), which means that this range may also be regarded as optimum for most agricultural soils. Under nearly neutral conditions, the effect of the pH is also particularly favourable for soil structure, i.e. high Ca^{2+} and Mg^{2+} content. In strongly acidic soils (pH < 4.5), high concentrations of Al^{3+} ions are found in the soil solution, which may damage plants (aluminum toxicity). Low pH also negatively affects soil bacteria (see below). Legumes are directly affected by this, because they live in symbiotic relationships with nitrogen-fixing bacteria (Sect. 3.7.4.1), which are impaired by acidic conditions.

pH affects many soil properties, and thus the growth conditions of plants and the living conditions and performance of soil organisms.

An increase in soil acidity (acidification) can be caused by a variety of factors. Widespread acidification and loss of bases has occurred in the upper soils of Central Europe in recent years, largely as the result of acid rain (Sect. 3.7.4.2). A decrease in pH can also result from development of acidic humus materials from litter, and as a result of the respiration of soil organisms, including soil roots (production of carbonic acid: $CO_2 + H_2O \leftrightarrow HCO_3^- + H^+$)

In agricultural soils, specific fertilizers (e.g. ammonium sulfate, liquid manure, superphosphate) can also lead to a decline in pH.

The extent of the effects of these on soil pH depends primarily on lime content, cation-exchange capacity, and base saturation on the exchange sites. The higher these are, the better is the buffer capacity of the soil, i.e., the soil's ability to bind H^+ ions and thus counteract soil acidification. This ability is, in turn, affected by the alkali content of the parent material, which is particularly high for carbonate rocks (calcite and dolomite). By application of calcium and/or magnesium carbonate, the buffering capacity of the soil can be improved, in other words the pH of acidic soils can be increased.

4.3.5 Soil Organisms

All the organisms living in the soil are known as the *Edaphon* (Greek: living in the soil). The most important function of the soil biological community is decomposition and transformation of dead remains of organisms in terrestrial ecosystems. These processes include the development of humus and the release of organic and inorganic compounds, including plant nutrients.

4.3.5.1 Composition and Function of the Soil Community

Traditionally, soil organisms are divided into two main groups:

1. The **soil flora** primarily consists of bacteria and fungi, which are the smallest organisms of the soil community and are also termed microflora. The soil flora also includes algae and lichens, and, in the broadest sense, subterranean plant organs.
2. The **soil fauna** is composed of the animals in the soil community, which are subdivided according to body size:

   - The microfauna primarily includes the protozoa (single-celled animals). Their most important groups in the soil are amoebae, ciliates, and flagellates.
   - The mesofauna comprises the smallest representatives of the soil arthropods, of which the most important groups are the Acari (mites) and Collembola (springtails). In addition, Nematodes (roundworms or eelworms) and the Enchytraeidae belonging to the Annelida (segmented worms) are also included in the mesofauna.
   - The macrofauna especially includes the Lumbricidae (earthworms), some insects, Chilopoda (centipedes), Diplopoda (millipedes), Isopoda (woodlice), some Arachnida (spiders), and Gastropods (snails and slugs).
   - The term megafauna is primarily applied to mammals, for example voles (Arvicolidae) and moles (Talpidae), which interact with the soil by construction of burrows and tunnels.

Little is known about the diversity of the Edaphon. It can be assumed that in most ecosystems the number of species living in the soil is several times the number of species found above-ground. This is primarily because of the species diversity of bacteria and fungi that live in the soil. Currently, probably only 10% of all species of the soil community have been described (Hawksworth and Mound 1991).

Beside classification according to size, the organisms of the soil community can also be divided into functional groups, which are defined by their position in the food web, in other words by their source of food. The major groups are, thus, detritivores, bacterivores (feed on bacteria), fungivores (feed on fungi), phytophages (primarily feeders on roots), predators, and omnivores. Figure 4.16 shows the food web of an agricultural soil, on the basis of this classification.

The most important resource on which the soil food web builds is dead organic material (detritus), which is primarily made up of plant residues (above-ground litter and dead root materials). Under the action of different groups of detritivores, the process of the transformation of organic materials is as follows:

- **Wash-out phase**: First, the readily soluble molecules, primarily sugars, are washed out by precipitation. These can then be used directly by microorganisms.
- **Break-down phase**: Soil animals, for example earthworms, roundworms, and springtails transform the remaining material by breaking it down into smaller pieces, ingesting it, and excreting most of it.

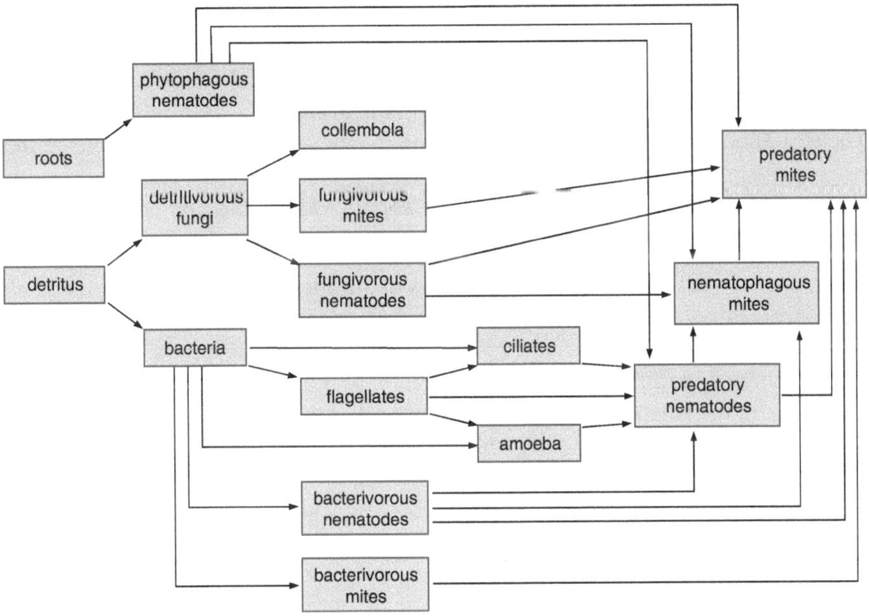

Fig. 4.16 Food web of the soil biological community in an intensively cultivated field of winter wheat in the Netherlands (Based on Moore and De Ruiter 1991)

- **Microbial phase**: The size-reduced and transformed residues are readily susceptible to the enzymes of fungi and bacteria and are used by them as a source of energy and nutrients. Thus, some of the material is decomposed into inorganic products in a relatively short period of time (between months and a few years). This process is called **mineralization**. In addition to CO_2 and water, diverse mineral nutrients are released in the process, and are then available to plants. Another portion of the materials (e.g. lignin, cellulose, resins, waxes, chitin) is decomposed more slowly and is transformed by the microorganisms into new organic intermediate products. Such **humic substances** can remain in the soil for hundreds of years.

The rate and extent of decomposition of organic substances is affected by, in addition to the abiotic conditions (including temperature, moisture, pH), primarily the composition of the materials themselves. An important factor is the carbon-to-nitrogen ratio of the material. The lower the **C/N ratio**, the faster is the decomposition, because then the organisms have a relatively large quantity of nitrogen compounds available for food. The litter of herbaceous plants has a C/N ratio of approximately 20–30:1. The autumn leaf litter of many trees often has a ratio of approximately 40–60:1; for straw it is 60–100:1. The rate of decomposition decreases in the same order. Coniferous litter is, because of its high resins content, very hard to decompose.

4.3.5.2 Factors Affecting the Soil Community

Most of the biomass of soil organisms is usually fungi and bacteria. Among the representatives of the soil fauna earthworms and protozoa usually predominate. The total biomass, and the proportions of the individual groups of organisms in the community, can, however, vary substantially in different soils. In addition to climatic conditions, the following factors strongly affect the composition of the soil community:

Pore spaces in the soil. Most representatives of the microflora are found in soil pores as small as 1–25 μm in diameter, and many species of microflora and mesofauna require pores with a diameter of 25–100 μm (1 μm = micrometre = 10^{-6} m = one thousandth of a millimetre). The composition of the soil community shifts with a decline in the pore size of a soil, to the benefit of the organisms with small body sizes, and therefore varies with soil type.

Water and air content of the soil. The availability of water is a precondition for the existence and activity of soil organisms. They also require oxygen, however, so water and air, in balanced quantities, must be present in the soil. The ideal conditions for microflora are when approximately 60% of the pore volume is filled by water. Many organisms (e.g. bacteria, protozoa, and nematodes) require a water film on the soil particles to be able to move. Extreme deviations from the balanced conditions (water saturation or drought) have a negative effect on the entire soil community.

Soil pH. Bacteria biomass is usually highest in neutral or slightly acidic soils and declines with more acidic or alkaline pH values. The reason for this is primarily that the activity of many bacterial enzymes is dependent on pH. This relationship is usually not evident for fungi and the soil fauna. Some of those species reach their highest density in relatively acidic sites whereas others reach this level in sites with neutral pH (Fig. 4.17).

Vegetation. The composition of the vegetation of a site determines the quantity and quality (e.g. C/N ratio) of the litter. This material serves as the food resource of the detritivores and thus affects the structure of the soil food web. For many soil organisms, the living biomass of plant roots is also important. This is true not only for plant feeders but also for fungi (mycorrhiza, Box 3.2) and bacteria, which can use root excretions (especially amino acids and sugars) as a food source (cf. Sect. 3.7.4.1). In addition, the vegetation has indirect effects on the environmental conditions of the soil, which can include effects on water balance (e.g. by shading, and water use of the plants) and on pH.

In agroecosystems, the soil organisms and their activities are also affected by the following factors:

Fertilization. The nutrient content of agroecosystems is directed by natural processes in only a limited way. Because a large part of the produced biomass is removed with harvest, the quantity of organic material that originates from plants is usually less than that found in natural ecosystems. The nutrient supply of plants is, instead,

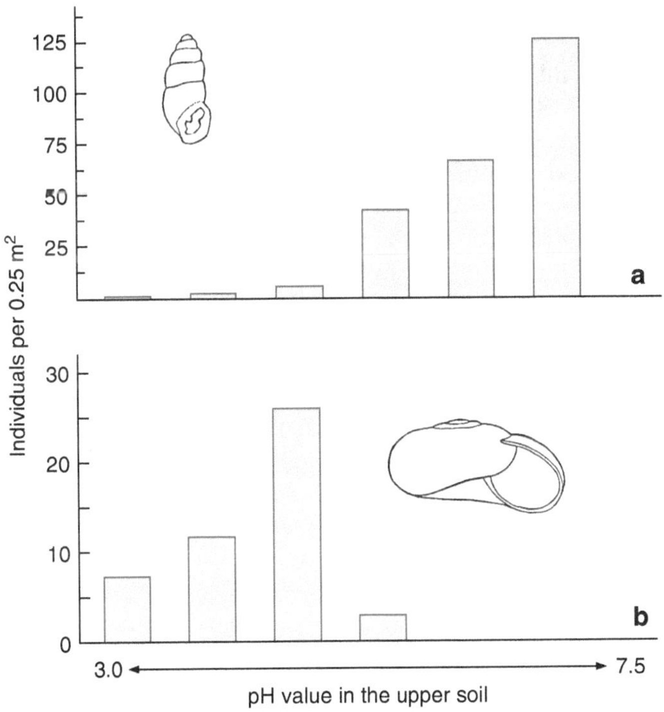

Fig. 4.17 Relationships between the pH of the upper soil and the number of individuals of two snail species in woodlands of southwest Germany. (**a**) *Carychium tridentatum*, (**b**) *Nesovitrea hammonis* (Based on Martin and Sommer 2004; drawings by Ch. Allgaier)

met with mineral or organic fertilizers. The qualitatively and quantitatively different resource supply of agroecosystems, in comparison with natural systems, can have diverse and complex effects on the biomass and composition of the soil community, on the abundance of individual groups of organisms, and on the structure of the soil food web. No broad generalizations can be made about the extent of such changes; they are affected not only by the type and quantity of fertilizer used but also by the individual abiotic site conditions.

Soil cultivation (Box 4.2). Ploughing causes inversion of the soil, which changes the soil structure and the soil microclimate conditions (temperature and moisture). These factors affect the organisms of the soil community in different ways. Negative effects result for many soil animals found in the soil (e.g. earthworms and some insect larvae) or on the soil's surface (e.g. ground beetles and some species of spider). On the other hand, the activities of the microorganisms and, thereby, the decomposition of humus are promoted by ploughing. In the long term, this type of soil cultivation therefore leads to a loss of organic matter in the upper soil. As a result of regular ploughing a plough pan may also form in the soil. The associated decline in pore volume, with changes in soil water content, also affect the living conditions of soil organisms.

Box 4.2 Soil Cultivation

Conventional tillage involves ploughing, whereby compact layers of the upper soil are broken up and turned upside down. Ploughing improves plant root penetration, water infiltration and storage capacity, and soil aeration. Ploughing is also a means of incorporating harvest residues and contributing to weed management (Sect. 5.1.3). After this so-called primary cultivation, the seed bed is prepared in which the soil material is broken down into smaller pieces and levelled with a variety of equipment. On this prepared area, sowing is conducted. However, this process is accompanied by several disadvantages, primarily negative effects on the soil fauna, loss of humus, increased risk of erosion, and development of a ploughpan.

As an alternative to conventional soil cultivation, **conservation tillage** can be used to avoid or minimize the disadvantages mentioned. In contrast with the effect of the plough, conservation tillage does not turn harvest and green manure residues into the soil, instead these residues are mixed with the soil close to the surface. This is achieved by use of equipment for soil-loosening (e.g. a cultivator), rather than soil-turning. Thus, the structure of the soil is maintained, in other words "conserved".

An additional form of tillage in agroecosystems is **direct seeding or no-till**, in which there is no soil tillage. Seed bed preparation and mechanical weeding are also omitted. Sowing solely involves creation of a furrow in the soil. In direct seed systems, the soil is permanently covered with plant residues, which can hinder the even distribution of seeds.

Conservation tillage, and to a greater extent direct seeding, have positive effects on some properties of the soil. In particular these include conservation or increase of the organic matter content of the soil, avoidance of soil compaction, and reduction of soil losses via erosion. Furthermore, with both methods the costs of machinery use and energy, and labour requirements, are lower than with conventional soil tillage. Because no-till and conservation tillage methods are only rarely accompanied by declines in yield, they are increasingly being used.

Pesticides and herbicides. The chemical and biological substances used to protect crops from pests, phytopathogens, or weeds not only affect the target organisms but can also often directly or indirectly damage species of the soil biological community.

4.3.5.3 The Importance of Humus in Agroecosystems

In principle, it is possible to replace the soil, as rooting space for plants, with a synthetic substrate (e.g. rock wool or coconut fibres) and grow such plants as tomatoes, cucumbers, or courgettes in a nutrient solution that is produced to meet the requirements of the plants. This soil-free method of production is used in greenhouses where, under controlled temperature and light conditions, production can be conducted irrespective of season. Complete de-coupling of plant production from natural nutrient cycles, including the delivery of organic materials and their transformation (the release of nutrients and the building of humus) by soil organisms, should not be the intention of farming, nor is it achievable. A nutrient-rich, entirely mineral soil is only of limited use for agricultural purposes because particular functions, which depend on the delivery of organic materials and the activity of soil organisms, are not, or only partially, fulfilled. Humus, which in agricultural soils originates from harvest residues (roots, straw, leaves) and, eventually, organic fertilizers (green manure, manure, liquid manure, compost), is of crucial importance in these functions and to **soil fertility**, which means the ability of the soil to facilitate plant growth and produce yields, i.e. the productivity of a site. In addition to organic substance, soil fertility is also determined by diverse abiotic factors (e.g. parent material, climate, weathering, water balance). Depending on the quality and quantity of humus, other soil properties that are important in agroecosystems are, in turn, affected. These include root penetration, ease of tillage, field capacity, rate of infiltration, aeration, and soil erodibility. Conditions suitable for high agricultural productivity are found in a soil which has approximately 45% mineral components, 5% organic substances, and a pore volume of approximately 50%.

Important agricultural measures for conservation and promotion of the humus content of an agricultural soil are:

- providing the soil with a sufficient supply of organic substances to compensate for humus loss;
- balanced and need-based application of mineral and organic fertilizers, pesticides, and herbicides;
- site-adapted crop rotation, as diverse as possible; and
- site-adapted and appropriate soil tillage that avoids soil compaction and erosion.

4.3.6 Soil Erosion

Soil erosion is removal of soil materials by water or wind. In principle, soil erosion is a natural process; it may, however, be strengthened or initiated in particular sites by some types of land use.

The extent and the form of **water erosion** (Fig. 4.18) depend primarily on:

- the intensity, frequency, and duration of rainfall. With increasing precipitation and energy of the raindrops, the effect on the soil surface increases (rain-splash erosion).
- the water-absorption capacity of the soil, which is affected by the infiltration rate, the water transfer capacity, and the current storage capacity of the soil (in other

Fig. 4.18 Symptoms of erosion in a savanna landscape

words the pore volume filled with air). Compacted zones in the upper and lower soil (e.g. ploughpan) and high water saturation of the soil increase the surface run-off and thus also increase the quantity of transported soil materials.

- the slope angle and the topography. The steeper and longer the slope, the higher the velocity of the surface runoff water, as a result of which increasingly large fragments may be transported. This can cause rill erosion (small channels) and, eventually, gully erosion (larger channels, too big to be removed by tillage). Measures to reduce water erosion on slopes include establishment of vegetation strips or dams along the contour line (a line that connects points of equal elevation on the surface of a slope) or terracing of the slope.
- the soil properties. Soil with high proportions of medium-sized grains (fine sand and silt) can be transported by water with relatively low velocities. Another factor is the aggregate stability of the upper soil, which is higher the better the organic and mineral components of the soil are mixed and the more strongly the fine mineral grains (sand, silt, clay) are stabilized by iron oxide.
- the soil cover. Vegetation is extremely important in the prevention of erosion, if coverage is sufficient. To achieve effective erosion prevention, agroecosystems should have at least 70% of the soil surface covered with plants (crops, cover crops, etc.). If coverage is less, for example immediately after sowing and when row crops (e.g. maize) are grown, soil protection can be achieved by conservation tillage, direct seeding, and mulching.

The extent of **wind erosion** is affected by wind speed, soil properties, and soil coverage, but in contrast with water erosion, less so by topography. Soils that are most strongly affected by wind erosion include dry, finely sandy and silty soils, with low vegetation cover. Wind erosion is of greatest importance in deserts and arid areas, in the wet–dry tropics, and in the winter–humid subtropics (Chap. 7) during the dry season. In the Ice Ages, wind erosion led to the deposition of loess (Sect. 3.5.2). Under humid climate conditions, wind erosion is of minor importance in the loss of soil.

In the last few decades, almost one third of agricultural land worldwide has been lost to wind and water erosion. The losses increase by more than 10 million hectares per year, which amounts to 75 billion tons of soil. Because nutrients are also lost with the soil, significant decreases in yield may result. In most agricultural sites, the losses range between 13 and 40 tonnes of upper soil per hectare per year. These are far higher than the natural rate of soil development, which is approximately 0.5–1 tonnes per hectare per year (Pimentel et al. 1995; Pimentel and Kounang 1998).

4.4 Crop-Associated Flora

In addition to the cultivated crop plants, other plant species usually exist in agroecosystems; these, as a whole, constitute the **crop-associated flora**, which consists primarily of wild plants, i.e., naturally occurring species that find suitable living conditions in agroecosystems. Most of these plants are typically pioneer species of open sites. Approximately three quarters of all wild plant species of the agricultural fields of Central Europe are annuals, which means they complete a full life-cycle from germination to seed formation within 1 year and usually have high growth rates (dry mass production per unit time). Not only wild plants, but also crop plants can be a component of the associated flora. This is especially true for crop-rotation systems, in which crop plants emerge from the previous stage of the rotation (e.g. canola from the previous stage emerging in wheat).

Crop-associated plants that cause more damage than they provide benefits in a crop stand, are termed **weeds**.

Beneficial properties of particular associated plants include protection of the upper soil from erosion (Sect. 4.3.6) or promotion of antagonists of pests (Sect. 5.2.4.5).

4.4.1 Competition Between Crops and Weeds

The damaging effects of weeds on crop plants are primarily because of interspecific competition for resources, namely light, nutrients, and water. As a result, the growth and development of crops and, finally, yields are negatively affected.

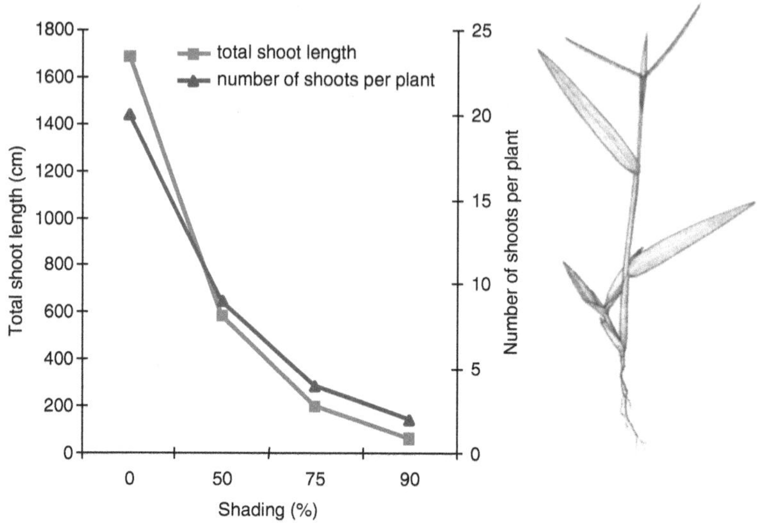

Fig. 4.19 Effect of shading on the number of vegetative shoots per plant and the resulting total shoot length of *Paspalum conjugatum* (greenhouse experiment of 80 days)

4.4.1.1 Competition for Light

Lack of light leads to a reduced rate of photosynthesis and thus to reduced primary production. Within a plant stand, competition for light is the result of shading. Its effect is primarily determined by the growth characteristics and growth form of the species in competition. In agroecosystems, the highest yield losses because of light competition are caused by weeds which are capable of out-growing the crops and thus shadow the crop plants.

Crop plants can also cast shade, however, which can result in the suppression of weed growth (Fig. 4.19). The growth form of crop plants, and their size and the position of their leaves, both affect the extent of self-shading and the effect on other plants.

Further factors of substantial importance in light competition are plant density and the spatial distribution of crops. Ideal conditions for suppression of weed growth by shading are close and even spacing between crops. An increase in planting density is, however, limited by intraspecific competition of the crops for light, water, and nutrients. With crops for which relatively wide spacing or row distances must be maintained because of harvest technology (e.g. maize, potato, sugar beet), it is not possible to achieve a complete shading of the soil surface, at least not during the early stages of development.

An example of a broad-leaved weed with a high growth rate is lamb's quarters (*Chenopodium album*, Fig. 4.20), the most common weed species found in Central European sugar beet production and also of great importance for many other crops. Another property of wild plants that enables them to compete for light against crop plants is a climbing growth habit. Important climbing weeds include stickyweed

Fig. 4.20 Lamb's quarters (*Chenopodium album*)

Fig. 4.21 Stickyweed (*Galium aparine*)

(*Galium aparine*; Fig. 4.21), climbing bindweed (*Fallopia convolvulus*), and vetches (*Vicia* species).

4.4.1.2 Competition for Nutrients

In agroecosystems, competition of plants for nutrients is not only determined by natural soil fertility, but also by fertilization. If and to what extent the competitive relationships between crops and associated plants are affected by these factors depends not only on the quantity of fertilizer applied and plant density, but also on the properties of the different species. Not all plant species have the same ability to efficiently utilize a high supply of nutrients (Sect. 3.3.1). Among weeds, there are species that can gain an advantage in growth and competition over crop plants or other associated species as a result of nitrogen fertilization. Examples are sticky weed (*Galium aparine*), twitch grass (*Alopecurus myosuroides*), and common wild

Fig. 4.22 Wild oat (*Avena fatua*)

oat (*Avena fatua*; Fig. 4.22). Nitrogen fertilization can lead to a decline in the yield of wheat when *A. fatua* covers more than 1.6% of the total area of the plant stand (Carlson and Hill 1985). There are, however, weeds upon which nitrogen fertilization has negative effects. These include primarily legumes, which are represented among the associated flora by the vetches.

4.4.1.3 Competition for Water

The importance of water as a factor of competition in agroecosystems is determined primarily by the quantity and distribution of precipitation during the growing period; it is, therefore, closely related with the prevailing climate conditions. Other important factors are the water-storage capacity of the soil and the ability of the plants to utilize the water stores. Species that have a large root system or a tap root can extract water from a greater depth than those with only a shallow root system. This is of particular importance when the upper soil layers dry out during temporary lack of precipitation. Primarily young plants, which, because of their weakly developed root systems, are dependent on water close to the soil surface, have disadvantages in such situations against deep-rooted weed species.

If and to what extent yield losses occur as a result of competition with weeds depends not only on resource availability, but also on the ability of the respective crop and weed species to utilize these resources.

4.4.1.4 Combined Competitive Effects

In many situations, the results of competition between crops and associated plants depend on the combined effects of different factors:

- Lack of light not only affects shoot growth but also affects development of the root system, because fewer photosynthetic products are available. This in turn has negative effects on the ability of the plant to utilize water and nutrients and therefore on its competitive ability with regard to these resources.
- Lack of water leads to a decline in photosynthetic performance because the plant contracts or closes stomata to prevent high transpiration loss; it can, therefore, take up less CO_2. Disadvantages also result from this in competition for other resources.
- Increased nutrient availability can affect the competition for light between plants in a stand. Nitrogen fertilization promotes the development of more above-ground biomass, leading to more shading by the plants. This can, on the one hand, lead to greater competitive success of the crop against weeds but, on the other hand, if they cast more shade on each other it can also lead to greater intraspecific competition between crop plants.
- Another effect of fertilization can emerge when the crop-associated plants also benefit from the greater availability of nutrients and thus produce more biomass despite more shading. In this case, a decrease in nutrient competition has a greater effect than an increase in light competition.

4.4.1.5 Particular Situations of Competition

In an agroecosystem, more than one associated plant species is usually present, which results in diverse and complex competitive effects on the crop species. In such a situation, it is hardly possible to determine to what extent the individual species contribute to the observed effects. Experiments have shown that particular associated species have very different effects on yield losses, depending on the crop species.

Example: The stiff ryegrass (*Lolium rigidum*; Fig. 4.23), which originates from the Mediterranean region, is an important weed in a variety of crops of southern Australia. Field experiments have shown how the presence of this species affects the yields of eight crop species. At a density of 300 *L. rigidum* plants m^{-2}, the cereal species oats, rye, and triticale had the highest competitive ability, which means that these crops maintained acceptable grain yields while at the same time suppressing growth and seed production by the weed. In contrast, the grain legumes tested, lupin and field pea, were poor competitors and produced no yields under the same conditions. These results show that some crop species have greater competitive ability than others against particular weed species. Depending on the variety and weather conditions, competition-dependent yield losses can vary substantially from year to year for each species (Table 4.4). In this experiment, it was also shown

Fig. 4.23 Stiff ryegrass (*Lolium rigidum*)

Table 4.4 Effect of *Lolium rigidum* (300 plants m^{-2}) on the yields of different crop species in southern Australia

Crop	Yield reduction (%)
Oats	2–14
Rye	14–20
Triticale[a]	5–24
Canola	9–30
Wheat	22–40
Barley	10–55
Pea	100
Lupin	100

Source: Based on Lemerle et al. 1995

[a]Triticale is a cereal breed that resulted from the crossing of wheat (*Triticum aestivum*) and rye (*Secale cereale*). Its name is a combination of *Triticum* and *Secale*

that yield losses for the crop plants resulting from *L. rigidum* were related to competition for light and for nutrients including nitrogen, phosphorus, and potassium (Lemerle et al. 1995).

Most cereal species and varieties are relatively competitive against weeds, whereas the yield losses caused by competition can be very high for field vegetables. This is true for carrots (*Daucus carota*) and onions (*Allium cepa*), for which, without weed management, only very low yields often achieved (Fig. 4.24). The reason for this is the slow rate of germination of these species, which is a

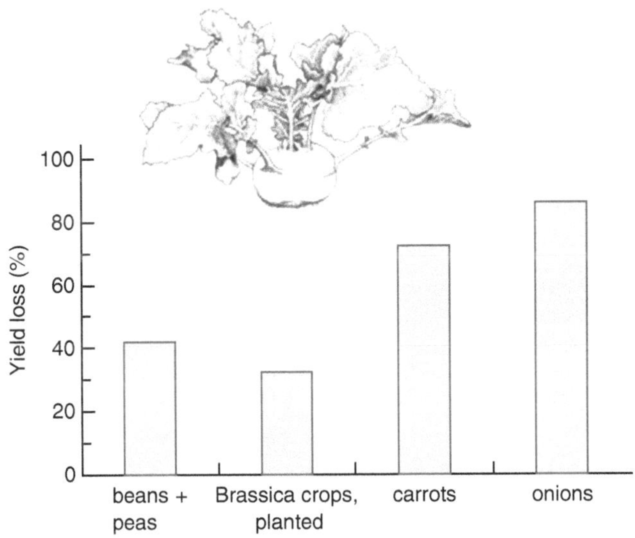

Fig. 4.24 Average yield losses for different vegetable species without weed management (Based on Laber 1999). An example of a *Brassica oleracea* variety is kohlrabi (drawing)

disadvantage in competition for light. In contrast, a high longitudinal growth rate is usually observed for cereal species but not for most of the broadleaved wild plant species. However, especially for winter cereals, grass species that predominantly germinate in autumn and develop synchronously with the cereal plants can act as weeds. However, the presence of such species is not necessarily related with high yield losses, even at high densities.

Example: In more than 100 individual experiments at different sites in southwest Germany, the competitive relationship between winter wheat and twitch grass was examined. With increasing twitch grass density (up to approximately 400 plants m^{-2}) a continuous decline in the grain yield of wheat was observed. However, with a further increase in weed density, there was no change in the average decrease in yield, in other words it remained constant at approximately 22% (Fig. 4.25). This can be explained by the fact that with higher plant density intraspecific competition increases and that with the threshold of approximately 400 plants m^{-2} the maximum biomass production per unit area was reached. As a consequence, both twitch grass and wheat plants achieved a smaller size. The ideal plant density for wheat under weed-free conditions is also in the range of 400–600 plants per m^2. However, there can be significant deviations from the averages presented in Fig. 4.25, depending on the site and the time of germination of the twitch grass. With the highest weed densities, the smallest yield loss was 16% and the highest was 50% (Koch 1994).

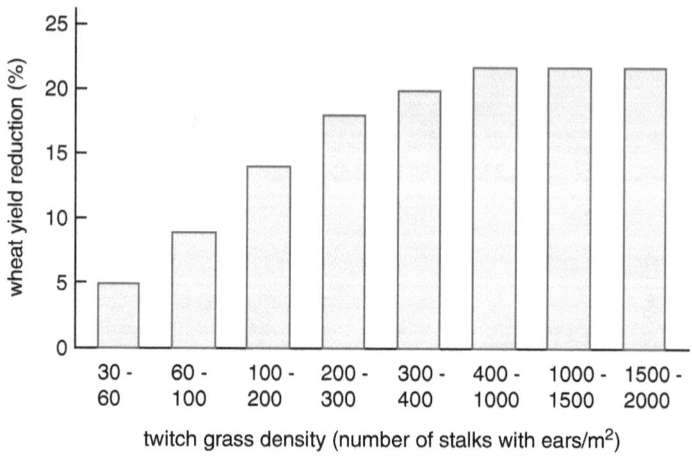

Fig. 4.25 Yield losses of winter wheat (as a percentage of the yield achieved under weed-free conditions) in relation to the density of twitch grass (Based on Koch 1994)

4.4.2 Allelopathy

Allelopathy is a specific type of interaction between plants that is distinguished from competition mechanisms for light, nutrients, and water. Allelopathy is the effect on one plant species of chemicals released by another. This usually has a negative effect as a result of inhibition of germination, growth, or development. The chemical substances, the so-called allelochemicals, can be released into the air (e.g. ethylene) or can reach the soil by being released from the roots, washed down from above-ground plant parts, or released during decomposition of plant material. Allelopathy can be of importance in agroecosystems in various ways (Buschmann 2003):

- Crop plants can be harmed by the allelopathic substances released by specific associated plants. The substances often originate from dead plant material. Examples are creeping thistle (*Cirsium arvense*) and common chickweed (*Stellaria media*), which have negative effects on barley and many other crop species.
- There are, however, also crop plants which inhibit the development of associated plants by allelopathy. Examples include sunflower, onion, cassava, rice, alfalfa, and tobacco, and also species that can be established as intercrops or undersown crops, for example black mustard (*Brassica nigra*) and specific clover species. Growth of weeds can also be inhibited by incorporation of allelopathically active plant materials into the soil, or by application of natural substances that can be derived from suitable plant species.
- One form of allelopathy is autotoxicity, in which the released substances exert a negative effect on plants of the same species. This explains at least some cases of the phenomenon of autointolerance (Sect. 2.4.1.2).

4.4.3 *Parasitic Flowering Plants*

Some flowering plant species have a parasitic life strategy, which means they draw resources from other plant species and thereby harm these host plants. Two groups can be identified on the basis of the resources utilized by the parasitic flowering plant:

1. **Hemiparasites** carry out photosynthesis, which means that they can produce their own carbohydrates and primarily draw water and nutrients from their host. Examples include mistletoe (*Viscum album*), different species of figwort (*Scrophulariaceae*), and field cow-wheat (*Melampyrum arvense*) and rattle (*Rhinanthus* species).
2. **Holoparasites** do not have chlorophyll and are, therefore, also dependent on the organic compounds of their host. Examples of this group are broomrape (*Orobanche* species) and dodder (*Cuscuta* species).

Not all parasitic flowering plant species can be definitively attributed to either one or the other group. For example, species of the genus *Striga* are capable of performing photosynthesis, but nevertheless draw carbohydrates from their host.

In addition, a differentiation can be made between shoot parasites (e.g. *Viscum* and *Cuscuta*) and root parasites (e.g. *Orobanche* and *Striga*). Germination of the seeds of root parasites occurs only in soil close to the host plant and is initiated by substances released by the host plant roots. A germination tube grows in the direction of the host plant and forms a contact organ (haustorium) after becoming attached to the root; this establishes a connection with the transport tissues of the host. This interaction shows that allelopathy is not only inhibitory but can also have supporting effects on plants of another species.

A variety of parasitic flowering plants attack agricultural crop species. Among the *Orobanche* species, the lesser broomrape (*Orobanche minor*) was still a widespread parasite of clover species at the beginning of the twentieth century in Central Europe, despite being severely repressed by intensification of agriculture. Other *Orobanche* species, which are primarily found in the Mediterranean region (e.g. *O. crenata*; Fig. 4.26), attack a variety of crop species, including grain and fodder legumes, sunflower, tomato, potato, and tobacco. The branched broomrape (*Orobanche ramosa*) is currently of increasing importance in Central Europe. This species has a broad range of hosts and attacks tobacco, canola, hemp, and tomato.

The *Striga* species are primarily found in the wet-dry tropics (Sect. 7.2.1.2). In the savannah regions of Africa, and in parts of Asia; they are the greatest problem for cereal production because they parasitize maize, sorghum, pearl millet, and rice. In recent decades the incidence of *Striga* has increased substantially in many of the regions affected. One reason for this is increasing reduction in the duration of the fallow period and the related greater survival of *Striga* seeds in the soil.

Fig. 4.26 Crenate broomrape (*Orobanche crenata*)

4.5 Phytophages

Almost tissues of living plants can serve as food for phytophagous organisms. The largest group of consumers of wild plants and crops are the insects, which are also the most species-rich class of the animal kingdom (Table 4.5). Other important groups of invertebrates with phytophagous representatives are the mites (Acari) and the roundworms (Nematodes).

4.5.1 Trophic Groups of Phytophages

Phytophages can be classified into groups on the basis of the plant parts that they consume and on the different strategies they use to acquire food. Species or their development stages that feed on vegetative plant parts (stem, leaves, or root) are called **herbivores**. Other phytophages, for whom there is no overall term, utilize products that are formed in the inflorescence of the plant in the context of reproduction. These products primarily include nectar, pollen, seeds, and fruits. For most phytophagous species, however, the feeding strategy of the larval stages is different from that of the adults.

Table 4.5 The seven largest orders of insects, number of currently known species, and proportion of phytophagous species

Order	Number of species	Phytophagous species (%)
Coleoptera (beetles)	349,000	35
Lepidoptera (moths and butterflies)	119,000	100
Diptera (flies)	119,000	30
Hymenoptera (wasps, bees, sawflies, ants)	95,000	11
Hemiptera (cicadas, aphids, planthoppers, leafhoppers, mealy bugs)	59,000	91
Orthoptera (grasshoppers, crickets, locusts)	20,000	100
Thysanoptera (thrips)	5,000	90

The actual number of insect species is certainly much higher. Some estimates suggest there are more than ten million insect species worldwide (Based on Schoonhoven et al. 1998)

4.5.1.1 Ectophagous Consumers of Plants

Herbivores that feed on surfaces of plants without penetrating the plant, are ectophagous consumers. On the basis of feeding strategy, two groups can be identified:

1. **Ectophagous tissue consumers** are species that remove plant parts while feeding, and therefore usually cause visible damage to the tissue, for example edge injuries, holes, or strips on the leaves, the last by removal of the epidermis.

   Important ectophagous consumers on aboveground plant parts include caterpillars (Fig. 4.27), grasshoppers and crickets (Orthoptera), and the larva and/or adult stages of many species of beetle (Coleoptera), including primarily the leaf beetles (Chrysomelidae; Fig. 4.28). In Central Europe, slugs of the keelback (Limacidae) and arion (Arionidae) families are important ectophagous consumers on crop plants. For roots, the larvae of specific weevil or snout beetle species (Curculionidae) and larvae of scarab beetles (Scarabaeidae) are of importance.
2. **Sap suckers** are species that feed on the sap of plants. On the basis of the plant tissues from which they draw their food, usually with their piercing-sucking mouth parts, three groups can be identified:

   (a) **Phloem suckers** tap into the plant tissue responsible for transport of photosynthetic products (phloem). The phloem sap is rich in sugar but poor in nitrogen compounds. Phloem suckers must therefore consume large quantities of sap to meet their amino acid needs. Excess sap is excreted as droplets, the so-called **honeydew**. The most important groups of phloem suckers are aphids (Aphididae; Fig. 4.29), scale insects (Coccoidea; Fig. 4.30), larvae of the whiteflies (Aleyrodidae), and jumping plant lice (Psylloidea). Many representatives of the cicadas (Auchenorrhyncha) are also phloem suckers (Fig. 4.31).

   (b) **Xylem suckers** use the tissues that transport water, inorganic nutrients, and stored substances from the roots to the aboveground plant parts (xylem).

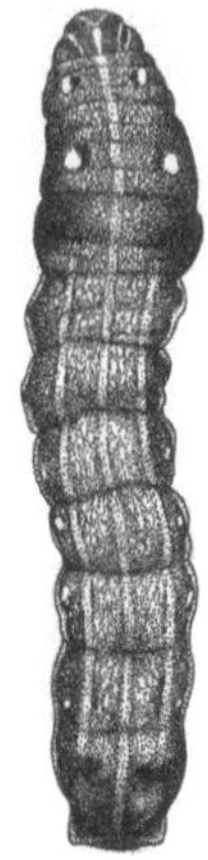

Fig. 4.27 Caterpillar of the oriental leafworm moth (*Spodoptera litura*), which attacks numerous crop plants in tropical Asia

Fig. 4.28 The bean leaf beetle (*Ceratoma trifurcata*) feeds on soybean, common bean and other leguminous plants

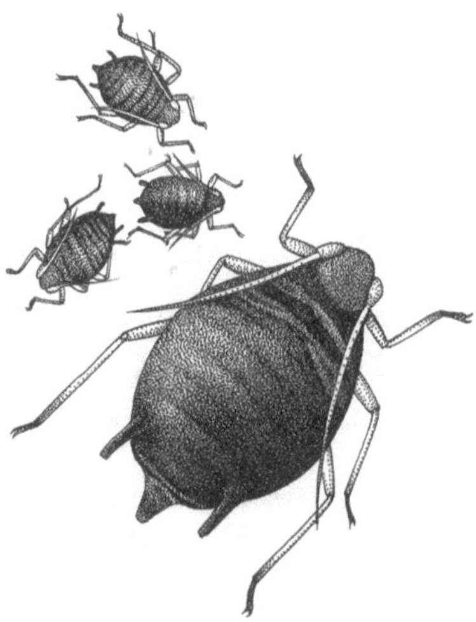

Fig. 4.29 Black bean aphid (*Aphis fabae*)—wingless female and nymphs

Similar to the phloem suckers, they primarily extract amino acids from the fluids they take from the plant. Xylem suckers include primarily particular groups or species of the cicadas.

(c) **Cell fluid suckers** tap into the epidermis and parenchyma cells. They include spider mites (Tetranychidae; Fig. 4.32), many of the thrips

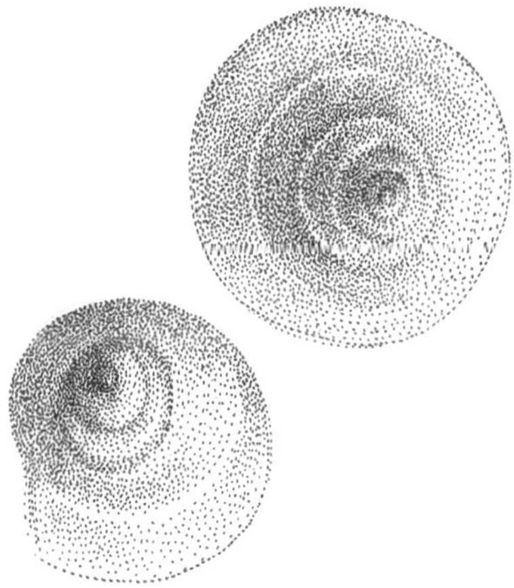

Fig. 4.30 San Jose scale (*Quadraspidiotus pernicosus*) attacks numerous fruit tree species. The females form a wax-like scale, covering the soft body

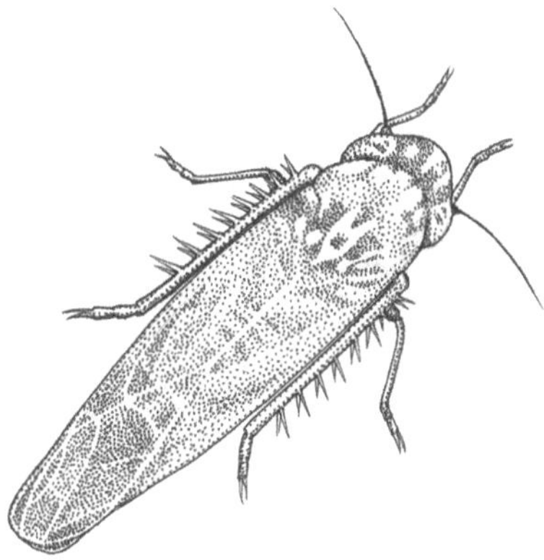

Fig. 4.31 The potato leaf hopper (*Empoasca fabae*) feeds on the phloem sap of a variety of crop plants, for example soybean and potato

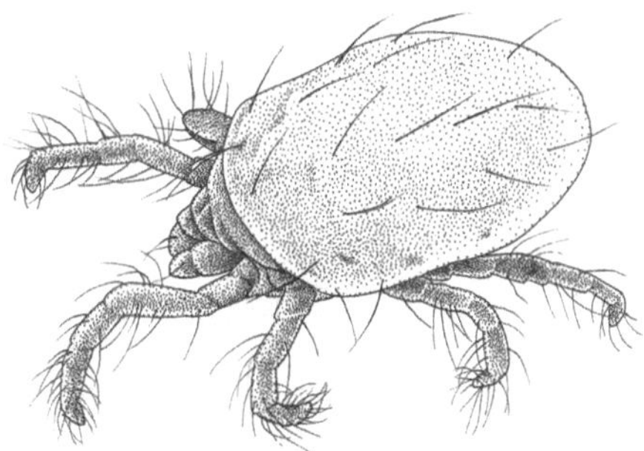

Fig. 4.32 Spider mite of the genus *Tetranychus*

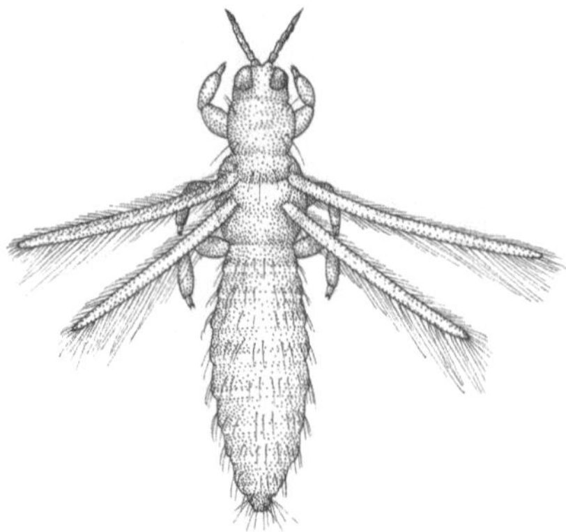

Fig. 4.33 Tobacco thrips (*Frankliniella fusca*) primarily attack tobacco and peanuts in North America

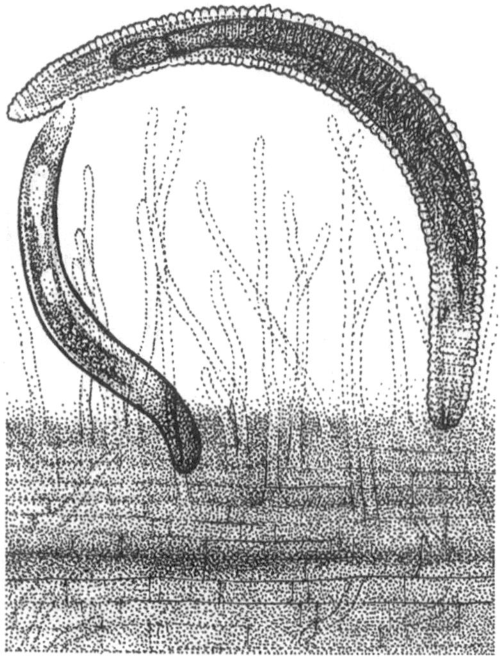

Fig. 4.34 Ring nematodes (genus *Criconemella*) obtain their food from the cells of the rhizodermis

(Thysanoptera, Fig. 4.33), some of the leafhoppers (Cicadellidae), and some nematode species. The last of these are found in the soil and obtain their food from cells of the rhizodermis and cell layers deeper in the root (Fig. 4.34). A special group are sedentary root nematodes, which colonize a site as larvae and remain there for the rest of their lives. By means of released exudate they cause the local destruction of cell walls, which ensures a permanent supply of cell fluid.

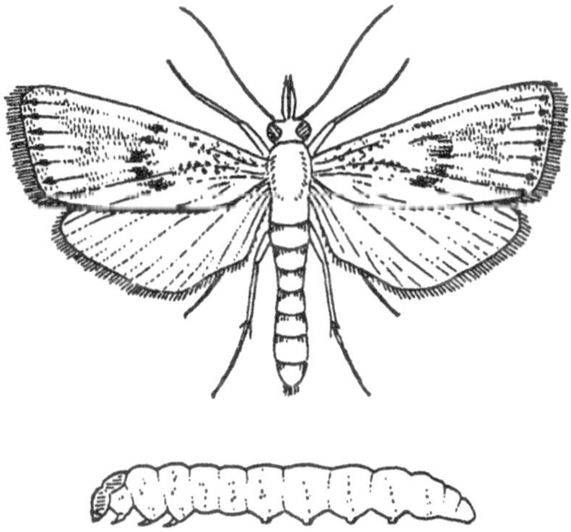

Fig. 4.35 Dark-headed rice stem borer (*Chilo polychrysus*)—imago and larva (Drawing by W. Lang)

4.5.1.2 Endophagous Consumers of Plants

Herbivores that penetrate the plant for feeding, and pass through some or all of their life stages there, are called endophagous consumers. They can be classified into groups on the basis of their feeding strategies:

1. **Leaf miners** are primarily insect larvae that feed on the cells of the spongy and/or palisade parenchyma of the leaves, so epidermal cells are not usually damaged. The feeding passages of the leaf miners are called tunnels or mines. They vary in form and pattern, depending on the species responsible. Most leaf miners are fly and butterfly larvae, but some are sawfly and beetle larvae. Nematodes can also enter leaves, via stomata.
2. **Shoot, stem, and wood borers** enter the respective plant parts and usually feed on that tissue. Shoot and stem borers are of particular importance for cereal crops. Young shoots are the primary food sources of the larvae of several fly species. Important stem borers in cereal stalks are the caterpillars of the snout moths (Pyralidae; Fig. 4.35) and owlet moths (Noctuidae). Caterpillars of several families of butterfly, and the larvae of snout beetles, are found in the stems of herbaceous plant species. Woody plant parts are mainly penetrated by the larvae of some butterflies and beetles. The latter group includes longhorn beetles (Cerambycidae) and bark beetles (Scolytidae).
3. **Root borers** consist primarily of groups of nematodes that colonise below-ground plant organs, including roots, tubers, rhizomes, and bulbs. From there, some species enter the shoot and even the ovules of the plant. The larvae of click beetles (Elateridae) and of some butterflies and flies are root borers.
4. **Gall formers** cause local growth anomalies in plant tissues, called galls, which can be found on a variety of organs Gall formers include insects, mites, nematodes, and microorganisms that penetrate the plant and initiate formation

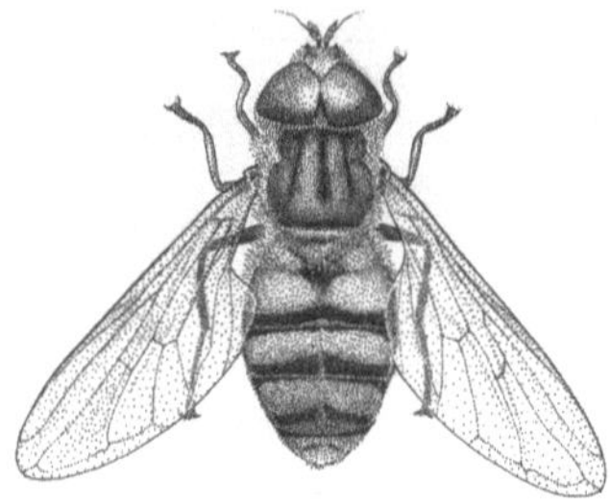

Fig. 4.36 Hoverfly (family Syrphidae)

of the galls. This occurs by release of substances that stimulate the plant to accelerate cell production or cell growth, which may be associated with increased levels of specific plant products (including phytohormones). The mechanisms of these interactions are not understood in detail. The gall tissue provides the organism with food, but also protection from enemies, all within a controlled environment. The largest groups among the gall-forming insects are the gall midges or gall gnats (Cecidomyiidae) and the gall wasps (Cynipidae). There are also gall formers among the aphids, but only very few among the beetles and butterflies. Among the mites, the only representatives are the gall mites (Eriophydae) with a body size of only 0.2 mm.

4.5.1.3 Consumers Related to Flowers, Seeds and Fruits

Many species of the flowering plants (Spermatophyta) exploit animals for pollination (Sect. 4.5.2) and for distribution of their seeds. To initiate this interaction with the animals, the plants offer them food produced for this purpose. Such relationships are usually beneficial for all the species involved and are therefore **mutualistic** (Sect. 3.1.3). In contrast, consumption of seeds is detrimental to the plant because potential offspring are thereby eliminated from the population.

1. **Flower visitors** (Anthophiles) usually visit the flower in search of food. Plants that rely on animals for pollination must attract these animals, a process which occurs primarily through the colour, smell, and form of the flowers. Most zoogamous plant species offer nectar, produced in special glandular tissues (nectaries), to flower visitors. Often, some of the pollen produced by the plant, and sometimes oils (lipids), also serve as reward for pollen transfer. The most important flower visitors and pollinators include the Apidae (bees and bumblebees), many butterflies and flies (Fig. 4.36), and some beetles and thrips. In the tropics, some bat and bird species are also important.
2. **Frugivores** in the narrow sense primarily include birds and mammals which ingest the fleshy hulls of seeds of some flowering plants. The seeds also are usually ingested and later excreted, and thus distributed. Most plant species in the tropical rainforests are dependent on this type of seed dispersal. In contrast, it is uncommon in plants of dry areas. In the broader sense of the term, many invertebrate animals also are frugivores, but do not usually contribute to the

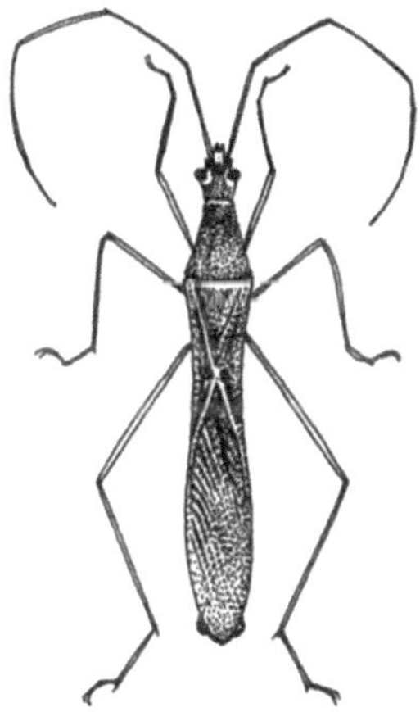

Fig. 4.37 Paddy bugs (Genus *Leptocorisa*) suck on milk-ripe rice kernels

dispersal of seeds. This is true for many of the true bugs (Hemiptera), which suck fruit juices, and for caterpillars of many butterfly species that develop in fruits. The larvae of many fly and beetles species, and some mites, scale insects, whiteflies, and jumping plant lice (psyllids), also feed on fruits.

3. **Granivores** are not only important pests of grain and beans during storage but can also occur on the plant where they feed on ripening seeds or the flower ovary and seed structures. Such consumers include species of the snout beetles (Curculionidae), seed beetles (Bruchidae), and caterpillars of several butterfly species. Some bug species pierce milk ripe seeds of graminoids (Fig. 4.37). Seeds lying on the ground are used by ants and rodents, among others.

4.5.2 Host Plant Range of the Phytophages

None of the species of phytophages feeds on all the plant species available in their environment; instead each uses a specific selection of these. Host plants species can, depending on the species of phytophage, range from one to several hundred, and this is used as a criterion for classification of phytophages into three groups:

1. **Monophagous** species feed on one plant species only, or on very few, closely related species within one genus. Examples are the yellow rice stemborer (*Scirpophaga incertulas*), whose larvae only develop in rice plants (genus *Oryza*) and the olive fruit fly (*Bactocera oleae*) whose larvae are only found in the fruits of the olive tree.
2. **Oligophagous** species have a host plant range limited entirely or predominantly to species of a specific plant family. Examples are the Colorado potato beetle (*Leptinotarsa decemlineata*) which feeds on approximately 14 plant species, all from the family Solonaceae (nightshades).

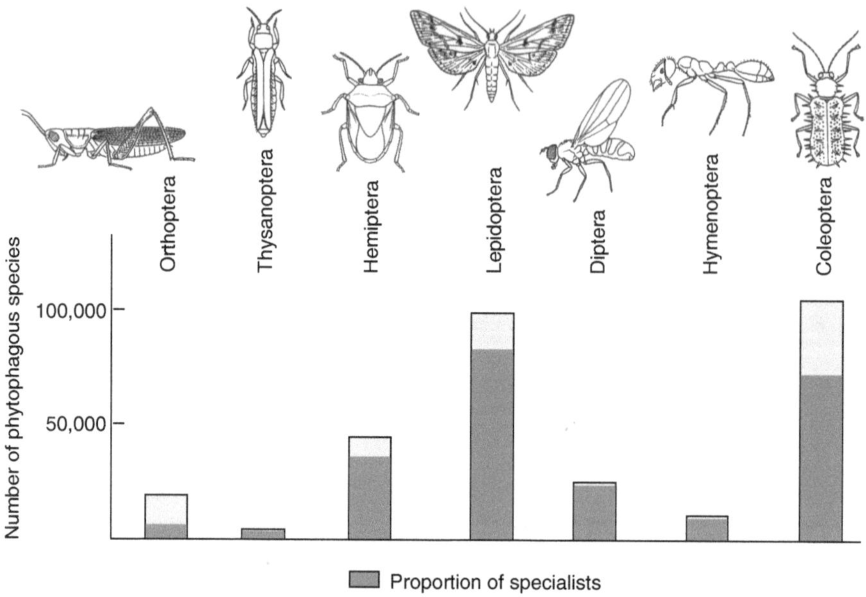

Fig. 4.38 Proportion of specialists among the phytophagous species of several orders of insects (Based on Bernays and Chapman 1994; drawings by W. Lang)

3. **Polyphagous** species obtain their food from plant species of several families. Extreme and agroecologically important examples are:

   - the two-spotted spider mite (*Tetranychus urticae*), which is found on approximately 900 plant species;
   - the whitefly *Bemisia tabaci*, which attacks 500 plants species of 74 families; and
   - the desert locust (*Schistocerca gregaria*), which feeds on over 400 plants species.

Because it is often difficult to definitively distinguish between monophagous and oligophagous species, together they are termed **specialist feeders (specialists)**. These are defined as phytophages that feed on a limited number of plant species with particular common characteristics, which are often found within one plant family. These specialists contrast with the polyphagous species, which are termed **generalist feeders (generalists)**. These feed on a broad range of plant species from more than one family.

It is highly likely that no more than 10% of all phytophagous insects utilize plants from more than three families as a food source (Bernays and Graham 1988). Approximately 60% of all leaf-sucking cicada species of England, and 70% of all species that mine the leaves of trees and shrubs feed on only a single plant species (Claridge and Wilson 1981). In almost all large insect orders most species are specialists (Fig. 4.38). The caterpillars of the butterfly species found worldwide as pests of crop plants are also mostly specialists (Fig. 4.39).

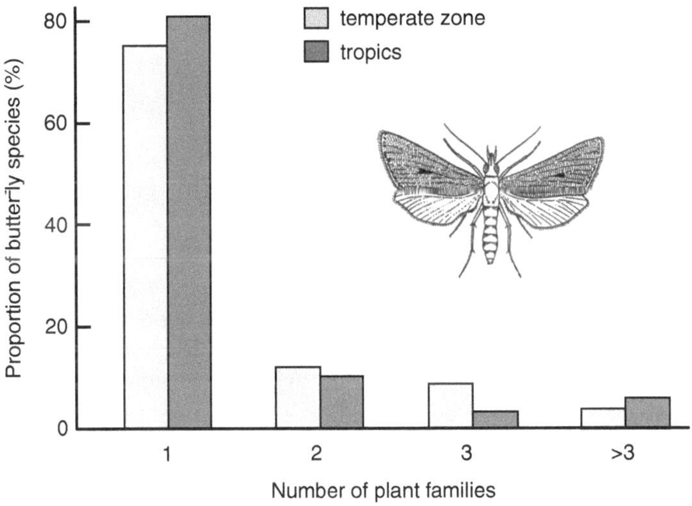

Fig. 4.39 Proportion of specialists among the pest caterpillars in agroecosystems of the temperate zone and the tropics (Based on Bernays and Graham 1988). The illustration shows the imago of the yellow rice stemborer (*Scripophaga incertulas*), whose larvae are specialized on the genus *Oryza* (cultivated varieties and wild species of rice; drawing by W. Lang)

More than three quarters of all phytophagous insect species of wild and cultivated plants are specialists.

The reasons for the different host plant preferences of the phytophages are to be found in the morphological and biochemical characteristics of the plants and are related to the defence and resistance mechanisms of the plants (Sect. 4.5.5).

4.5.3 Crop Pollination

One of the most important ecosystem services to agriculture (Sect. 3.6) is the pollination of crops by animals, especially insects. Approximately one third (35%) of the world food production depends on biotic pollination. Animal-pollinated crop species include vegetables, fruits, oil crops, pulses, and nuts (whereas all cereals of the Poaceae family are wind-pollinated). For European crops, Williams (1994) concluded that production of 84% of 264 crop species depends at least to some extent on animal pollination. With regard to the tropics, a list of 1,330 crop species including their potential pollinating taxa, indicates that approximately 70% of tropical crops benefit from animal pollination (Roubik 1995).

Fig. 4.40 Domestic honey bee (*Apis mellifera*; drawing by Ch. Allgaier)

One of the most important crop pollinators in many regions of the world is the domestic honey bee *Apis mellifera* (Fig. 4.40). Honey bee colonies have declined substantially during recent decades, however. It is estimated that colonies decreased by more than 50% between 1950 and 2007 in the United States and by 25% since 1985 in Europe. It is generally assumed there is no single reason for this development, rather a combination of different factors, which may include (a) infection with the parasitic mite *Varroa destructor*, pathogens, for example *Nosema* spp., and several viruses, all of which have almost global distribution, (b) direct and indirect effects of insecticides and other agrochemicals, (c) lack of continuous food resources in homogenous landscapes, (d) effects of weather extremes related to climate change, and (e) lack of genetic diversity and vitality.

Furthermore, domestic honeybees are not appropriate pollinators for many tropical crops, and bee keeping is uncommon or difficult in many tropical regions. Instead, several tropical crops (e.g., cocoa, passion fruit, dragon fruit, vanilla, oil palm) depend on pollination by naturally occurring insects, for example wild bees (Apoidea, including social bees, for example bumblebees and stingless bees, and a variety of families of solitary bees), butterflies, and moths (Lepidoptera) and hoverflies (Syrphidae).

For these reasons there is increasing focus on the importance of naturally occurring insects to crop pollination. Many studies have dealt with the question of if, or under which conditions, wild pollinators can compensate for the reduction in honey bee populations. Major overall results have indeed shown that wild pollinators in agricultural regions can be even more important than domesticated bees in pollination of crops. However, species numbers, abundance, and success of wild pollinators are highly dependent on the availability of natural or semi-natural habitats in the surrounding environment, for example area or proportion of uncultivated land, diversity of wild flower resources, and overall landscape diversity. In temperate regions, for example North America and Europe, the most important habitats for wild pollinators include semi-natural open lands, for example fallow fields, marginal grasslands, field and forest margins and riparian habitats. In tropical regions, however, natural forests have been shown to be an essential habitat for wild bees, which are important pollinators for field and plantation crops, for example fruit trees and coffee.

Examples: Winfree et al. (2008) studied bee visits to different vegetable crops on 29 farmland sites in the northeast of the United States. Wild bees were the dominant

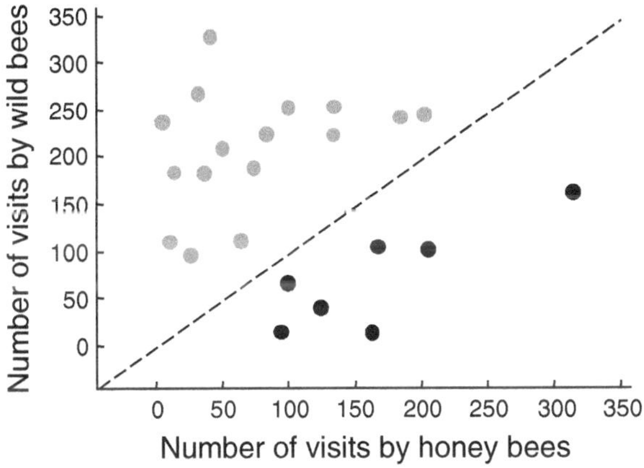

Fig. 4.41 Total number of visits to crop flowers by honeybees vs. wild bees. Each point is a study site. Points above the 45° line indicate that wild bee visits exceeded honeybee visits (Based on Winfree et al. 2008)

flower visitors for three of the four crops studied, including watermelon (*Citrullus lanatus*), bell pepper (*Capsicum annuum*), and tomato. Domesticated honey bees accounted for the other visits. Figure 4.41 shows the 1 year results for water melon visits. They also found that the wild bee communities (54 recorded species) visiting tomato flowers were distinct from those visiting watermelon flowers, indicating that a diverse community of natural pollinators is necessary for full pollination. Winfree et al. conclude from the landscape structure in their study area that high habitat heterogeneity may support high wild bee abundance even in landscapes with a low proportion of natural habitat overall.

Steffan-Dewenter et al. (2002) analysed the effects of landscape context on the distribution of wild bees (Hymenoptera: Apoidea) at multiple spatial scales in a region of Central Germany. They selected 15 landscape sectors and determined the percentage of semi-natural habitats and the diversity of habitat types at different spatial scales (radius 250–3,000 m). The percentage of semi-natural habitats varied between 1.4 and 28%. To observe bee visits, patches of potted flowering plants (four perennial and two annual species) were placed in the centre of each landscape sector in the same habitat type, a grassy field margin adjacent to cereal fields. Species richness and abundance of solitary wild bees was in close positive correlation with the percentage of semi-natural habitats at small scales up to a radius of 750 m (Fig. 4.42), whereas bumblebees and honey bees did not respond to landscape context at these scales.

Klein et al. (2003) analysed the effects of wild bee diversity and abundance on the fruit set of coffee (*Coffea arabica*) in Indonesia, where it is traditionally grown under a canopy of shade trees. For 24 such agroforestry sites, differing in shade and forest distance, they showed that fruit set of self-fertilizing coffee is highly variable and related to bee pollination, and that fruit set could be predicted from the number of flower-visiting bee species. That is, diversity, not abundance, explained variation in

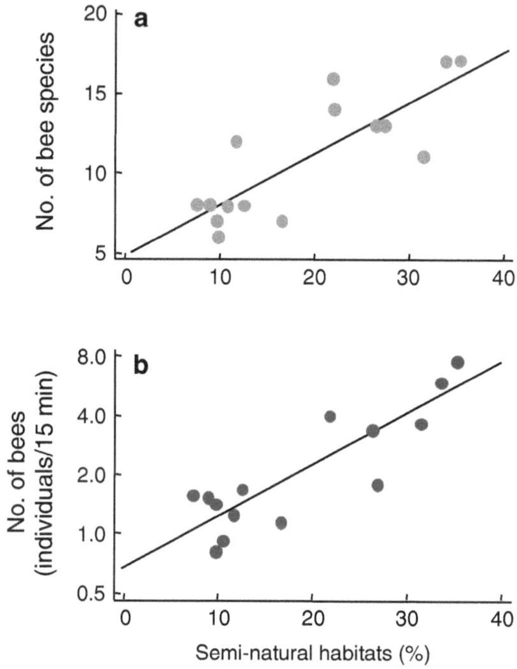

Fig. 4.42 Relationship between species richness (**a**) and abundance (**b**) of flower-visiting solitary wild bees on experimentally exposed potted plant species and the proportion of semi-natural habitats within a radius of 750 m (Based on Steffan-Dewenter et al. 2002)

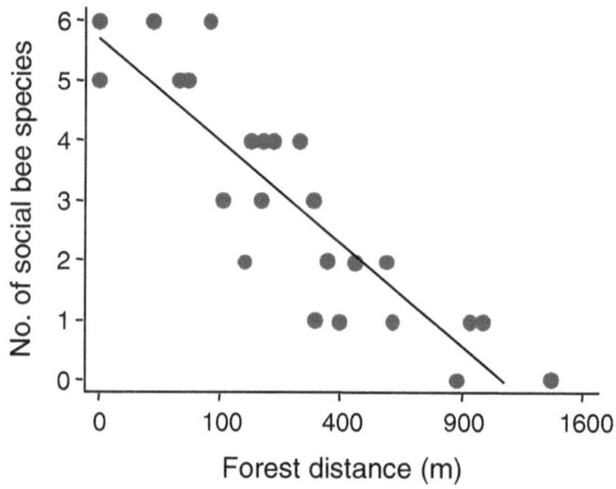

Fig. 4.43 Number of social bee species in relation to forest distance in a fragmented tropical landscape of Sulawesi, Indonesia (Based on Klein et al. 2003)

fruit set, so collective involvement of a species-rich bee community was important for pollination success. Further experiments showed that single flower visits from rare solitary species led to higher fruit set than for abundant social species. Pollinator diversity was affected by two habitat properties indicative of guild-specific nesting requirements: the diversity of social bees decreased with forest distance (Fig. 4.43), whereas the diversity of solitary bees increased with light intensity of the agroforestry systems. Klein et al. point out that conservation of rainforest adjacent to adequately managed agroforestry systems could improve the yields of farmers.

4.5.4 *Pests and Their Effects on Yield*

In agroecosystems, pests are phytophages that cause yield losses of crop plants. However, phytophages cannot be classified as damaging, useful, or neither on the basis of their species identity alone. Rather, the circumstances determine whether a species is a pest. A crucial factor is the number of individuals present, which in turn is affected by a variety of environmental factors (e.g. weather, crop type, natural enemies). How large must a specific population of phytophages be to be regarded as damaging? For individual species, this is evaluated by using the **economic threshold**. The economic threshold is the density of a phytophagous species (per plant, per unit area, or per trap) at which control treatment will provide an economic return. Below this threshold, yield losses may occur, but the costs required to prevent them are higher than the profit from the yield difference.

Yield losses usually correlate positively (but not necessarily linearly) with the population density of the pest. The damage threshold can be affected by a variety of factors, of which the age of the plant and its resource supply are the most important. In assessment of population density, specific factors must also be considered, including the position of traps (e.g. yellow pan or sticky traps) in or near the field, and the time of day when the recording is made.

Examples: The Bavarian State Institute for Agriculture (Bayerische Landesanstalt für Landwirtschaft 2000) has set damage thresholds for several phytophagous species of rapeseed. For the cabbage stem weevil (*Ceutorrhynchus napi*) it is between 10 and 15 individuals per yellow pan trap within 3 days. This species invades the crop after the end of February and remains in the fields for approximately 4 weeks. For the pollen beetle, *Emligethes aeneus*, the damage threshold is six individuals per plant for winter rape and three individuals per plant for summer rape. Counting should be conducted within 10 m of the field's edge. For the cabbage aphid, *Brevicoryne brassicae*, the threshold is 150 individuals on 25 young rapeseed plants. For later growth stages of the plant, no threshold is given.

Individual crop plant species and varieties are not subject to one pest species only, but usually to a more or less complex community of phytophagous species, so yield losses are often not caused by one species alone. As an example, Table 4.6 shows a selection of phytophagous species that can be found on wheat.

4.5.4.1 Plant Compensation for Feeding Damage

Damage to plants does not always result in reductions in fitness or yield losses (cf. Box 3.1). Under some conditions, both wild and cultivated plants are capable of partially or completely compensating for the biomass and energy losses caused by phytophages. An essential factor that affects this ability is the **development stage** of the plant. In early development stages, plants have a greater capacity for compensation than in later stages.

Table 4.6 Examples of consumers or pests of wheat in Central Europe

Consumer	Plant part consumed
Ectophagous consumers of the plant	
Field slugs (*Deroceras* species)	Leaves of seedlings
Corn ground beetle (*Zabrus tenebrionides*), larva	Leaves of seedlings
Cereal leaf beetle (*Oulema melanopus*)	Leaves (strips)
Barley flea beetle (*Phyllotreta vittula*)	Leaves (strips)
Turnip moth (*Agrotis segetum*), caterpillar	Leaves (edges)
Frit fly (*Oscinella frit*), larva of the spring generation	Young plant (growing point)
Wheat bulb fly (*Delia coarctata*), larva	Young plant (growing point)
Hessian fly (*Mayetiola destructor*), larva	Stem
Crane flies or daddy long legs (*Tipula* species), larva	Roots and shoot base
Click beetle (*Agriotes lineatus*), larva	Roots and shoot base
Rose-grain aphid (*Metopolophium dirhodum*)	Phloem sap
Grain aphid (*Sitobion avenae*)	Phloem sap
Bird cherry-oat aphid (*Rhopalosiphum padi*)	Phloem sap
Barley Thrips (*Limothrips denticornus*)	Epidermis cells (cell fluids)
Endophagous consumers of the plant	
Leafminer (*Hydrellia griseola*), larva	Leaves
Wheat stem sawfly (*Cephus pygmaeus*), larva	Stem
Lesion nematodes (*Pratylenchus* species)	Roots
Consumers of reproductive organs	
Wheat bug (*Eurygaster maura*)	Ripening seeds (milk-stage)
Bishop's mitre shield bug (*Aelia acuminata*)	Ripening seeds (milk-stage)
Corn ground beetle (*Zabrus tenebrionides*), adults	Ripening seeds
Rustic shoulder-knot (*Apamea sordens*), caterpillar	Ripening seeds
Frit fly (*Oscinella frit*), larva of the summer generation	Ripening seeds
Lemon wheat blossom midge (*Contarinia tritrici*), larva	Flower ovary

Examples: For Brussels sprouts, artificial removal of 60% of the leaves 3 weeks after planting led to yield losses of less than 10%. The vulnerable stage of the plants is reached after 6 weeks, at which point leaf loss of 60% results in a yield loss of approximately 40%. For plants closer to the stage of full development, after approximately 15 weeks, the yield was no longer affected by loss of leaves (Fig. 4.44; Wit 1982). Similar relationships between tissue loss, growth stage of the plant, and yield have also been found for other crop plants (e.g. potato, soybean, rice, cotton) and wild plants.

Not only development stage but also **resource availability** determines if and by how much attack by phytophages affects the fitness and yield of plants. Compensation for a loss of tissue can only occur if the plants are sufficiently provided with nutrients, water, and light. Competition also affects resource availability (Sect. 4.4.1).

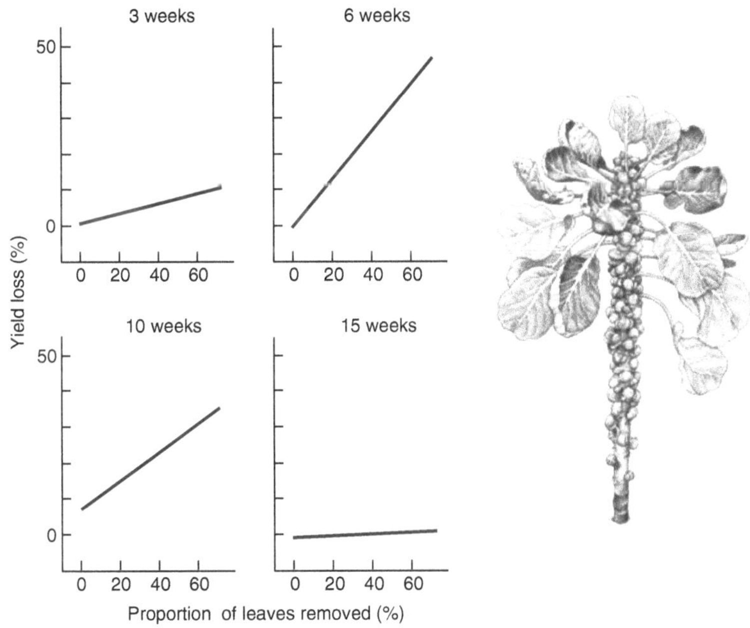

Fig. 4.44 Relationships between the proportion of leaves removed in different growth stages (weeks after planting) and yield loss for Brussels sprouts (Based on Wit 1982)

4.5.4.2 Resource Availability and Pests

A good supply of nutrients and water results in ideal growth and development conditions not only for plants but also for pests. Of great importance for the phytophages is the nitrogen content of the plants.

Animals have a substantially greater protein demand than plants because they require proteins for biosynthesis of their tissues, for which plants primarily use carbon compounds. Plant dry mass contains between 0.03 and 7% nitrogen. The highest concentrations (>3%) are found in young growing tissues and in seeds. The nitrogen content of the food of phytophages must contain at least 1–3% nitrogen to ensure normal development of the animal (Mattson 1980). Individual leaf-feeding insects often react to tissues with a low nitrogen content by increasing food intake, which leads to increased damage to the plants. In wild and cultivated plants that are well supplied with nitrogen, the phytophages that feed on them usually have a shorter development period or are found in higher densities than on plants with lower nitrogen content. An increase in the yield of crop plants by fertilization can, at least in part, be negated by simultaneous promotion of pest populations.

Examples: For tomato plants a close relationship is evident between nitrogen fertilization and development of the Colorado potato beetle. The better the plants are supplied with nitrogen, the more larvae reach the adult stage (Hunt et al. 1992; Fig. 4.45).

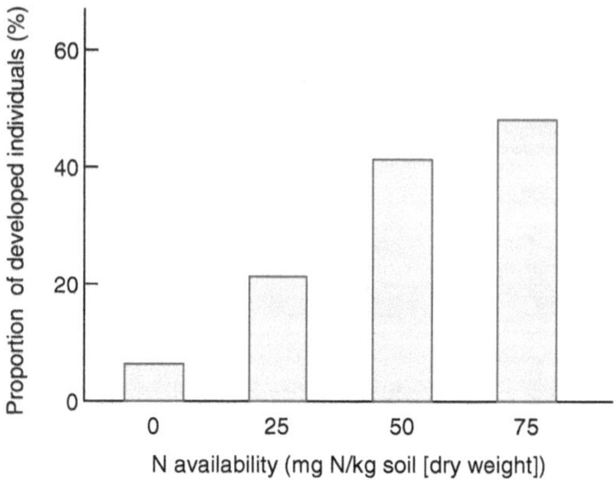

Fig. 4.45 Proportion of larvae of the Colorado potato beetle (*Leptinotarsa decemlineata*) that developed to the adult stage on tomato plants with different nitrogen supply (Based on Hunt et al. 1992)

The females of the whitefly species *Trialeurodes vaporariorum* prefer to lay their eggs on nitrogen-fertilized tomato plants, and in higher numbers, which consequently results in greater pest incidence (Jauset et al. 1998).

The caterpillars of the two butterfly species, *Sesamia calamistis* (Noctuidae) and *Eldana saccharina* (Pyralidae) are important stem and spike boring pests on maize in West Africa. Nitrogen fertilization leads to a greater incidence of pests and greater damage to the plants. However, the positive effect of fertilization on plant growth outweighs the negative effects of greater pest incidence. Pest infestation has been observed to result in yield loss of 20% for unfertilized plants but this was reduced to 11% for fertilized plants (Sétamou et al. 1995).

Yield and fitness of a (crop) plant is affected by complex relationships between availability of resources, ability to compensate for pest damage, and pest density.

4.5.5 *Plant Defences Against Herbivores*

Plants are usually able to resist attacks by phytophages, especially herbivores, and pathogens. For the ability of plants to defend themselves against their enemies, or limit the effects of damage, the terms **plant defence** and **plant resistance** are used. These terms refer to genetic traits of a wild or cultivated plant to sustain a lower level of damage, or a lower loss of fitness, than another plant of the same species without these genetic traits.

The term resistance is not consistently defined and its use for herbivores is different from its use for phytopathogens (Sect. 4.6.4). The term also refers to properties of plants related to specific abiotic factors (e.g. salt resistance or herbicide resistance). However, the term *tolerance* is often used in the same way. Tolerance also has meaning in relation to herbivores and phytopathogens, but is defined differently for each of the two groups of pests (compare Sect. 4.6.4).

In the interactions between plants and herbivores, the "resistance" is a superior term used in three different ways:

1. **Antixenosis** refers to the ability of the plant to prevent or to limit the establishment of herbivores. Thus it refers to attributes that can effect the behaviour of the animals and force them to seek an alternative host plant (for food or oviposition).
2. **Antibiosis** is the ability of the plant to have a negative effect on the metabolism or on the survival, growth, or development of herbivores when they start to feed on the plant.
3. **Tolerance** is the ability of the plant to maintain a relatively high growth rate or fitness, despite the occurrence of damage. In contrast with the two other forms of resistance against herbivores, this form does not affect the enemy. Tolerance is, therefore, not measured on the basis of the effect but rather on that of the result.

Resistance is a relative property, which means it must be examined in relation to other plants with different genetic attributes; it can range from low to high.

Examples: Some varieties of rice are resistant to attacks of *Nephotettix virescens* to different extents. This leafhopper species is an important rice pest in Southeast Asia and can transmit the tungro virus when piercing the plant tissue. This, in turn, can cause high yield losses in rice. Figure 4.46 shows the mortality of the nymphs of *N. virescens* on six different rice varieties. Two of are not regarded as resistant, because mortality of the animals is only between 7 and 11%. Two other varieties may be regarded as moderately resistant, because they cause mortality of 28 and 34%. On the other two varieties, approximately two thirds of the nymphs die (64 and 66%). Because of this rather obvious effect they are regarded as resistant to *N. virescens*. On these plants, the leafhoppers primarily tap into the xylem, the sap of which is, apparently, a less suitable food source than the phloem sap normally utilized (Myint et al. 1986).

The ability of a plant to repel or to harm herbivores is a result of specific morphological, physiological, and biochemical traits, which are also referred to as **resistance factors**.

4.5.5.1 Morphological Traits

Plant tissues have a variety of structures which are adaptations to different environmental conditions. For plants growing on dry sites, structures that reduce water loss are of primary importance. These include outer epidermal walls reinforced with cellulose or cutin, wax layers, and deposits or thick coverings of plant hairs to reduce evaporation. However, many of such morphological traits also function in defence against herbivores.

Trichomes. Plant hairs (trichomes) are mainly single-cell structures which emerge from epidermis cells and occur in a broad range of forms and sizes on different plant parts. Special forms include glandular hairs, which excrete specific substances. In

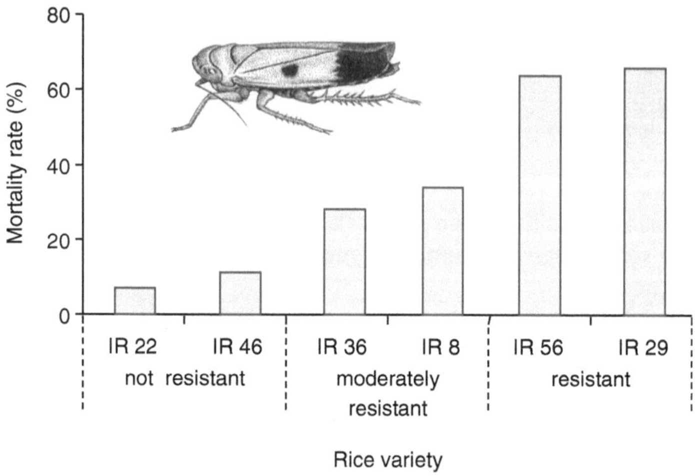

Fig. 4.46 Mortality of nymphs of the leafhopper *Nephotettix virescens* after feeding on different rice varieties. "IR" indicates breeding of the variety by the International Rice Research Institute (IRRI) in the Philippines (Based on Myint et al. 1986)

Fig. 4.47 Cereal leaf beetle (*Oulema melanopus*)

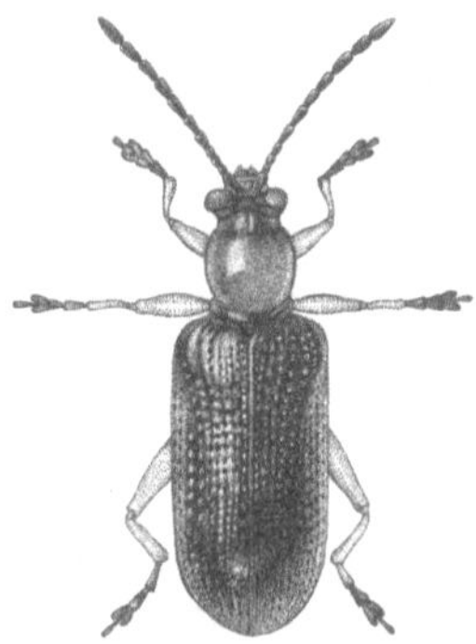

many cases, it has been shown that the presence of trichomes can inhibit herbivore feeding, which leads to increased mortality of the animals.

Examples: Both the larvae and the adults of the cereal leaf beetle (*Oulema melanopus*; Fig. 4.47) feed on the leaves of cereal species (e.g. wheat and barley). The larvae remove the cell layers from the upper surface of the leaves down to the lower epidermis, producing strip-shaped windows. The density of the trichomes on the leaves, which differ for the different wheat varieties, affects the development of the young larvae. Their survival (Fig. 4.48) and weight is reduced with increasing numbers of plant hairs, because the hairs make it difficult for the animals to reach the epidermis of the leaves. The trichomes themselves are unsuitable as food because of their high cellulose and lignin content (Schillinger and Gallun 1968).

The young caterpillars of different species of butterflies are similarly harmed by the leaves of hairy cotton varieties. Sucking insects, for example aphids and

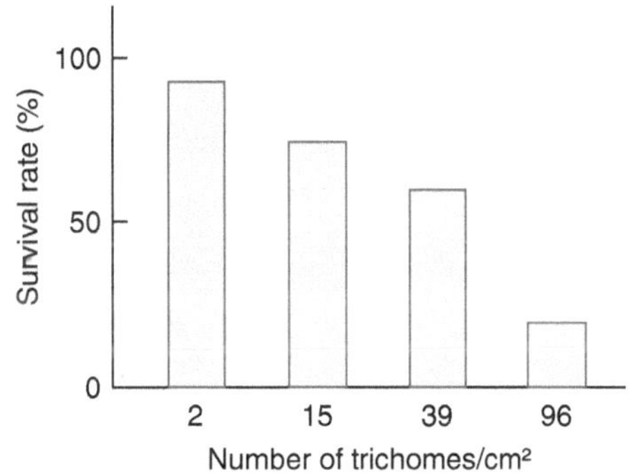

Fig. 4.48 Survival of larvae of the cereal leaf beetle on leaves of wheat varieties with different trichome densities (Based on Schillinger and Gallun 1968)

leafhoppers, are also often negatively affected by the hairy varieties of host plants (e.g. field bean, soybean, wheat). This is because the trichomes hinder the animals from tapping into the plant tissue with their mouth parts. Glandular hairs are especially widespread among the Solanaceae (e.g. tomato). They release sticky substances to which the animals adhere, or substances that serve in biochemical defence.

However, having trichomes does not always ensure effective protection against herbivores. Some species are not affected by the trichomes; yet others can even benefit from them. Spider mites, which construct webs on plants, often find more suitable conditions for this on surfaces with hairs than on surfaces without hairs (e.g., varieties of sugar cane and cotton). Infestation with spider mites may, therefore, actually be higher for hairy varieties. In these cases, the form and density of trichomes may also be important, and not all spider mite species react in the same way. Even within a herbivore species, variable effects of trichomes may be found, depending on the stage of development of the individuals.

Waxes. Plant surfaces are usually coated with a layer of wax. Waxes are esters of a long-chain alcohol and a fatty acid. However, the waxes on the cuticles of plants contain a variety of additional hydrocarbons, for example alkanes, fatty alcohols, and other substances, which vary quantitatively and qualitatively depending on the plant species.

The first contact between the plant and its enemy usually occurs at the wax layer. The wax layer can act as a mechanical barrier that inhibits the herbivore's attack on the plant cell. However, of greater importance to the behaviour of most animals is often the biochemical composition of the wax, which is perceived by the herbivore via chemoreceptors on contact with the plant. This assessment can determine whether the herbivore starts feeding or whether it leaves the plant.

Tissue structures. Some physical properties of plant tissues affect resistance. Cell walls reinforced with lignin at different sites of the plant are usually important. These make it more difficult for the stem borers to enter the plant and for sap suckers to tap into the transport tissues. A higher silicic acid or silicate content of the cell walls of the epidermis of grasses (e.g. rice, sorghum) can affect these interactions in a similar manner.

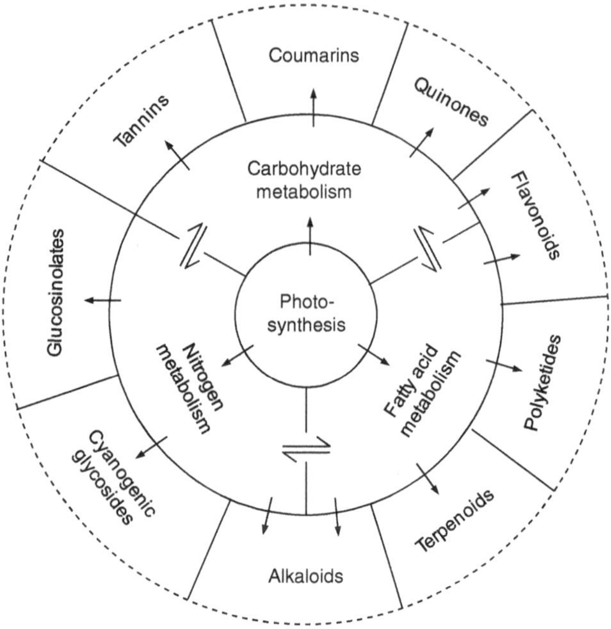

Fig. 4.49 Biochemical derivation of different groups of secondary metabolites from the products of the primary metabolism of the plant (Based on Hartmann 1996)

4.5.5.2 Secondary Plant Metabolites

The metabolism of plants includes several biochemical processes that produce so-called primary metabolites, which include carbohydrates, amino acids, nucleotides, lipids and organic acids. In principle, production of these substances occurs via the same pathways for all plants. In addition, different primary metabolites serve as the basis for numerous additional substances, known as **secondary plant metabolites** (Fig. 4.49). These compounds are not directly involved in the normal growth, development, or production of a plant. Production of most of these compounds is restricted to a particular plant species, or a narrow set of species (genera or families). In other words, the respective taxa produce a characteristic array of secondary metabolites by which they can be distinguished from other taxa. In the following section, the most important secondary metabolites will be discussed.

1. **Nitrogenous compounds**. These comprise a wide range of substances that are derived from amino acid metabolism and include several groups:

   Alkaloids contain at least one N atom, which is usually found in a heterocyclic ring. At least 10,000 such compounds are known in plants. These include many stimulants, drugs, and spices (Table 4.7).

   Glucosinolates contain both sulfur and nitrogen; approximately 120 of these compounds are primarily found among the Brassicaceae (mustard family). When plant tissues are damaged, glucosinolates are hydrolysed by the enzyme myrosinase, which is found in the cell spatially separated from these

Table 4.7 Examples of wild and domestic plants that are used as stimulants, drugs, or spices because of their content of particular alkaloids

Plant species	Alkaloid
Coffee (*Coffea species*)	Caffeine
Tea (*Camelia sinensis*)	Caffeine
Cocoa (*Theobroma cacao*)	Theobromine, caffeine
Kola nut (*Cola nitida*)	Caffeine
Tobacco (*Nicotiana species*)	Nicotine
Marijuana (*Cannabis sativa*)	Cannabinol
Coca (*Erythroxylon coca*)	Cocaine
Opium poppy (*Papaver somniferum*)	Morphine, codeine
Peyote (*Lophophora williamsii*)	Mescaline
Areca palm (*Areca catechu*)	Arecoline
Khat (*Catha edulis*)	Cathine, cathinone
Black pepper (*Piper nigrum*)	Piperine
Nutmeg (*Myristica fragrans*)	Myristicin
Paprika or pepper (*Capsicum* species)	Capsaicin

compounds. In this way isothiocyanates (mustard oils) and other reaction products, for example thiocyanates and nitriles, are released. Such substances are responsible for the typical taste of mustard, cress, horseradish, and cabbage vegetables.

Cyanogenic glycosides are compounds from which hydrogen cyanide is released when plant tissues are damaged. The approximately 75 known representatives of this group of compounds include amygdalin, which is found in the seeds of different *Prunus* species (e.g. apricot, sweet and sour cherry, bitter almond, plum, and peach), and linamarin in cassava (*Manihot esculenta*) and beans (*Phaseolus* species).

Non-protein amino acids are amino acids other than the approximately 20 amino acids which occur naturally as components of proteins. Of approximately 900 known compounds, many are found in the seeds of legumes (Fabaceae), for example L-Dopa in *Mucuna* beans.

2. **Phenolics.** Biosynthetically, these compounds are derived from carbohydrates and contain an aromatic ring with at least one hydroxy group in their basic structure. The phenolics found in plants are divided into more than a dozen large groups, of which only a few examples are discussed here:

Flavonoids found in plants comprise approximately 4,000 compounds, including anthocyanins and flavonols. Among these are many natural colourants (pigments) of flowers and fruits of numerous plant species.

Tannins are phenolic polymers found in the bark and leaves of many woody plants, and also in wine and tea, for example. Because of their protein-binding properties, they are also suitable for tanning animal hides and producing leather.

Lignins are also phenolic polymers and have great importance among the vascular plants as structural components.

Coumarins, which comprise approximately 500 known compounds, are primarily found in umbellifers (Apiaceae), legumes, and many grass species. In the living plant, coumarins occur as glycosides, i.e. bound do a sugar. They are only released during drying and are responsible for the typical smell of dry hay.

3. **Terpenoids**. The members of this group of compounds are made up of five carbon atoms, the so-called isoprene unit. These approximately 22,000 known substances are derived from the biosynthesis of fatty acids. Their classification is based on the number of carbon atoms, specifically on the number of C_5 units.

Hemiterpenes (C_5) and **monoterpenes** (C_{10}) comprise a large number of usually volatile substances, many of which are found as components of essential oils. They are responsible for the characteristic smell of plant species which are used primarily as sources of essential oils and aroma compounds and as spices. Examples include menthol from the peppermint plant (*Mentha piperita*), linalol from coriander (*Coriandrum sativum*), and thymol from thyme (*Thymus vulgaris*). The monoterpenes include the pyrethrins of the flowers of *Chrysanthemum* species, which can be used as natural insecticides (Sect. 5.2.1.2).

Sesquiterpenes (C_{15}) are also components of essential oils, but the phytohormone abscisic acid also belongs to this group. It is found in plants and promotes leaf and fruit fall, among other things.

Diterpenes (C_{20}) are important components of resins, for example abietic acid in conifers. They also comprise the gibberellins which, as phytohormones, promote seed germination and flower formation.

Triterpenes (C_{30}) consist of a range of different compounds. Many of these are found in waxes and resins. Another group are the cucurbitacins, which are found as bitter substances in the cucurbits (Cucurbitaceae). The triterpene saponins include, for example, ginsenosides in ginger (*Panax ginseng*) root and azadirachtin of the Neem tree (*Azadirachta indica*) which has an insecticidal property (Sect. 5.2.1.2).

Tetraterpenes (C_{40}) are the carotinoids found in the chloroplasts of plants. They are responsible for the yellow or red colour of the leaves, flowers, and fruits of many species. Examples include lycopene (responsible for the red colour of tomatoes and rose hips) and zeaxanthin (responsible for the yellow colour of maize kernels).

Polyterpenes ($C_{>40}$) include the important compound caoutchouc (natural rubber). It is found as an emulsion in the latex of the rubber tree (*Hevea brasiliensis*; Fig. 4.50) and in other plants of the spurge family (Euphorbiaceae).

The Importance of Secondary Metabolites for Herbivores

Why do plants produce this array of secondary metabolites? Only a few compounds, for example flower pigments and plant hormones, seem to fulfil important functions. The vast majority of the secondary metabolites are not involved in the growth and

Fig. 4.50 Extraction of latex from the trunk of the rubber tree (*Hevea brasiliensis*)

development of the plants. However, most of these compounds most probably justify their existence in that they serve to protect plants from the negative effects of other organisms. This assumption is supported by the fact that secondary metabolites have negative effects on the metabolism of herbivores:

- Alkaloids affect a broad range of physiological processes in the organisms of consumers. They can inhibit the synthesis of DNA, RNA, and proteins, membrane transport, and the effect of enzymes, or block the receptors of nerve tissues.
- Isothiocyanate and cyanide primarily impair cell respiration by inhibition of the associated enzymes.
- Many non-protein amino acids are very similar to some protein-forming amino acids. They are therefore sometimes accidentally built into proteins during biosynthesis and, as a result, enzyme function is impaired.
- Tannins form complexes with proteins. As a result, protein digestion and the availability of amino acids in the food to consumer organisms are limited. Plant materials with more than 2% tannin content are not suitable as food for herbivores, whether insects or vertebrates.
- Many of the volatile terpenoids and the components of waxes have a deterring effect on herbivores. Pyrethrin has a neurotoxic effect on insects, and other terpenoids affect hormone balance.

Although plants can have different effects on their herbivorous enemies by production of secondary metabolites, they are not usually protected against attacks by this mechanism. Herbivores and other animals have families of enzymes in their digestive tracts which are called mixed-function oxidases (MFOs). Relatively small amounts of the damaging substances taken up in food can be rendered ineffective by these enzymes, primarily by two mechanisms:

- toxic compounds are broken down into harmless components with enzymes and channelled into metabolism, or

- the toxic compounds maintain their basic structure but are chemically bound and then excreted or stored in body tissues.

Furthermore, there are large differences in the ability of herbivores to tolerate high concentrations of secondary metabolites in their food. Specialist feeders are, because of special physiological adaptations, able to make large quantities of toxic substances ineffective. This ability is usually the result of specific enzymes and, therefore, usually restricted to single compounds or to a very limited group of secondary metabolites. Generalist feeders, in contrast, are dependent on food that contains only a small concentration of the toxic substances.

Example: Allylglucosinolate is a mustard oil found in many species of the Brassicaceae (e.g. in *Brassica oleracea*). Because even low concentrations of this compound are acutely toxic to caterpillars of the swallowtail butterfly species, *Papilio polyxenes*, members of the mustard family do not serve as a food source for this species. *P. polyxenes* feeds exclusively on plants that contain furanocoumarins (primarily found in representatives of the Apiaceae). The caterpillars of the small white butterfly (*Pieris rapae*) are not affected even by artificially high concentrations of allylglucosinolate, because this species feeds exclusively on Brassicaceae and is adapted to cope with this compound. Development of the caterpillar of the owlet moth *Spodoptera eridania*, a generalist, is inhibited by moderately high quantities of allylglucosinolate, which means this species tolerates higher quantities of this substance in its food than *P. polyxenes*, but less than *P. rapae* (Blau et al. 1978).

Food Selection by Herbivores

The selection of food by herbivores is primarily determined by chemical compounds in the plants, which may either stimulate or deter feeding. How the individual species react to particular substances depends on whether they are generalist or specialist feeders.

- Generalists are predominantly affected by the primary plant metabolites, which serve the herbivores as food sources (e.g. sugars) and stimulate feeding. The host plant range of generalists is, in effect, determined by the absence of feeding-inhibiting substances (especially secondary metabolites). It is often not the plant species that determines on palatability but rather the development stage of the plant, because the concentration of many compounds in plant tissues can change with the development of the plant. Thus, shoots of some grass species contain high concentrations of substances that inhibit the feeding of grasshoppers whereas the same plant in a later stage of development exerts a positive effect on the feeding of these insects.
- Specialists in their search for food are enticed by the characteristic secondary metabolites of the plants to which they are adapted. Thus, for the specialists, substances that act as inhibitors for generalists act as feeding stimulants. Such substances therefore also serve as identifying features for the feeding plants and determine the host plant range of the specialist species.

4.5.5.3 Direct Induced Defence

The structural and chemical barriers of a plant before contact with an attacking herbivore are its constitutive defence (or resistance) mechanisms. However, plants are also able to react on an attack by organisms, in other words to initiate additional defence which is deployed only when protection is needed. A physiological reaction of the plant to herbivore feeding, which leads to reduction in the attractiveness to the attacker of the plant or the attacked parts of the plant, is called **direct induced defence** (or direct induced resistance). Types and effects of such mechanisms are considered in the following sections.

Another type of induced defence is called indirect induced defence. It includes plant traits that by themselves do not affect the susceptibility of host plants, but can serve as attractants to natural enemies of the attacking insect. Such interactions are considered in Sect. 5.2.4.3.

Direct Induced Increase of Secondary Plant Metabolites

Mechanisms of direct induced defence are based either on increased production of specific substances that are already present in the tissue or on synthesis of new compounds. Such changes can occur within a few hours or over several days. Herbivore feeding often leads to an increased concentration of secondary metabolites in the damaged tissue, from which negative effects can result for generalist feeders.

Example. The caterpillars of the cabbage looper (*Trichoplusia ni*) are generalists and occasionally feed on wild parsnip (*Pastinaca sativa*). The plants react to damage of their tissues by increasing production of furanocoumarins (Fig. 4.51). The development of the caterpillars that feed on previously damaged leaves is severely impaired by the elevated levels of secondary metabolites. The individuals achieved significantly lower weight than those that ingested undamaged leaf material (Zangerl 1990).

Direct Induced Production of Inhibitors of Insect Digestive Enzymes

In addition to secondary metabolites, plants also produce several proteins and polypeptides which have negative effects on herbivores by use of primary metabolic pathways. One of most important groups of such compounds includes inhibitors of insect digestive enzymes, especially the **protease inhibitors**. These inhibit the activity of the enzymes that break down proteins (the proteases) in the digestive system of consumers. In the plant kingdom, such protease inhibitors are widespread and not limited to particular taxa, as are the secondary metabolites. Most of the protease inhibitors found in plants affect serin-proteases, including trypsin and chymotrypsin. These are important enzymes in protein breakdown (proteolysis) in insects, mammals, and many other groups of animals. The protease

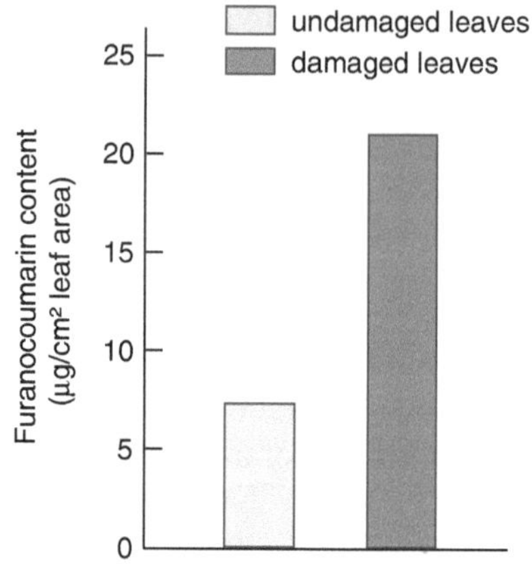

Fig. 4.51 Herbivore feeding leads to increased levels of furanocoumarins in the leaves of wild parsnip, *Pastinaca sativa* (Based on Zangerl 1990)

inhibitors thus affect protein digestion by herbivores, which consequently impairs growth and development.

The importance of protease inhibitors in direct defence against herbivores was discovered by Green and Ryan (1972). They found that the amount of a particular protein with protease-inhibiting properties increased fourfold in young tomato plants in response to attack by Colorado potato beetles. This change occurred not only in the leaves which had been damaged but next day in other, undamaged leaves of the same plant. The induced production of protease inhibitors is, thus, a **systemic defence**, in other words a reaction that not only affects the damaged tissues, but the entire plant. Other experiments have demonstrated that ingestion of protease inhibitors inhibits the development of herbivores.

Example: Caterpillars of the butterfly *Spodoptera exigua*, a generalist species, were fed with tomato leaves that contained different levels of protease inhibitors. A significant inverse relationship was evident between inhibitor concentration and body weight of the caterpillars (Broadway et al. 1986; Fig. 4.52).

Other types of inducible defensive proteins include proteases with direct toxicity, for example lectins, that reversibly bind to specific mono or oligosaccharides and polyphenol oxidases. These are anti-nutritive enzymes that catalyze the oxidation of phenolics to quinones, reducing the nutritive value of the wounded plant (Chen 2008).

Not surface structures but also the production of different substances gives plants the ability to defend themselves against herbivores. Overall, the biochemical mechanisms of defence are far more important than morphological defences.

Mechanisms of Direct Induced Defence

Induction of a direct defence mechanism in plants not only results from herbivore feeding, but also from attack by pathogens (Sect. 4.6.4.2), from artificial damage to

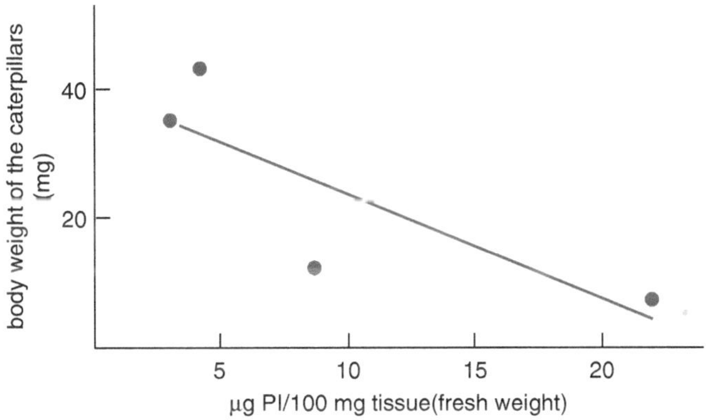

Fig. 4.52 Relationship between the body weight of caterpillars of the butterfly *Spodoptera exigua* and the concentration of protease inhibitors in the tomato leaves of their food (Based on Broadway et al. 1986)

the tissues, and from the external application of specific substances. The manner in which the plants react to such effects depends on the inducing substances, which are called **elicitors**. Such substances can be released by the plants themselves (e.g. as a consequence of artificial damage to tissues by removal of leaf parts), but usually originate from the organism causing the damage (animal or pathogen). Several chemically very different substances identified in herbivorous insects are found in the oral secretions of the animals and act as elicitors. It is already known that such substances include specific proteins and peptides and fatty acid amino acid conjugates. These substances come into contact with the damaged plant tissues or receptors during feeding and, as a consequence, production of the respective defence substances is initiated via specific metabolic pathways. Attack by herbivores usually elicits rapid accumulation of jasmonic acid, as a result of activation of constitutively expressed biosynthetic enzymes (Bonaventure and Baldwin 2010). Little is known in detail on the course of these processes. The genes responsible for the production of the protease inhibitors are activated by a variety of signal pathways which in each case involve such phytohormones as salicylic acid, abscisic acid, and ethylene.

Artificial application of jasmonic acid to the leaf surfaces of, for example, tomato (Fig. 4.53) and potato leads to similar reactions in the plants as herbivore feeding. The development of insects that feed on the treated plants is thereby impaired (Fig. 4.54). However, jasmonic acid itself has no toxic effect on the animals (Thaler et al. 2001).

Specificity of Direct Induced Defence

The physiological reactions initiated in the plants by the elicitors are complex and consist of quantitative changes in numerous substances within in the different tissues.

Fig. 4.53 Tomato plant (*Solanum lycopersicum*)

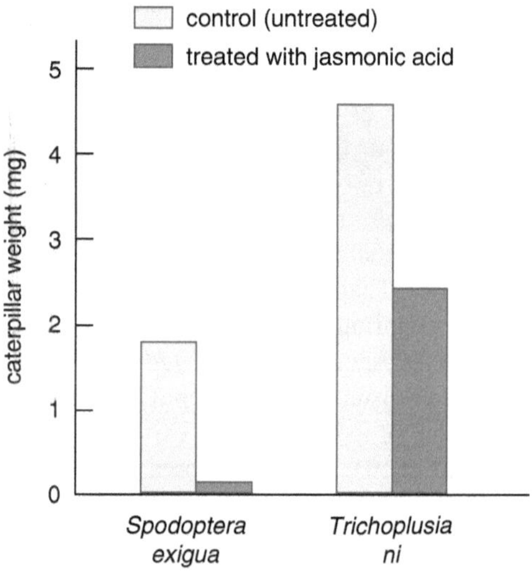

Fig. 4.54 Body weight of caterpillars of two butterfly species on tomato plants that were sprayed with jasmonic acid (approximately 0.1 mg per plant) and on untreated control plants, on which the animals fed (Based on Thaler et al. 2001)

Example: Feeding of caterpillars of the owlet moth *Helicoverpa zea* initiates three physiological changes in tomato plants: (1) a systemic increase in the concentrations of jasmonic acid, protease inhibitors, some proteins, the alkaloid tomatin, and the enzyme polyphenoloxidase, (2) an increase in the concentration of some enzymes (lipoxygenase, phenylalanine-ammonium-lyase, and peroxidase) in the damaged leaves, and (3) a decrease in the concentration of the two phenolic compounds chlorogenic acid and rutin (Thaler et al. 2001).

As a result of such processes, it can be assumed that, as a result of the defences induced in the plants, different effects are exerted on the herbivore attackers, which are not caused by the effect of one substance alone. This raises the question of the specificity of induced defences and includes two different aspects:

1. **Specificity of the plant response**. As far as known, plants not only react differently, depending on the form of damage to their tissues, but also produce, depending on the herbivore species that causes the damage, a different range of induced substances.

   Example: Feeding of the caterpillars of *Helicoverpa zea* on the leaves of tomato increased concentrations of protease inhibitors and the enzyme pholyphenoloxidase in the plant. The phloem-sucking aphid species *Macrosiphum euphorbiae* induces no increase in the synthesis of protease inhibitors in the tomato plants, but initiates increased activity of the enzymes peroxidase and lipoxygenase (Stout et al. 1998).
2. **Specificity of the effects**. The effects of the induced defences are usually relatively unspecific, i.e. individual herbivore species can be affected by the reactions that were initiated by other species. Conversely, herbivores can initiate the synthesis of substances that have negative effects on other organisms (e.g. phytopathogenic fungi).

   Example: Feeding of the caterpillars of *Helicoverpa zea* on the leaves of tomato induces the synthesis of substances that have a negative effect on four other organisms including an aphid species (*Macrosiphum euphorbiae*), a spider mite species (*Tetranychus urticae*), caterpillars of *Spodoptera exigua*, and the phytopathogenic bacterium *Pseudomonas syringae* (Stout et al. 1998).

4.5.5.4 Effectiveness of Plant Defences

The previous sections have shown that plants have numerous mechanisms to defend themselves against feeding enemies. Many herbivores are, however, able to utilize at least a specific range of plant species, perhaps at specific stages of their development, as food sources, by use of a variety of strategies. The following overview shows how both the plants and the herbivores manage to secure the survival of their populations.

1. The morphological structures of plant surfaces can act as an initial barrier to some feeding enemies and prevent or impede their attacks on the plant tissues. These defence mechanisms can affect both specialists and generalists, but do not usually provide complete protection against herbivores.
2. The second level of constitutive defences is the presence of secondary metabolites. In so far as these are present in some development stages and tissues in sufficient concentrations, they prevent the feeding activities of generalists or at least strongly limit the extent of damage. However, specialists adapted to the respective secondary metabolites are not restricted in their feeding.

Fig. 4.55 Tobacco plant (*Nicotiana tabacum*)

3. An induced increase in the concentration of secondary metabolites can deter some generalists and, sometimes, specialists also, if the concentration of such substances tolerated by the herbivores is exceeded in the food.

   Example: Caterpillars of the tobacco hornworm (*Manduca sexta*) feed exclusively on the Solanaceae and appear as pests on tobacco (Fig. 4.55). The animals can tolerate a some nicotine in their leaf food and have the ability to excrete ingested nicotine in the unchanged form. However, when the concentration of this substance in the leaves exceeds approximately 0.5%, the animals consume less food and their growth is impaired. Baldwin (1988) showed this by artificially increasing uptake of nicotine by tissues of plant leaves by use of a solution of the compound.
4. With the induced synthesis of protease inhibitors and other proteins, a level is reached at which the feeding activity of specialists, which are not or only slightly affected by the other mechanisms, are impaired.

   Example: The cabbage stem flea beetle (*Phsylliodes chrysocephala*) is, as a Brassicaceae specialist, adapted to the glucosinolates of its host plant. Feeding experiments with artificial foods have shown that the rate of feeding increases with increasing glucosinlate content. When canola plants were damaged by herbivores or treated with jasmonic acid, significantly higher concentrations of glucosinolates were found within a few hours. It is to be expected that by this means the feeding behaviour of the beetles would also be positively affected. However, there was no increase in the rate of feeding on the leaves of plants with increased glucosinolate content induced by application of the jasmonic acid. It can, therefore, be assumed that, in addition to the increase in the concentration of secondary metabolites, other mechanisms of induced defence are initiated

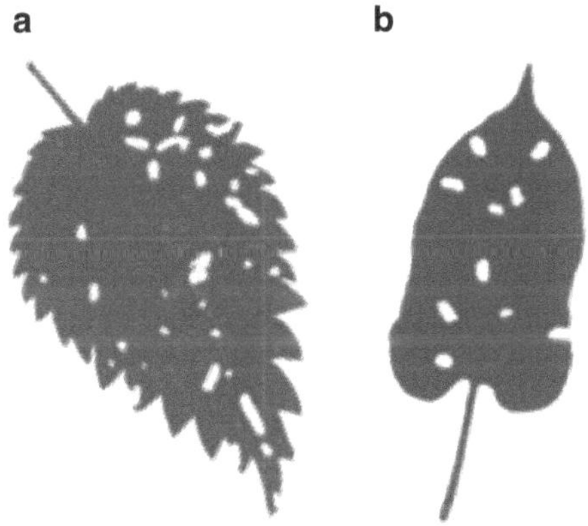

Fig. 4.56 Leaves of two wild plants from their natural sites, damaged by herbivores. (**a**) Stinging nettle (*Urtica dioica*) from southern Germany, (**b**) *Merremia umbellata* (Convolvulaceae) from northern Thailand

which are targeted at specialists and which counteract the feeding-stimulation effect of the glucosinolates (Bartlet et al. 1999).

Although the mechanisms of induced defence do not provide the plants with comprehensive protection against the attacks of herbivores, they can have an important effect on the distribution and extent of feeding damage, and on the feeding behaviour of the animals. For wild plants, at least, herbivorous arthropods usually cause only minor loss in biomass. This loss usually amounts to only 5–10% of the leaves, which can usually be found on any plant in nature (Fig. 4.56).

Feeding is, furthermore, rarely concentrated on a particular area of the plant but is, rather, relatively evenly distributed. This is the case both for a particular leaf and between leaves. According to the hypothesis of Edwards and Wratten (1983), such patterns are essentially a result of the effects of induced defence by the plants: in reaction to the feeding activities of a herbivore, specific compounds are produced in higher concentrations, which leads to a decline in the food quality of the tissues. A consumer is, therefore, forced to stop food intake and search for another site on the tissue or plant. In extreme cases the herbivore has to give up a particular plant as food source and move to another plant. This situation occurs especially when the action of the induced defence is systemic. In any case, the result is spatial distribution of the consumers and the areas of feeding damage, from which the plant draws a relative advantage. The loss of entire leaves is worse than minor damage to several single leaves, which can, e.g., become important in competition with other plants for light.

In many production systems it has been shown that phytomass loss of plants resulting from herbivore feeding is often far higher than 5–10% when no effective pest management is carried out. Several factors may be responsible for this:

- Fertilization and irrigation not only promote growth of the plant but also result in improved food quality for herbivores (Sect. 4.5.4.2).
- For many cultivated species and varieties, breeding measures have led to a change in the physiological properties which were accompanied by a decrease in resistance to pests (Sect. 5.2.2).

- Cultivated plants are usually grown in monocultures, which facilitates the ability of the herbivores to find their hosts and promotes the development of high pest densities.
- In uniform plant stands, the conditions for existence of many predators and parasitoids which appear as antagonists to the pests are limited (Sect. 5.2.4.5).
- Insecticides usually act not only on the pests but also have negative effects on their antagonists. The density of some pests is, sometimes, not reduced by use of insecticides but rather remains unchanged or even increases (Sect. 5.2.1.4).

4.6 Phytopathogens

The microorganisms which cause plant diseases (phytopathogens) are essentially viruses, bacteria (including phytoplasmas, a special group of bacteria that are primarily characterized by the lack of a cell wall), and fungi. They utilize substances and energy from the plants for their own development which, similar to the feeding activities of herbivores, can lead to a decline in fitness and yield. The development of a plant disease, the pathogenesis, has several phases:

1. In the first stage, attack of a pathogen, i.e. its contact with the plant, occurs; this, in phytopathological terminology, is called **inoculation**. An inoculum is thus a pathogen that can reproduce and may be viruses, bacteria cells, or fungal spores.
2. In the second phase, the **infection** occurs, which means the pathogen enters the plant. Because undamaged plants are protected by dermatic tissues (epidermis, rhizodermis), the pathogen must find a way by which it can enter the plant or a plant cell. Depending on the type of pathogen, this occurs via plant wounds, which are most often caused by other factors, via natural openings in the plant (e.g. stomata), or by production of enzymes, by means of which the cuticle and cell walls can be dissolved.
3. During further progress of the pathogenesis, the causative agent spreads from the site of infection into the plant. Emergence of the disease is accompanied by a variety of biochemical and physiological changes in plant metabolism, which may be the direct result of activities of the pathogen (e.g. the effect of toxins) or which may result from the plant's defences against the pathogen (Sect. 4.6.4). External signs of the disease are histological or anatomical modifications, which are termed **symptoms**. These include numerous indications that are usually typical of a particular pathogen. The most important symptoms include wilting, changes in colour, death of cells or tissues, and changes in the growth of the plant or of plant parts.
4. In the last phase of the pathogenesis, reproduction of the pathogen usually occurs.

Table 4.8 Examples of economically relevant viral diseases of crop plants

Disease	Vectors	Affected crops (examples)
Beet necrotic yellow vein virus	*Polymyxa betae* (fungus)	Sugar beet
Barley yellow dwarf virus	Wheat aphids	Barley, oats, wheat
Potato leafroll virus	Aphids	Potato
Tobacco mosaic virus	None	Tobacco
Sugar beet savoy disease	*Piesma quadrata* (beet leaf bug)	Sugar beet, kale
Plum pox virus	Aphids, leafhoppers	Plum, peach
Cucumber mosaic virus	Aphids	Cucumber, pumpkins, melons
Pea venation mosaic	Aphids	Pea, bean, alfalfa

4.6.1 Viruses

Although viruses are often described as microorganisms, they are, in fact, not living organisms in the strict sense. They do not have metabolism of their own and have only one type of nucleic acid (either DNA or RNA), whereas true organisms possess both types. Viruses can only reproduce themselves in living organisms and are dependent on the nucleic and amino acids of their host for this. Phytopathogenic viruses do not have the ability to independently enter an intact plant or its cells but can, instead, only do this via wounds in the tissue that are usually the result of other organisms. Simultaneously, these other organisms often serve the viruses as **vectors** through which they may be transported from one plant to another. The most important vectors include sap-sucking insects (especially aphids, leafhoppers, and shield bugs), nematodes, mites, and phytopathogenic fungi. Inside the plant, the virus can spread with the aid of viral transport proteins via the plasmodesma (i.e., cytoplasmatic bridges between the cells, which are formed during cell reproduction).

The diseases caused by viruses (Table 4.8) are accompanied by a variety of changes in the plant which, especially, include the stimulation of nucleic acid and protein synthesis, but photosynthetic processes and the phytohormone balance may also be affected. Symptoms of viral pathogens include lesions (dead plant tissue), discolouration, and reduced or altered plant growth.

4.6.2 Bacteria

Phytopathogenic bacteria utilize the metabolic products of the plant for their own metabolism and reproduction. However, they are not dependent on living plants for their survival, but instead can also feed on organic substances outside the plant. Most bacteria enter a plant via the stomata, lenticels, wounds, or root hairs, which do not have a cuticle. Inside the plant, they move through the intercellular spaces or spread via the transport tissues (phloem, xylem).

By use of enzymes, many bacteria are able to dissolve cell walls, the middle lamella, and finally the plasmalemma inside the plant to utilize the cell contents as substrate. Bacterial infections often affect the metabolic process of the plant as a result of the effect of toxins. Most diseases caused by bacteria result in the dying of plant parts or in growth anomalies (Table 4.9).

Table 4.9 Examples of economically relevant bacterial diseases and pathogens of crops

Disease	Bacterial agent	Affected crops (examples)
Fire blight	*Erwinia amylovora*	Apple, pear, quince
Black rot	*Xanthomonas campestris*	*Brassica* species
Ring rot	*Corynebacterium sepedonicum*	Potato
Blackleg rot	*Erwinia carotovora*	Potato
Crown gall	*Agrobacterium tumefaciens*	Sugar beet
Halo blight	*Pseudomonas phaseolicola*	*Phaseolus* beans
Angular leaf spot	*Pseudomonas tabaci/amygdali*	Tobacco

4.6.3 Fungi

All fungi are heterotrophic and, therefore, dependent on the presence of organic compounds which they can obtain from a variety of substrates. On this basis they can be classified into several groups:

- obligatory parasites can only feed on living organisms,
- facultative parasites can feed on both living tissues and dead organic matter (detritus),
- perthophytes attack living organisms and feed on tissue they destroy,
- detritivores do not feed on living plants, but act as saprophages.

Many phytopathogenic fungi can, in contrast with viruses and bacteria, not only enter plants via natural openings and wounds, but also through intact plant surfaces.

Colonization of the host plant (inoculation) begins with adhesion of a spore followed by germination of the spore and the development of a germination tube. This forms an organ (appressorium) that attaches to and penetrates cells and develops into an infection hypha. Depending on the species of fungus, this then either enters the plant via openings (e.g. stomata) or penetrates an epidermal cell directly. In the latter case, the cuticle, middle lamella, and cell wall must be overcome by use of appropriate enzymes (primarily cutinase, pectinase, and cellulase). In the obligatory parasites, the infection hyphae develop into a so-called haustorium through which the fungi can draw nutrients from the plant cells (Fig. 4.57).

The fungus does not usually spread throughout the plant but is restricted to a more or less large area of affected tissue. As a result, cells usually die. Table 4.10 shows some examples of fungal diseases.

4.6.4 Plant Defences Against Pathogens

In contrast with herbivore interactions, no differentiation is made between antixenosis, antibiosis, and tolerance in the resistance of plants to pathogens

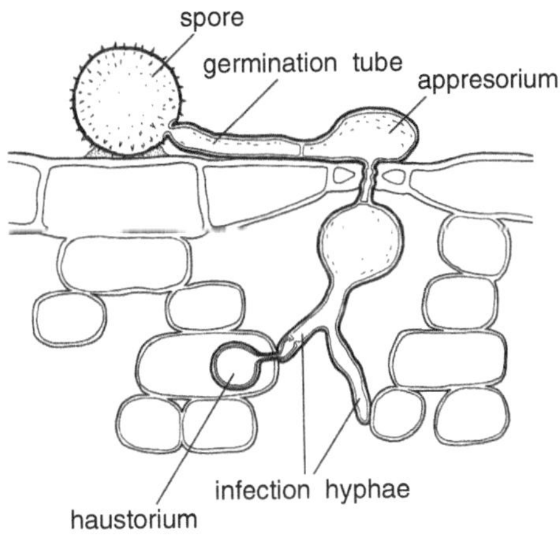

Fig. 4.57 Schematic representation of the infection structure of a rust fungus (Order Uredinales) which enters a leaf via the stomata (Drawing by Ch. Allgaier)

Table 4.10 Examples of economically relevant fungal diseases and pathogens of crops

Disease	Fungal pathogen	Affected crop plants (examples)
Powdery mildew	*Erysiphe (Blumeria) graminis*	Wheat, barley, rye
Stripe rust	*Puccinia striiformis*	Wheat, barley
Ergot	*Claviceps purpurea*	Rye, wheat, barley
Common smut	*Ustilago zeae*	Maize
Bakanae	*Fusarium fujikuroi*	Rice
Bean rust	*Uromyces appendiculatus*	*Phaseolus* beans
Late or potato blight	*Phytophthora infestans*	Potato
Blackleg disease	*Phoma lingam*	*Brassica* plants
Beetroot and mangold black leg and dry rot	*Pleospora bjoerlingii*	Sugar beet
Apple scab	*Venturia inaequalis*	Apple
Black sigatoka	*Mycosphaerella fijiensis*	Banana

(cf. Sect. 4.5.5). Instead, a differentiation is made between two extreme forms of the plant–pathogen relationship:

- **Incompatibility** refers to the situation in which the plant can successfully defend itself against a potential agent of disease, in other words when the plant can prevent reproduction of the pathogen. This is the case when the plant has morphological, physiological, or biochemical barriers that cannot be overcome by a particular pathogen. A plant with such properties is described as being resistant to the respective pathogen.
- **Compatibility** refers to the situation in which a pathogen succeeds in entering the plant, and reproduces. In such a case, the plant is susceptible, which is evident from the occurrence of disease symptoms (e.g. yellowing, wilting, lesions).

There is, however, a continuum between resistance and susceptibility. In other words, just as with the herbivores, gradual differences in the degree of resistance exist (cf. Sect. 4.5.5). These are based primarily on the plant's ability to defend itself against pathogens and the speed with which the plant reacts.

Different from the interactions between plants and herbivores, **tolerance** is not seen as a form of resistance to phytopathogens, but has a different meaning. For pathogens, tolerance is defined as the ability of a plant to endure infection by the disease agent without suffering substantial damage or loss in fitness. Thus, in contrast with resistance, infestation occurs but no symptoms of disease emerge (passive tolerance) or compensation for the damage occurs (active tolerance).

As for plant defences against herbivores, it is possible to distinguish between constitutive and induced mechanisms in the types of resistance of the plants against pathogens.

4.6.4.1 Constitutive Mechanisms

The mechanisms of constitutive resistance are based on the morphological characteristics of the plant and on the chemical compounds it has available before contact with the pathogen occurs. The most important mechanical protection against a pathogen is an intact surface, which viruses and bacteria, at least, cannot overcome. The wax layer is particularly important in defence against phytopathogenic fungi, because this layer has lower wettability owing to its hydrophobic properties, and therefore impairs germination of spores.

Secondary metabolites (Sect. 4.5.5.2) are also important in defence against pathogens. A variety of substances (primarily phenolic compounds) are released by plants on to the surfaces of leaves, roots, and seeds, where they have an effect on microorganisms. These consist of, for example, inactivation of enzymes with which fungi attack plant dermal tissues or inhibition of spore germination and the growth of mycelia. Additional secondary metabolites are located in plant cells and have a toxic effect on the disease agent after the tissue has been damaged. In infected areas formation of such substances often increases.

4.6.4.2 Induced Mechanisms

The reaction of plants in the defence against pathogens is initiated by elicitors (Sect. 4.5.5.3), which originate from the attacking pathogen or from the damaged dermal tissues of the plant. The reaction, which is initiated by a signal pathway, can be limited to the affected area (as occurs with histogenic reactions, the hypersensitive reaction, and with production of phytoalexins), or can involve the entire plant (systemic resistance).

Histogenic Reactions

Reactions of the plant to the attacking disease agent originating in tissues (histogenic) are primarily found in the cell wall. Papillae are often formed as a defence against fungi, the infective hyphae of which enter the wall of the epidermis cells. This involves the deposition of substances (e.g. phenolics, callose, silicic acid) from the cell on the inside of the cell wall. The papillae are particularly resistant to fungal enzymes. However, fungi often enter the plant faster than it can form such a barrier, so comprehensive and effective protection is not achieved by formation of papillae.

Another often observed histogenic reaction of plants is increased production of lignin. As a result, not only is mechanical resistance increased but possibly also a barrier for materials is built up. This impairs permeation of the enzymes and toxins of the attacker and also reduces the attacker's supply of water and nutrients.

Hypersensitive Reaction

A phenomenon which often occurs after attack of a plant by a pathogen is the death of cells at the site of infection. This process can occur within minutes or hours and is not initiated by the pathogen but rather by the plant itself. As a result of this hypersensitive reaction the attacker is isolated by the plant before it can spread further. The dead tissue is visible as dot-shaped lesions. Production of highly reactive, damaging compounds of oxygen, including superoxide (O_2^-), the hydroxyl radical (OH^-), and hydrogen peroxide (H_2O_2), is probably responsible for initiation of hypersensitive cell death. Production of phytoalexins has also been observed in association with this.

Production of Phytoalexins

Phytoalexins are secondary metabolites (mainly terpenoids and phenolics) that are produced after the plant has been attacked by a pathogen. They have unspecific antimicrobial effects (primarily on fungi). Their synthesis only occurs in the tissues of affected areas. Currently more than 350 such compounds are known, from approximately 30 plant families, of which approximately one third are in the Fabaceae family (legumes). Examples of phytoalexins are gossypol in cotton, pisatin in pea, and rishitin in potato and tomato.

Systemic Resistance

For many plant species it has been shown that defence reactions initiated in response to the first infection by a pathogen lead to increased resistance of the plant to subsequent attacks of pathogens. This "immunization" is termed **acquired systemic resistance** and can be regarded as a special form of induced systemic

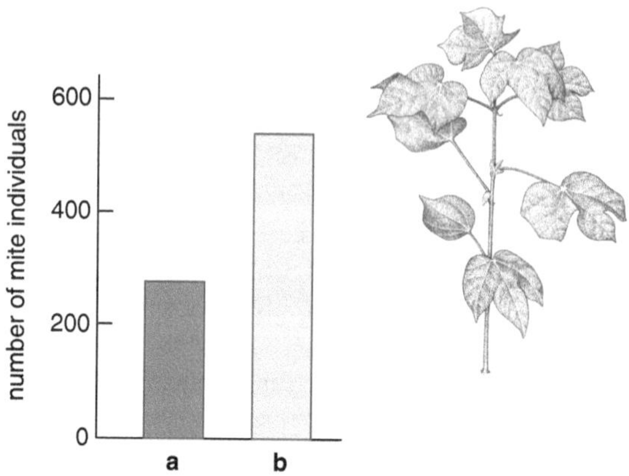

Fig. 4.58 Number of spider mites on cotton plants that (**a**) were attacked in the seedling stage by *Verticilium* wilt, and (**b**) that were not infected (Based on Karban et al. 1987)

resistance or defence (Sect. 4.5.5.3). It has been demonstrated as occurring in cucumber plants, which are normally susceptible to the fungus *Colleotrichum lagenarium*. When the leaves of plants are infected by this pathogen the first time, typical disease symptoms are visible. However, attacks by the fungus are successfully repelled by leaves that develop later on the same plant (Conrath and Kauss 2000). In this case, as in others, it was found that the acquired systemic resistance applied not only to the pathogen that caused the first infection, but also to a variety of other pathogens, for example viruses and bacteria. This also occurs when the situation is reversed, i.e., an initial infection with viruses or bacteria provides resistance against fungi. The acquired systemic resistance is thus a very unspecific defence mechanism which is effective not only against phytopathogens but may also be effective against herbivores.

Example: Incidence of spider mites (*Tetranychus urticae*) on cotton shoots previously infected with the fungus *Verticilium dahliae* (the agent of *Verticilium* wilt) was lower than on plants with no previous contact with the fungus (Fig. 4.58). When the order was reversed, initial attack by mites was also observed to reduce the susceptibility of the cotton plants to the fungus (Karban et al. 1987).

The phenomenon of acquired systemic resistance is based on defence processes that include the synthesis of so-called PR proteins (PR = pathogenesis related). These include compounds that may be found in plants after a first infestation and, at least in part, have a direct function in defence. For example, some PR proteins attack the cell walls of some fungi or are involved in defence mechanisms as enzymes. It has also been shown that different defence reactions of the plant (e.g., histogenic changes and the hypersensitive reaction) occur faster and to a greater extent after systemic resistance has been acquired.

Salicylic acid has a specific function in the systemic resistance to pathogens. Its concentration in plant tissues increases substantially after microbial infection. Salicylic acid acts as a signal substance, is transported in the phloem, and is involved in the production of PR proteins. Artificial application of salicylic acid to a leaf surface can initiate these processes. In this an effect occurs similar to that which results from jasmonic acid treatment against herbivorous insects (Sect. 4.5.5.3). Compounds that are structurally similar to salicylic acid are substantially more effective because they can enter tissue more easily. One of these is benzothiadiazol, which is used as an active ingredient in synthetic formulations used to initiate systemic resistance. It does not have a direct affect on the pathogenic organism, but instead takes on the role of salicylic acid in the signal pathway. Benzothiadiazol is used, for example, as preventative protection in the treatment of wheat against powdery mildew (*Erysphe graminis*) or of lettuce against downy mildew (*Bremia lactuca*). Disease infestation is not always completely prevented but is at least limited or delayed.

References

Baldwin IT (1988) Short-term damage-induced increases in tobacco alkaloid protect plant. Oecologia 75:367–370

Bartlet E, Kiddle G, Martin A, Williams I, Wallsgrove R (1999) Wound-induced increases in the indolyl glucosinolate content of oilseed rape and their effect on subsequent herbivory. Entomol Exp Appl 91:163–167

Bayerische Landesanstalt für Landwirtschaft (2000) Integrierter Pflanzenschutz. Rapsschädlinge. www.Lfl.Bayern.de/Publikationen

Bernays EA, Chapman RF (1994) Host-plant selection by phytophagous insects. Chapman and Hall, New York

Bernays EA, Graham M (1988) On the evolution of host specificity in phytophagous arthropods. Ecology 69:886–892

Blau PA, Feeny P, Contardo L (1978) Allylglucosinolate and herbivorous caterpillars: a contrast in toxicity and tolerance. Science 200:1296–1298

Bonaventure G, Baldwin IT (2010) New insights into early biochemical activation of jasmonic acid biosynthesis in leaves. Plant Signal Behav 5:287–289

Broadway RM, Duffey SS, Pearce G, Ryan CA (1986) Plant proteinase inhibitors: a defense against herbivorous insects? Entomol Exp Appl 41:33–38

Buschmann H (2003) Die Bedeutung der Allelopathie für die Landwirtschaft. Praxis der Naturwissenschaften (Biologie) 4(52):6–11

Carlson HL, Hill JE (1985) Wild oat (*Avena fatua*) competition with spring wheat: effects of nitrogen fertilization. Weed Sci 34:29–33

Chen MS (2008) Inducible direct plant defense against insect herbivores: a review. Insect Sci 15:101–114

Claridge MF, Wilson MR (1981) Host plant associations, diversity and species-area relationships of mesophyll-feeding leafhoppers of trees and shrubs in Britain. Ecol Entomol 6:217–238

Conrath U, Kauss H (2000) Die systemisch erworbene Resistenz von Pflanzen gegen Krankheitserreger. Biologie in unserer Zeit 31:202–208

Daubenmire RF (1974) Plants and environment, 3rd edn. Wiley, New York

Edwards PJ, Wratten SD (1983) Wound induced defences in plants and their consequences for patterns of insect grazing. Oecologia 59:88–93

FAO-UNESCO (1988) Soil map of the world. World soil resources report 60, FAO, Rome

Gliessman SR (2000) Agroecology. Lewis Publishers, Boca Raton

Green TR, Ryan CA (1972) Wound-induced proteinase inhibitor in plant leaves: a possible defense mechanism against insects. Science 175:776–777

Grenz J (2004) Using a simulation model to evaluate parasitic weed management strategies: case study of the association *Vicia faba – Orobanche crenata*. Agroecology series 7, Apia Verlag, Laubach
Hartmann T (1996) Diversity and variability of plant secondary metabolism: a mechanistic view. Entomol Exp Appl 80:177–188
Hawksworth DL, Mound IA (1991) Biodiversity databases: the crucial significance of collections. In: Hawksworth DL (ed) The biodiversity of microorganisms and invertebrates: its role in sustainable agriculture. CAB International, Wallingford, pp 17–29
Hunt DW, Drury CF, Maw HE (1992) Influence of nitrogen on the performance of Colorado potato beetle (Coleoptera: Chrysomelidae) on tomato. Environ Entomol 21:817–821
Jauset AM, Sarasúa MJ, Avilla J, Albajes R (1998) The impact of nitrogen fertilization of tomato on feeding site selection and oviposition by *Trialeurodes vaporariorum*. Entomol Exp Appl 86:175–182
Karban R, Adamchak R, Schnathorst WC (1987) Induced resistance and interspecific competition between spider mites and vascular wilt fungus. Science 235:678–680
Klein A-M, Steffan-Dewenter I, Tscharntke T (2003) Fruit set of highland coffee increases with the diversity of pollinating bees. Proc Royal Soc Lond B 270:955–961
Koch W (1994) Grundzüge des Pflanzenschutzes mit dem Versuch einer Bewertung. PLITS 12(4), W. & S. Koch Verlag, Stuttgart
Laber H (1999) Effizienz mechanischer Unkrautregulationsmaßnahmen im Freilandgemüsebau. Dissertation, Fachbereich Gartenbau, Universität Hannover
Lemerle D, Verbeek B, Coombes N (1995) Losses in grain yield of winter crops from *Lolium rigidum* competition depend on crop species, cultivar and season. Weed Res 35:503–509
Martin K (1994) Struktur und Nahrungsnetze aquatischer Reisfeld-Biozönosen im traditionellen System der Ifugao. Plits 12(5), W. und S. Koch Verlag, Stuttgart
Martin K, Sommer M (2004) Relationships between land snail assemblage patterns and soil properties in temperate-humid forest ecosystems. J Biogeogr 31:531–545
Mattson WJ (1980) Herbivory in relation to plant nitrogen content. Ann Rev Ecol Syst 11:119–161
Moore JC, De Ruiter PC (1991) Temporal and spatial heterogeneity of trophic interactions within below-ground food webs. Agric Ecosyst Environ 34:371–397
Myint MM, Rapusas HR, Heinrichs EA (1986) Integration of varietal resistance and predation for the management of *Nephotettix virescens* (Homoptera: Cicadellidae) populations on rice. Crop Prot 5:259–265
Pimentel D, Kounang N (1998) Ecology of soil erosion in ecosystems. Ecosystems 1:416–426
Pimentel D, Harvey C, Resosudarmu P, Sinclair K, Kurz D, Mcnair M, Crist S, Sphritz L, Fitton L, Saffouri R, Blair R (1995) Environmental and economic costs of soil erosion and conservation benefits. Science 267:1117–1123
Roubik DW (1995) Pollination of cultivated plants in the tropics. FAO Bulletin 118
Scheffer F, Schachtschabel P (2002) Lehrbuch der Bodenkunde, 15th edn. Spektrum Akademischer Verlag, Heidelberg
Schillinger JA, Gallun RL (1968) Leaf pubescence of wheat as a deterrent to the cereal leaf beetle, *Oulema melanopus*. Ann Entomol Soc Am 61:900–903
Schoonhoven LM, Jermy T, Van Loon JJ (1998) Insect-plant biology. From physiology to evolution. Chapman & Hall, London
Sétamou M, Schulthess F, Bosque-Pérez NA, Thomas-Odjo A (1995) The effects of stem and cob borers on maize subjected to different nitrogen treatments. Entomol Exp Appl 77:205–210
Siebert S, Döll P (2010) Quantifying blue and green virtual water contents in global crop production as well as potential production losses without irrigation. J Hydrol 384:198–217
Steffan-Dewenter I, Münzenberg U, Bürger C, Thies C, Tscharntke T (2002) Scale-dependent effects of landscape context on three pollinator guilds. Ecology 83:1421–1432
Stout MJ, Workman KV, Bostock RM, Duffey SS (1998) Specificity of induced resistance in the tomato, *Lycopersicon esculentum*. Oecologia 113:74–91

Thaler JS, Stout MJ, Karban R, Duffey SS (2001) Jasmonate-mediated induced plant resistance affects a community of herbivores. Ecol Entomol 26:312–324

Walter H, Breckle S-W (1999) Vegetation und Klimazonen, 7th edn. UTB, Eugen Ulmer Verlag, Stuttgart

Williams IH (1994) The dependences of crop production within the European Union on pollination by honey bees. Agric Zool Rev 6:229–257

Winfree R, Williams NM, Gaines H, Ascher JS, Kremen C (2008) Wild bee pollinators provide the majority of crop visitation across land use gradients in New Jersey and Pennsylvania. J Appl Ecol 45:793–802

Wit AK (1982) The relation between artificial defoliation and yield in Brussels sprouts as a method to assess the quantitative damage induced by leaf-eating insects. Zeitschrift für angewandte Entomologie 94:425–431

Zangerl AR (1990) Furanocumarin induction in wild parsnip: evidence for an induced defense against herbivores. Ecology 71:1926–1932

Chapter 5
Management of Unwanted Organisms

5.1 Weed Management

Suppression of the growth of weeds in agroecosystems can be accomplished in numerous ways. The measures used are either intended to give the crop a favourable starting position in the competition with other species or to keep the crop stand free from weeds, as far as possible.

5.1.1 *Herbicides*

Herbicides are chemical compounds that serve for management or elimination of unwanted plants. In the nineteenth century, such inorganic compounds as copper(II) sulfate, iron(III) sulfate, or sodium chlorate were first used to combat weeds. Today's herbicides are mostly organic compounds from a variety of chemical groups. Currently, there are more than 200 substances with approximately 20 different mechanisms of action. In most cases, the herbicides impair the metabolism of the plant by inhibiting enzyme activity. The most important processes in which herbicides interfere are:

- photosynthesis (e.g. by impairing electron transport in the photosystems);
- carotenoid synthesis (by preventing the production of carotenoids, which are important as protective pigments of chlorophyll);
- cell respiration (by impairing the acquisition of energy from stored organic compounds);
- fatty acid metabolism (e.g. by impairing the formation of waxes and cutins on the plant's surface and thus disturbing the plant's water balance);
- amino acid synthesis (by which the formation of various proteins and enzymes is prevented);

K. Martin and J. Sauerborn, *Agroecology*, DOI 10.1007/978-94-007-5917-6_5,

- cell growth (by herbicides which correspond in their effect to the natural plant hormone auxin and lead to uncontrolled growth of the plant); and
- cell division (by disturbance of the processes of mitosis).

Herbicides are usually applied in the liquid form and are taken up by the plant either through the roots (soil herbicides) or through the leaves (leaf herbicides). Herbicides can be further differentiated into contact herbicides, which are only effective at the sites at which they come into contact with the plant, and systemic herbicides, which are transported inside the plant to the locations in which they are effective.

Another criterion used for classification of herbicides is the selectivity of their impact. Non-selective or broad-spectrum herbicides are effective against a wide range of different plant species. In agroecosystems, they are often applied before sowing or germination of the crop (the so-called pre-emergent herbicides). In contrast, selective herbicides are only effective against certain types of plants and can therefore be used against individual weeds in a crop stand without significantly damaging the crop (post-emergent herbicides). The active agents to be applied are selected according to the respective crop and weed species. The selectivity of a herbicide is often not only determined by the physiological and morphological characteristics and the stage of development of a plant, but also by the quantity of herbicide applied, the time of application, and different environmental factors (e.g. weather, soil properties).

As protection against the effects of specific herbicides, so-called safeners or antidotes were developed for several cereal species. They selectively protect crop plants from herbicide damage without reducing activity in target weed species. Safeners consist of additional ingredients in herbicides which are taken up by the crop and suppress the activity of the herbicides in different ways (e.g. by breakdown or deactivation of the active agent of the herbicide during crop metabolism). Safeners do not impair the desired effect in the target species (primarily grasses), because these have physiological properties different from those of the crop species.

5.1.2 Herbicide-Resistant Crops

Genetic engineering of crop plants is another means of creating biochemical selectivity against herbicides. By transfer of appropriate genes from another organism (primarily bacteria), it has been possible to bestow crop species with resistance to specific active agents. Such transgenic plants are capable of synthesizing enzymes that enable them to degrade the active ingredient in a herbicide, rendering it harmless.

Fig. 5.1 Common barnyard grass (*Echinochloa crus-galli*)

The most important crops with such properties are varieties of soybean, rapeseed, maize, and cotton resistant to glyphosate or glufosinate. Both of these substances are non-selective herbicides which impair the synthesis of amino acids in plants and can thus only be applied to stands of resistant varieties as post-emergent herbicides. Farmers can thereby easily control weeds during the entire growing season and have more flexibility in choosing times for spraying.

In chemical weed management of such crops, however, the farmer is dependent on herbicides that contain the compound to which the particular crop variety being grown is resistant. This relationship is called a herbicide resistance system, which consists of the resistant variety and the so-called complementary herbicide, which does not damage the respective crop variety. The important advantage of this method is that a non-selective herbicide can be applied to a crop stand and chemical or mechanical weed management is thus no longer necessary before sowing. It is also possible then to mostly or entirely give up soil cultivation because the disadvantages of no-till methods (conservation soil tillage and direct seeding; Box 4.2), in relation to weed management are eliminated. Whether the herbicide resistance system is generally a more environmentally friendly method of weed management than the application of selective herbicides cannot be estimated. There is always a threat that with the repeated application of herbicides with the same active mechanisms, which is necessarily the case in the herbicide resistance system, resistant populations of weeds will develop. Such weeds are already a problem in various production systems, having developed as result of continuous application of specific active compounds. Examples of weed species with herbicide-resistant populations include the slender meadow foxtail (*Alopecurus myosuriodes*) and the common windgrass (*Apera spica-venti*) in cereal production in Europe, common barnyard grass (*Echinochloa crus-galli*; Fig. 5.1), white goosefoot (*Chenopodium album*), and other species in corn and soybean production in North America.

5.1.3 Cultivation Measures

The growth and development conditions of weeds can be affected by a variety of measures in the design and management of cultivation systems. These include:

Crop rotation. The importance of individual weed species in an agroecosystem is essentially determined by the respective crop species, because the cultivated species and varieties differ in their competitive ability against different weed species. When production is conducted in the form of monocultures, then the populations of the respective problematic weeds can increase from year to year. For example, the repeated production of winter cereals promotes such grass species as foxtail grass and common windgrass, which primarily germinate in autumn. Such developments can be limited by appropriate crop rotation (e.g. by switching between cereals and root crops). Although such crop rotation promotes a diverse assemblage of weeds, dominance of specific weed species in one of the crop stands can be avoided.

Undersown crops. The stands of such plants can suppress the growth of weeds by shading (Sect. 4.4.1.1) and in some cases also as a result of allelopathic effects (Sect. 4.4.2).

Sowing date. For many crop species, the choice of the sowing date has a substantial effect on the competitive ability of the crop against some weed species.

Examples: With an early spring sowing date of summer cereals or beet, such weed species as common wild oat and sticky weed develop almost simultaneously with the crop. With a later sowing date, the germinating weeds can be eliminated during seed bed preparation.

An early sowing date for winter cereals in autumn favours the development of weed species that germinate during this time, but leads to a suppression of the growth of annual species that emerge in the following spring. At that time, the cereal stand is already so closed that the plants cast substantial shade.

The incidence of the parasitic weed *Orobanche crenata*, and the biomass produced on the host plant, are reduced with decreasing soil temperature (Fig. 5.2). This situation can be used in the production of such species as lentils, by moving the sowing date from October to December.

Soil cultivation. Reduction in the density of weeds can be achieved by mechanical soil cultivation with a variety of equipment (e.g. hoe, weeding harrow, cultivator, plough). Depending on the method, growing plants are cut into pieces, rooted out, or buried. This is primarily accomplished by the plough, whereby seeds are also transferred into lower soil layers. In many weed species, the germination can thus be reduced. However, previously buried seeds can also be brought back to the surface by ploughing.

Variety selection. The varieties of a crop species can vary in their ability to compete with weeds. This is based on characteristics which include height, leaf size, leaf position, and growth rate and are important in, primarily, competition for light (Sect. 4.4.1.1).

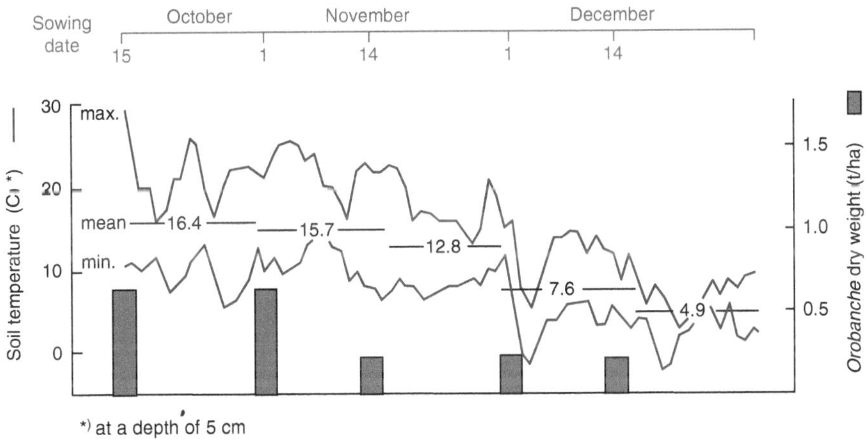

Fig. 5.2 Effect of sowing date of lentils and soil temperature on the dry mass production of *Orobanche crenata* in Tel Hadya, northwest Syria (Based on Sauerborn 1989)

5.1.4 Additional Methods of Weed Management

In addition to the methods of weed management described above, other measures exist that, depending primarily on technical and financial feasibility, can be applied as either complementary or alternative measures:

Seed cleaning. Often the seeds of weeds are gathered during harvesting of crops and can find their way back to the fields in the course of the next sowing. With modern methods of seed cleaning, this means of dispersal can be strongly limited, but this can also contribute to the loss of biodiversity of wild plants (Sect. 3.3.2).

Manual labour. Manual removal of weeds and use of simple tools (hoes) are the most labour-intensive methods of weed management. In small-scale agricultural systems in the tropics, these methods are often of great importance because of the lack of alternatives (cf. Sect. 2.4.1). Because of the high labour input, manual weed management can become a factor limiting the amount of land cultivated. Even in intensively managed agroecosystems, primarily in field vegetable production, manual weeding can make an important contribution to securing yields.

Thermal methods. By use of a variety of methods it is possible to manage weeds by temperature treatment. The methods used include flaming with tools that produce an open flame or release infrared radiation, as well as microwave treatment. Another technique is solarization, which is primarily successful in regions with high insolation (e.g. in the Mediterranean region). In this method, areas of cultivation are first irrigated and then covered with foil through which light can pass. This results in a substantial increase in the surface temperature of the soil and thus causes the seeds and shoots of many weed species to die off.

Application of thermal methods is limited in agroecosystems because of cost and effectiveness. However, flaming techniques can be an alternative to manual weed management, especially in field vegetable production.

Biological methods. In contrast with biological control of pests (Sect. 5.2.4), management of weeds in agroecosystems by use of organisms (phytophages and phytopathogens) is not currently significant. It is rarely possible to control weeds with bioherbicides, i.e. with phytopathogens applied to fields as spore suspensions. On the basis of the classical principles of biological control (Sect. 5.2.4.7), in many cases it has been possible to repress plant species on continents to which they were introduced from other continents (e.g. some grassland weeds and aquatic plants).

5.2 Pest Management

The most important objective of pest management in agroecosystems is the prevention of economically significant yield losses of crops. Means by which this is achieved primarily include application of chemical substances, cultivation of pest-resistant varieties, use of specific means of cultivation, and use of antagonists of the pests.

5.2.1 Chemical Pest Control

The most common type of management of animal pests in agroecosystems is the application of **pesticides**, i.e. chemical compounds that are designed to counter different groups of target organisms (Box 2.2). Pesticides can be classified as either synthetic or natural, the latter being derived from plants, animals, or microorganisms. The substances either act as contact poisons, i.e. they enter the organism through the body surface or as stomach poisons, the latter by ingestion during feeding. Because the insects are the largest and most important group among the animal pests, the following sections focus on insecticides.

5.2.1.1 Synthetic Insecticides

Today, almost all synthetic insecticides are organic compounds; inorganic compounds, for example arsenic or hydrogen cyanide (prussic acid) are no longer used or are banned. Most synthetic insecticides belong to the following groups of compounds:

Organochlorine compounds. These compounds impair the transmission of impulses in the nervous system by blocking the ion channels of the axon.

One of the first insecticides of this group, DDT (dichlorodiphenyltrichloroethane) was used during and after the second World War in the military hygiene sector against lice, fleas, and bed bugs, later also in the fight against malaria and its vector (*Anopheles* mosquitoes) and against agricultural pests. DDT is difficult to break down, can enter the food chain, and accumulates in the fat tissues of organisms. For these reasons, its production and use in have been prohibited in most countries. Other substances from this group are e.g. hexachlorocyclohexane (HCH) and endosulfan.

Organophosphates. These compounds (esters of phosphoric acid) include not only insecticides but also chemical weapons, for example the nerve gases tabun and sarin. They have neurotoxic effects by inhibiting the enzyme acetylcholinesterase and thus disrupting the transmission of nerve impulses. Most of the insecticides from this group, with some exceptions, for example parathion (E 605), have low toxicity to mammals but are highly toxic to aquatic organisms (fish and shellfish). Some of these substances are also taken up by plants and distributed within their transport systems. As a result of this systemic effect, they also work against sucking insects and mites. Other organophosphates enter the plant via leaf and stem tissues. Examples of this type of compound include chlorpyrifos, malathion, and phosphamidon.

Carbamates. This group consists of derivatives of carbamic acid, which have a neurotoxicological effect by inhibition of acetylcholinesterase, similar to the organophosphates. This group also includes substances that are taken up by plants and distributed in their tissues. Pirimicarb acts systemically via the roots and is regarded as safe for some aphid antagonists, for example ladybirds and green lacewing larvae. Other examples of carbamates are methiocarb and carbofuran.

Pyrethroids. These synthetic compounds are chemically related to the pyrethrins which occur naturally in the flowers of some *Chrysanthemum* species. The pyrethroids have a greater efficacy and lower light sensitivity than the natural plant substances and can be produced more easily. They are contact poisons that do not enter the plant and have a neurotoxicological effect similar to that of the organochlorine compounds. Active agents used in agriculture include deltamethrin, cypermethrin, and fenvalerate.

Chloronicotinyls. This group of substances was developed in the 1980s. As neurotoxins, they inhibit the acetylcholine receptors and are thus similar to nicotine in their effect. They act as contact poisons, but are also taken up by the plant and act against a wide range of insects. The most important substance is imidacloprid.

Benzoyl ureas. In contrast with most of the other synthetic insecticides, these compounds (e.g. diflubenzuron) are not neurotoxins, but instead block chitin biosynthesis and thus the development of arthropods. Because their effect is not immediately lethal to animals, the possibilities of use of these substances in agriculture are limited. They are primarily used against pests in the forestry sector.

5.2.1.2 Natural Insecticides

Substances derived from organisms that are used to manage pests largely include secondary plant metabolites (Sect. 4.5.5.2), which the plants produce to protect themselves from feeding enemies. Some of these plant-based insecticides have been used for thousands of years in the management of pests, but many have lost most of their importance since the development of synthetic insecticides. Worldwide, approximately 2,400 plants are known that can be used against pests, but only a few are produced commercially. Nevertheless, some of these species are of regional importance and can serve small-scale farmers in the tropics, especially, as an alternative to (expensive) synthetic insecticides. In the search for new substances, natural insecticides are also of interest to the chemical industry, because they can potentially be the source of initial compounds or models for more effective synthetic compounds. An example of this has already been given above—the pyrethrins were the basis for the development of the pyrethroids. Other examples of plant-based insecticides are:

Nicotine. Extracts from tobacco plants (*Nicotiana* species) have a neurotoxicological effect and were, before the development of DDT, among the most important agents in insect management. However, nicotine is also highly toxic to warm-blooded animals (endotherms) and other organisms, for example earthworms.

Quassin. This substance is derived from the South American tree species *Quassia amara*. Until the 1940s it was used to control aphids and caterpillar pests in Europe and North America.

Rotenone. This neurotoxin comprises a variety of substances found in tropical legumes of the genera *Derris*, *Lonchocarpus*, and *Tephrosia*. They act not only on insects but also on fish and other aquatic organisms.

Azadirachtin. This substance is found in the seeds, leaves, and bark, of the neem tree (*Azadirachta indica*), which originates from the Indian subcontinent. It is an antifeedant for many species of insects and also has an inhibitory effect on their metamorphosis. Neem leaves are traditionally used against storage pests. Neem extracts (primarily from the seeds) are also suitable for use in agroecosystems and are gaining increasing importance not only in tropical regions, but also in Europe and North America.

The most important natural insecticides originating from microorganisms are the toxins of the bacterium ***Bacillus thuringiensis* (Bt)**. For approximately 100 years it has been known that this soil-living species produces proteins with toxic effects on insects. After oral ingestion and enzymatic activation, these so called δ-endotoxins destroy the intestinal mucosa of insects, which subsequently die. More than 140 such proteins are known; they are produced by different subspecies of *B. thuringiensis* and each affects a specific group of insects. These primarily include butterflies, beetles, and diptera (flies and mosquitoes), but not all representatives within these orders. The Bt proteins are not dangerous to mammals and other vertebrates.

Bt formulations have been used as bioinsecticides since the 1930s and consist of dried bacteria spores or crystallized toxins. They are primarily used in maize, potato, fruit, and vegetable production. They are of greatest importance in organic agriculture. Another way of using Bt toxins is to transfer the respective genes into crop species (Sect. 5.2.2.1).

5.2.1.3 Pheromones

Substances that are released by organisms and have an attractive or repulsive effect on individuals of the same species are called **pheromones**. As sexual attractants, pheromones are important for numerous insect species. They are released by the females and have an attractive effect on the males. Many of these substances, which are usually specific to particular species, can be produced synthetically and used, e.g., in the management of pest caterpillar species in fruit and grape production. However, pheromones, usually applied in insect traps, are not usually suited to direct control, because the number of individuals caught usually makes up only a small portion of the total population. The traps primarily help to identify the start of the flight period of the pest butterflies and enable estimation of population density. Thus, they are used to determine the optimum time period and method with which to manage the pests. Against some species, for example the codling moth (*Cydia pomonella*) or the European grapevine moth (*Lobesia botrana*), sources of pheromones are placed among the crops to confuse the males. When concentrations of the attractive substances in the air are high, the males are no longer able to locate the females, whereupon reproduction, and thus the number of offspring, are reduced.

Another strategy which makes use of pheromones is the development of genetically modified crops that produce pheromones that are repellent to a pest. This approach has been realized in the development of a wheat variety which releases the sesquiterpene (*E*)-beta-farnesene, an alarm pheromone for many aphids. Such pheromones are released by aphids when they are attacked by their enemies, for example ladybirds, and causes them to disperse. This wheat variety was created by transformation of an (*E*)-beta-farnesene synthase gene from peppermint (*Mentha* × *piperita*) and is expected to repel aphids from the crop plant by emission of their alarm pheromone (Rothamsted Research 2012).

5.2.1.4 Unwanted Effects of Insecticides

With the exception of the Bt toxins, most synthetic and natural insecticides are unspecific, which means they usually affect a wide range of insect species. Furthermore, the neurotoxic substances not only affect insects and other arthropods, but almost all organisms with a nervous system. Nevertheless, it is possible to use insecticides in such a manner that not all species of the agricultural community are affected in the same way. Such selectivity is not usually based on fundamental changes in the mechanism of the applied substance but is essentially because of the following factors:

Dose-dependent effects. Some insects are more harmed than others by the same quantity of an insecticidal compound. Such differences are dependent on such factors as the size of the species, the development stage (egg, larva, pupa, adult), morphological characteristics, for example the structure of the cuticles, or physiological differences in the ability to detoxify harmful substances. By application of small quantities of the substances it is therefore possible to successfully manage some pests while at least some of the other insects survive. However, the latter can be harmed by sub-lethal effects. Sub-lethal insecticidal effects do not lead to the immediate death of the individuals but can, instead, reduce fertility, delay metamorphosis or impair metabolism and thus have a negative effect on the development of the population. Such disturbances can be found among predators and parasitoids when they feed on phytophages contaminated with insecticides.

Contact-dependent effects. Another way to limit the effects of insecticides to specific target groups is to avoid contact between the substances and some organisms. With treatments adjusted to the presence or the phases of activity of specific species or their developmental stages, it is possible to limit effects on organisms other than the pests. For example, pollinators, for example bees or parasitoids, can be protected when application of insecticides is avoided during the flowering stage of a crop. Effects on other organisms can be avoided, or at least reduced, by use of active agents that are taken up by the plant and primarily act as toxins when ingested by herbivores.

Use of methods to limit the impact range of insecticides requires detailed knowledge of the application of the products. Even humans can be harmed by inappropriate application of insecticides, for example by application of excessive quantities or by too frequent application. Approximately 90% of cases of illness or death that are caused worldwide by synthetic pesticides occur in developing countries. This can be primarily attributed to a lack of education and insufficient protection when working with the substances.

The insecticides and other pesticides applied to crop stands often end up in the environment in large quantities. As a result of water and wind, the substances also reach ecosystems far from the areas of application (Bedos et al. 2002).

Insecticide-Induced Pest Resurgence

There have been instances when the population density of some pests has not been reduced by application of an insecticide but has, instead, increased substantially. This phenomenon, known as insecticide-induced pest resurgence, has been observed in many production systems, for different pest species, and after application of different active agents. The causes of this are complex and not usually

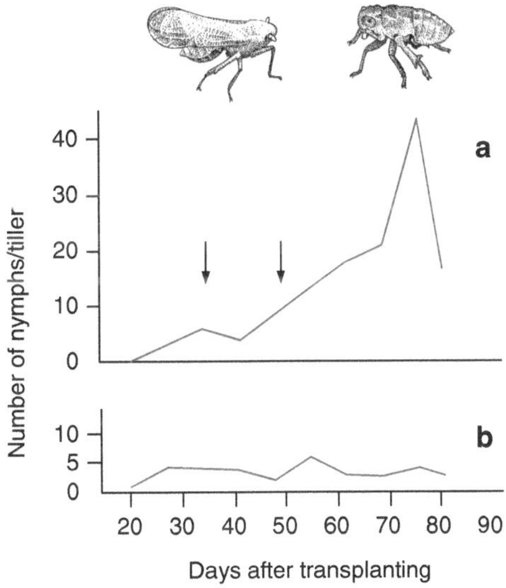

Fig. 5.3 Population development (number of nymphs) of the brown planthopper (*Nilaparvata lugens*) on rice plants that (**a**) were treated with the insecticide deltamethrin on two occasions (*arrows*) and (**b**) were not treated (Based on Heinrichs et al. 1982)

attributable to a single factor. The following mechanisms may be involved (Hardin et al. 1995):

- The applied substance, or the quantity applied, has a stronger effect on natural enemies of the pest than on the pest species itself. As a result, the effect of the antagonist on the pest population decreases, and the latter subsequently increases. In addition, other specific insecticidal effects, for example those affecting alternative prey species of the antagonist (Sect. 5.2.4.2) or competitors of the pest, can also be involved.
- In many cases, it has been shown that the insecticide affects the physiological processes of plants. Such changes can have a positive effect on the herbivores via the uptake of food and may improve their ability to detoxify insecticidal substances or increase their fecundity.

Example: For some rice varieties, the concentration of free nitrogen compounds in the transport tissues increased after application of the pyrethroid deltamethrin. As a result, the quality of food of the brown planthopper (*Nilaparvata lugens*) improved, and was believed to be the main reason for the observed resurgence of the pest (Heinrichs et al. 1982; Fig. 5.3).

Insecticide Resistance

Another factor resulting in reduced susceptibility of some insect pests is explained by insecticide resistance. In contrast with the mechanisms of insecticide-induced pest resurgence, insecticide resistance is an evolutionary process based on mutation

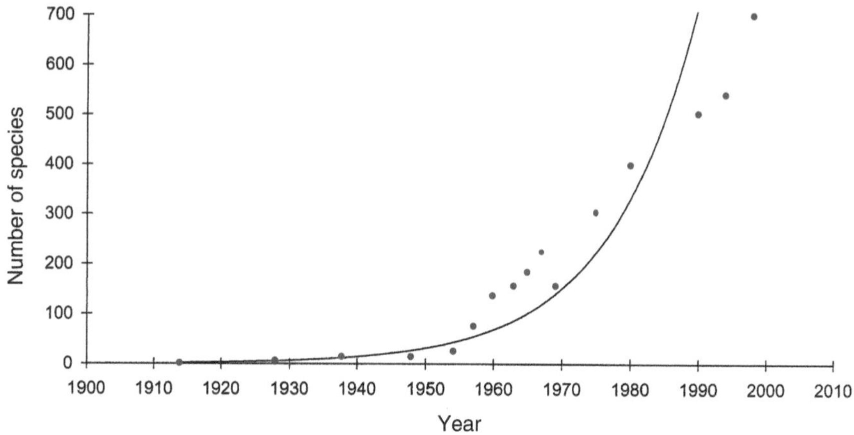

Fig. 5.4 Increase in the number of arthropod species resistant to at least one insecticide in the course of the twentieth century (Based on Thacker 2002)

and selection (cf. Box 2.1). Mutations can, in some individuals of a pest population, lead to the development of a resistance to a specific insecticide. These properties are generally based on biochemical mechanisms, for example the synthesis of specific enzymes with which the toxic substances are broken down, or on changes in the physiological characteristics of the organism as a result of which the toxic compounds no longer have their effects. The repeated application of insecticides acts as the selection factor through which susceptible individuals of the population become eliminated while the resistant individuals continue to exist and reproduce.

The number of arthropod species that have developed resistance to one or more insecticidal compounds has significantly increased in recent decades (Fig. 5.4) and is likely to continue to increase in the future.

Chemical control of such pests is then only possible with substances against which the respective populations have not yet developed resistance. In response to pesticide resistance, farmers often try to achieve success by increasing the applied quantities of already ineffective substances. However, the resistance that has developed in a population cannot be reversed. Development of insecticide resistance can be delayed by avoiding unnecessary insecticide applications and, if necessary, using the lowest concentrations possible. In other words, adopting the approach of integrated agriculture (Sect. 2.3.6) contributes to resistance management. This may include the application of other measures in pest management, which are the subject of the following sections.

Insecticides not only have toxic effects on specific organisms, but also affect the interactions of the species and other processes in the agricultural community in complex ways.

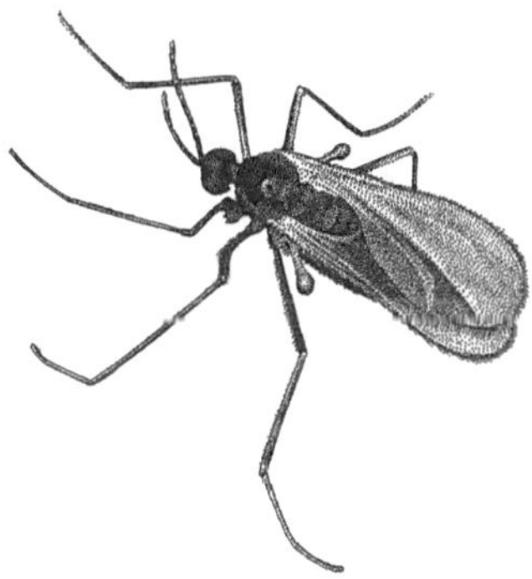

Fig. 5.5 Hessian fly or barley midge (*Mayetiola destructor*). The larvae of this species attack the wheat plant at the point of attachment of the leaf sheath to the stem

5.2.2 Pest-Resistant Crops

One way of managing pests without application of insecticides is the breeding and cultivation of resistant crop varieties. In the context of the definition of plant resistance against herbivores (Sect. 4.5.5), it has already been pointed out that the different mechanisms of plant defence are based on genetic properties of the plants. During the process of domestication and selection, many of the varieties of the different crop species grown today lost the genes that provide resistance, because the breeding was usually primarily directed toward high yields. The objective of **resistance breeding** is improvement of the genetic performance of a crop, to increase its ability to defend itself against the organisms that cause damage.

The phenomenon of pest resistance has been known for a long time. As early as 1792, the wheat variety "Underhill" was produced in North America which, in contrast with other varieties, is rarely attacked by the Hessian fly or barley midge (*Mayetiola destructor*; Fig. 5.5). However, the directed breeding of resistant varieties only began at the beginning of the twentieth century, with increasing knowledge of genetics.

In conventional resistance breeding, the objective is crossing of a desirable but pest or disease-susceptible plant variety with another variety that is a source of resistance. Often the respective resistance genes are found in old landraces or in wild forms from the centre of origin of the crop. By crossing, initially undesirable characteristics are also transferred into the breeding lines, which means that the previously selected characteristics of modern varieties (e.g. high yield) are partially lost. It is therefore necessary, by repeated back-crossing with the modern variety, to regain these characteristics, without losing the crossbred resistance.

Crossbreeding of a resistant variety can take up to 15 years and is undertaken with knowledge of the risk that the pest or disease can overcome the resistance in a shorter period than was required for the breeding.

5.2.2.1 Transgenic Pest-Resistant Crops

The insect resistance of transgenic crops (Sect. 2.3.4) is based, almost always, on the presence of genes of *Bacillus thuringiensis*. At the beginning of the 1980s, the first transfers of Bt-endotoxin genes into the genomes of plants succeeded. Subsequently, these plants were able to produce the toxins by themselves and were thus resistant to some herbivores. Compared with the application of insecticides, use of **transgenic Bt crops** has the following **advantages**:

- Bt plants have systemic protection that is present at all stages of growth and thus over the entire cropping period.
- Bt plants are also protected against some endophagous consumers (e.g. stem borers); this is difficult to achieve by application of many insecticides.
- Bt plants can be used against insects that have become resistant to some insecticides.
- Overall, the frequency and quantity of application of conventional insecticides can be substantially reduced by use of Bt plants, which reduces costs for the farmer and effects on the environment.

In addition to these advantages, cultivation of Bt crops is also accompanied by **problems and risks**:

- The relatively high specificity of Bt toxins is a disadvantage when economic damage is caused by different insect groups, which is usually the case. Pest species that are not affected by the Bt toxins (e.g. representatives of the true bugs, aphids, spider mites, and thrips) can still only be managed by use of insecticides.
- Insects may not only become resistant to synthetic insecticides, but also against Bt toxins, which has already occurred in some species as a result of application of Bt formulations. This process can be accelerated by use of transgenic Bt crops, because the herbivores are constantly subject to a selection pressure as a result of the systemic effect of the toxins. If insects acquire systemic resistance against transgenic Bt crops, they can also no longer be managed by use of Bt formulations. This would be particularly disadvantageous for organic agriculture, in which such products are used.
- Negative effects of Bt crops against non-target organisms cannot be definitively ruled out. They might affect animals (primarily antagonists of the pests; Sect. 5.2.4.1) and plants. By gene flow via pollen transfer, introgression of Bt genes to non-transgenic plants of the same species (e.g. maize) or to closely related wild forms might occur. In the extreme case, transgenic wild plants acting as weeds can therefore become more serious. As a result of their resistance against specific insects, these wild plants would have an advantage which would not only promote their dispersal but also harm insect species other than pests as a result of their toxic effects. By incorporation of harvest residues of Bt crops into the soil, microorganisms could also be affected. It is also possible that the genes of Bt toxins could be transferred to related *Bacillus* species.

Overall, the complexity of ecological systems is a substantial challenge to experiments to assess the risks, benefits, and inevitable uncertainties of genetically engineered plants. Collectively, current studies emphasize that these can vary spatially, temporally, and according to the trait and cultivar modified (Wolfenbarger and Phifer 2000).

In principle, genes other than those of *Bacillus thuringiensis* can be transferred to plants and provide them with increased resistance to insects. Most attempts are concentrated on the development of transgenic plants that synthesize proteins (e.g. protease inhibitors) that are harmful to the pests.

Example: A trypsin-inhibitor gene that is found in the winged bean (*Psophocarpus tetragonoloba*) has been produced synthetically and transferred into rice. In the transgenic plants, caterpillars of the stem borer *Chilo suppressalis* developed more slowly than in non-transgenic control plants (Mochizuki et al. 1999). In this case, and in many similar cases, the effects on the target organisms have, so far, been not sufficient enough to enable the use of such plants commercially.

5.2.3 Cultural Methods of Pest Management

By use of a variety of agricultural methods, the environmental conditions of the agroecosystem can be changed such that colonization by pests or the increase of current populations can be limited. Such measures include modifications in time (crop rotation and sowing date) and space (plant density, soil cultivation).

5.2.3.1 Crop Rotation

Most pest species are feeding specialists and feed on specific plant species only (cf. Sect. 4.5.2). By crop rotation in which a crop species that affected by pests is followed by another crop species that is not used by these pests, development of pest populations can be interrupted.

Example: A measure used to manage the western corn rootworm (*Diabrotica virgifera*) and the northern corn rootworm (*Diabrotica barberi*) in North America is annual rotation of maize and soybean. The larvae of these beetles feed on the maize roots and develop into adults at the end of the season. The females lay their eggs in the soil that same autumn and the larvae of the next generation then emerge in the following spring. The species are then confronted with soybean, on which they are unable to feed, instead of maize. However, in some areas *D. barbari* has adapted to the annual rotation by expanding the larval development stage to 2 years. Some populations of *D. virgifera* now also lay their eggs in soybean fields, which means that the larvae that emerge the following year can then feed on maize (O'Neal et al. 2002). In both cases, the regular rotation of crops as a method of reducing the population of the beetle became ineffective.

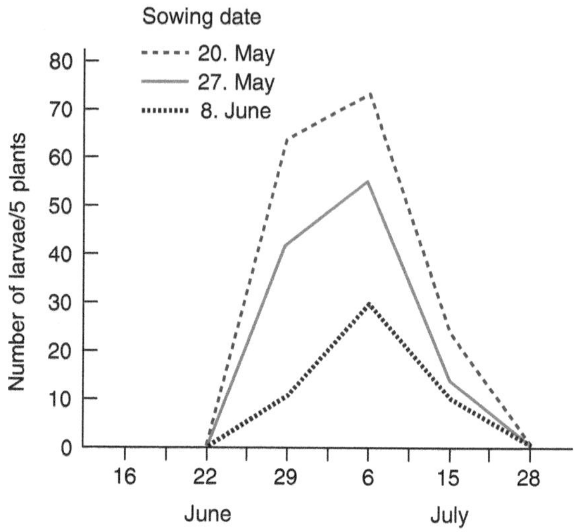

Fig. 5.6 Relationship between the sowing date of sunflowers in North Dakota and subsequent pest incidence of the larvae of the sunflower beetle *Zygogramma exclamationis* on the plants (Based on Charlet and Knodel 2003)

5.2.3.2 Sowing Date

The sowing or planting date of a crop can be chosen so that it does not coincide with the period of migration or with the highest density of a pest population. This measure is therefore based on minimizing the temporal coincidence (cf. Sect. 5.2.4.2) between the appearance of a pest species and the stage of crop development susceptible to the pest.

Examples: Larvae and adults of the sunflower beetle (*Zygogramma exlamationis*) specialize on sunflowers and feed on the plant leaves. The species is the major defoliating pest of sunflower in North America. A study conducted in North Dakota showed that numbers of adults and larvae decreased as planting date was delayed. Even a planting delay of only 1 week was sufficient to significantly reduce feeding on the sunflower plant and thus the extent of damage (Charlet and Knodel 2003; Fig. 5.6).

The females of the wheat stem sawfly (*Cephus cinctus*) lay their eggs in wheat stalks, on which the hatched larvae feed. The wheat plants are vulnerable only after stem elongation has begun and before grains begin to fill. A delay of the sowing date by 2–3 weeks in spring is sufficient in Montana (USA) in some years to completely avoid infestation of the larvae. At that time, the phase of oviposition has already passed (Morrill and Kushnak 1999).

5.2.3.3 Plant Density

The number of individuals of some pest species is affected by the number of plants per area and the distance between rows in the field. But, on manipulation of plant stand density, the yield is often affected more by intraspecific competition of the plants than by the pest incidence.

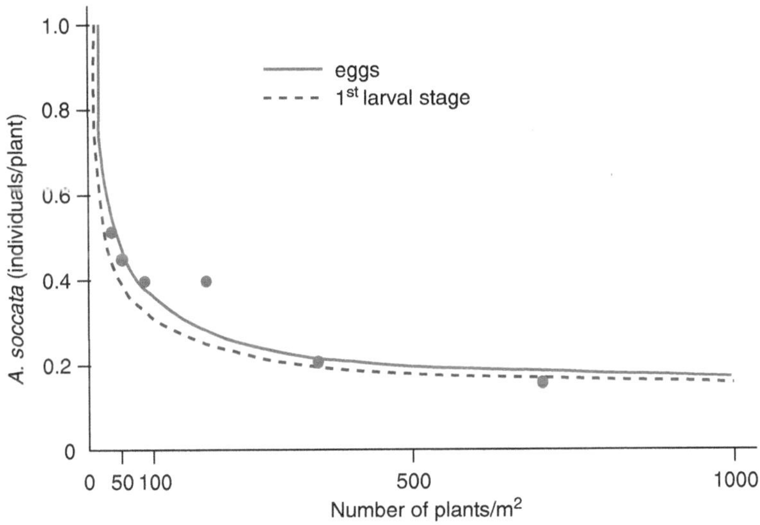

Fig. 5.7 Relationship between the number of individuals of different development stages of the sorghum shoot fly (*Atherigona soccata*) and plant density of sorghum plants in the field (Based on Delobel 1982)

Examples: The incidence of the sorghum shoot fly *Atherigona soccata* on sorghum decreases as plant density increases (Fig. 5.7). However, although an increase in the plant density reduced the damage caused by *A. soccata* an increase in yield was not achieved. The reason is intraspecific competition between the plants, which acts as a yield-limiting factor (Delobel 1982).

The opposite conditions were found in the cabbage root fly (*Delia radicum*), whose larvae feed on the roots of Brassicaceae. The higher the density of their host plants (different cabbage crops), the higher was the number of pupae in the soil (Finch et al. 1976; Fig. 5.8).

5.2.3.4 Soil Cultivation

In soil cultivation, soil-conserving practices (conservation tillage and direct seeding) can be distinguished from conventional ploughing (Box 4.2). Many pest species have development stages that are found in the upper layers of the soil and whose habitat is destroyed by ploughing. The pests then reach either the unprotected soil surface, where they may dry out and are threatened by predators (e.g. birds that often follow the plough), or deeper soil layers where they can no longer complete their development. Furthermore, soil management in agroecosystems may affect the soil arthropod fauna in different ways.

Examples: The number of eggs of cabbage flies (*Delia* species) and the root damage caused by their larvae in rapeseed was lower in ploughed fields than in no-till systems (Dosdall et al. 1998). On the other hand, some pest species occur at

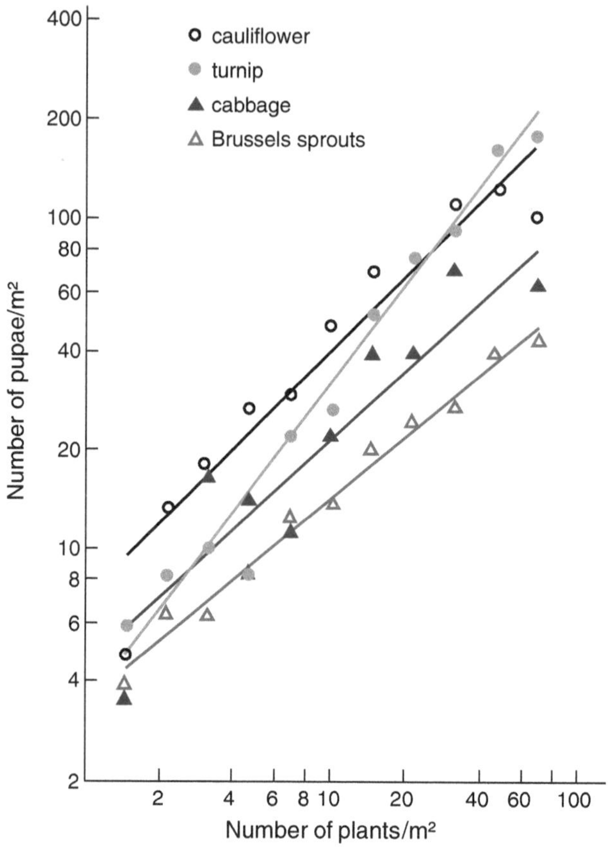

Fig. 5.8 Relationship between the number of pupae of the cabbage fly or maggot (*Delia radicum*) in the soil and the plant density of different Brassicas (Based on Finch et al. 1976)

lower densities on directly sown fields, which was found to be the case for the crucifer flea beetle (*Phyllotreta crucifera*) in rapeseed (Milbrath et al. 1995). Conventional ploughing had a negative effect on antagonists that live at the soil surface, primarily ground beetles and spiders (Holland and Reynolds 2003).

5.2.4 Biological Pest Control

The use of living organisms to limit the populations of phytophages, with the objective of reducing or preventing damage to specific plants, is called **biological pest control**. According to this definition, measures that are biological in the broadest sense, for example breeding for resistance, and application of natural substances and pheromones, are not regarded as biological pest control, because they are not based on the use of living organisms.

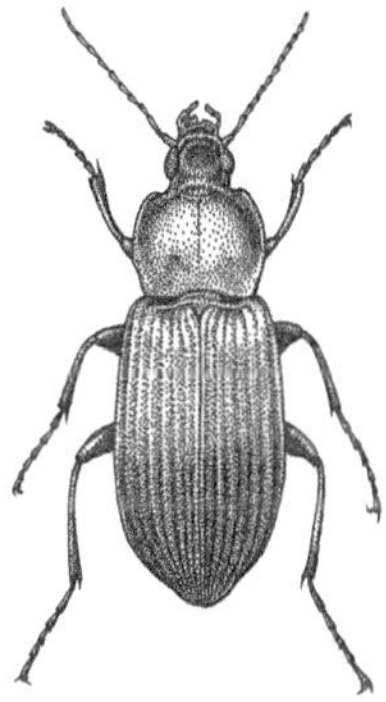

Fig. 5.9 *Poecilus cupreus* is a common ground beetle species in agroecosytems of Central Europe

5.2.4.1 Main Groups of Antagonists

The organisms that are of importance in biological pest control are primarily predators, parasitoids, and animal pathogens. They occur in agroecosystems as **natural enemies** of the pests and are termed **antagonists**. In specific cases, plants can also be used in biological pest management (Sect. 5.2.4.8). The most important groups of antagonists of phytophagous pests in agroecosystems are discussed in the following sections.

Ground Beetles (Carabidae)

In Europe there are approximately 2,700 ground beetle species, but only a few of these are found in agroecosystems (Fig. 5.9). In the fields of Central Europe, 20–30 species are found on average. In tropical cultivation landscapes, the diversity of the Carabidae is usually substantially lower. Most ground beetle species are polyphagous predators but some occasionally feed on plant material (e.g. seeds) also. Some species even occur as pests. The corn ground beetle (*Zabrus tenebriodes*), primarily found in eastern Central Europe, feeds on the ears and milk-ripe grains of cereals and grasses. Normally, however, ground beetles find their food and prey primarily on the soil surface and only occasionally visit plants. The larvae of ground beetles develop in the soil and are also mainly predators. In the temperate latitudes, one generation usually develops per year. Depending on the species, the larvae develop in summer and the adults overwinter, or vice versa.

Rove Beetles (Staphylinidae)

In Central European agroecosystems, the number of rove beetle species is often greater than the number of ground beetle species. The biology of both families is similar in many respects. The rove beetles also live primarily on the soil surface in fields, where most of the species are predators. Because most of the species are only

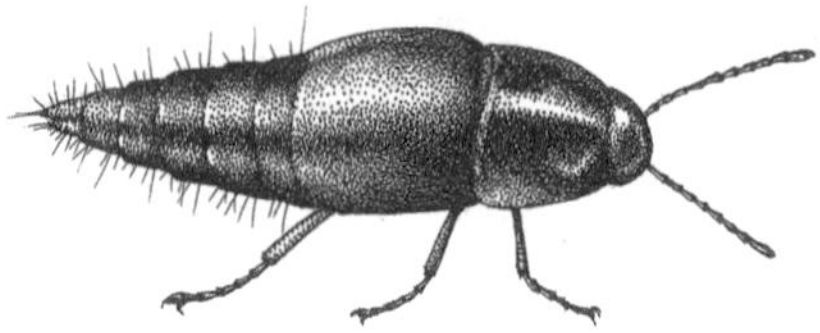

Fig. 5.10 Rove beetles of the genus *Tachyporus* climb on the stems of wheat and prey on aphids

Fig. 5.11 The 14-spotted ladybird beetle (*Propylaea quatuordecimpunctata*) with larva

a few millimetres in size, their range of prey includes correspondingly small organisms, for example springtails (Collembola), mites, nematodes, and some insects.

Species of the genus *Tachyporus* (Fig. 5.10) have a preference for aphids in their prey selection. In cereal stands, they are counted among the most important consumers of these pests, because the beetles are able to climb on the plant stalks. The larvae of the genus *Aleochara* are parasitoids that attack the pupae of different fly pests of the Brassicaceae, notably *Delia* species.

Ladybird Beetles (Coccinellidae)

The ladybirds are a family of beetles with approximately 5,000 species worldwide, many of which occur in agroecosystems (Fig. 5.11).

Approximately 90% of all species are predatory and primarily feed on sap-sucking herbivores, for example aphids, scale insects, whiteflies, nymphs of leafhoppers, and spider mites. Most of the Coccinellidae are more or less specialized on one of these groups. Many species have a preference for aphids and are therefore termed aphidophages. Often, particular species within the aphids are preferred. The seven-spotted ladybird (*Coccinella septempunctata*) primarily feeds on aphid species that attack cereals. The ladybird species *Semiadalia undecimnotata* has a preference for the black bean aphid (*Aphis fabae*), a pest of many vegetable crops. When there is a lack of prey, ladybirds also utilize nectar, honeydew, or pollen as a food source. The larvae of Coccinellidae usually have the same type of prey as the adults.

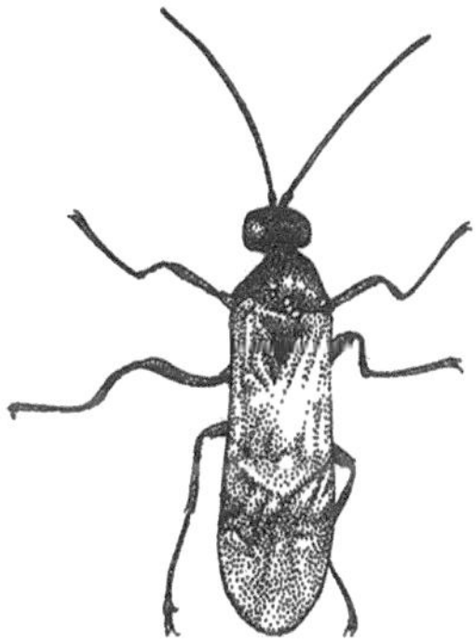

Fig. 5.12 Parasitoid wasp of the genus *Cotesia* from the family of the braconid wasps (Braconidae)

In the temperate latitudes, aphidophagous species develop one generation per year, usually in spring. Other species can reproduce several times per year. The aphidophagous species are characterized by high mobility.

Wasps (Apocrita)

The suborder wasps (Apocrita), which belongs to the order Hymenoptera (sawflies, wasps, bees, and ants) comprises numerous families, including many parasitoids and predators.

Parasitoid wasps. This group includes approximately 50,000 known species overall, which are assigned to at least 36 families (Fig. 5.12). The adults are predominantly small to tiny animals, some of which are even smaller than 0.3 mm. Most of the members of the Ichneumonidae (ichneumon wasps), however, reach a size of more than 5 mm. The larvae of the parasitoid wasps attack insects almost exclusively. The adults usually feed on nectar and pollen.

Predatory wasps. Important predator families of the Apocrita are the wasps (Vespidae) and the sphecid wasps (Sphecidae). The latter are primarily found in the tropics. In both families, prey animals (insects and spiders) primarily serve to feed the larva whereas the adults primarily feed on sugar-containing plant products (nectar, fruit). Whereas the Vespidae, which include the common wasp (*Paravespula vulgaris*) and the hornet (*Vespa crabo*), only use their stinger in self-defence, the sphecid wasps use it to paralyze their prey with poisons.

Ants (Formicidae)

Most ant species, which, similar to the wasps, belong to the Hymenoptera, are omnivores and feed on plant products, for example nectar and seeds, and on a wide range of living and dead arthropods and other invertebrates. The South and Central American leaf-cutter ants have another strategy, they practise fungal cultivation (Sect. 1.1). Some ant species also use the honeydew secreted by aphids and scale

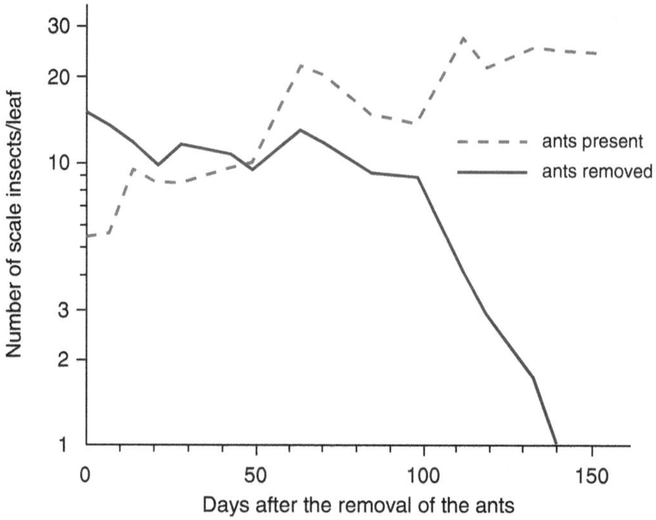

Fig. 5.13 Effect of the removal of the ant species *Pheidole megacephala* on the population of the scale insect species *Coccus viridis* on coffee shrubs of a plantation in Hawaii (Based on Reimer et al. 1993)

insects as a component of their diet, and so they also protect these animals from their enemies. In most agroecosystems of the temperate regions, ants are relatively rare. They reach their highest diversity in the tropics (cf. Table 3.1), where they can have quite different effect on agricultural systems. An example of a species acting as a predator and a serious pest is the red fire ant (*Solenopsis invicta*), an invasive species of North America originating from South America (details are given in Sect. 3.3.3).

The Asian weaver or green ants *Oecophylla smaragdina,* which inhabit trees, are traditionally used as antagonists of pests in fruit production in China and Vietnam. However, other ant species often have negative effects in such systems, because they utilize the scale insects as sources of honeydew and simultaneously eliminate their antagonists. In Australian citrus plantations, outbreaks of these pests were attributed to the presence of the ant *Iridomyrmex rufoniger* (James et al. 1999). In coffee plantations on Hawaii, the scale insect species *Coccus viridis* disappeared after the ants (*Phediole megacephala*) were experimentally removed from the plants (Reimer et al. 1993; Fig. 5.13).

True Flies (Diptera)

Some families of flies are predatory species that feed on other insects. Examples include the adults, and usually also the larvae, of the robber flies (Asilidae), several members of the gall midges or gall gnats (Cecidomyiidae), which feed on aphids and scale insects, and the larvae of many hoverflies (Syrphidae), which are more or

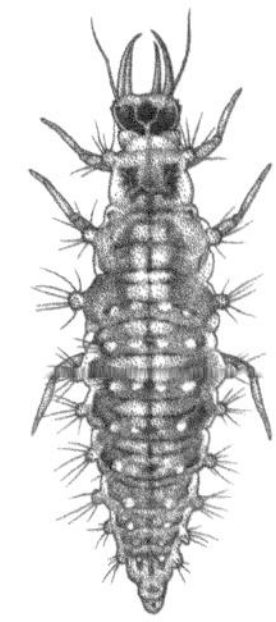

Fig. 5.14 Larva of a green lacewing of the genus *Chrysopa*

less specialized on aphids. Numerous species of different families of the Diptera are parasitoids. These include the big-headed flies (Pipunculidae) that attack leafhoppers and are important in rice production, amongst other systems.

Net-Winged Insects (Neuroptera)

From this order, the most important representatives are the green lacewings (Chrysopidae). Their larvae (Fig. 5.14) feed on smaller arthropods, which primarily include the aphids. The adults of these species are herbivores, predators, or omnivores.

True Bugs (Heteroptera)

Predatory true bugs include some representatives of the minute pirate bugs or flower bugs (Anthocoridae), capsid bugs (Miridae), damsel bugs (Nabidae), chinch bugs (Lygaeidae), and assassin bugs (Reduviidae). The last are primarily found in the tropics. Like many phytophagous bugs, the predatory representatives of the true bugs also have piercing-sucking mouthparts and feed on the body fluid of their respective prey.

Species of the genus *Nabis* (Nabidae) and *Geocoris* (Lygaidae; Fig. 5.15) are common predators in many crops in North America (e.g. cotton and soybean). Flower bugs of the genus *Orius* are used in the control of thrips in greenhouses. *Cyrtorhinus lividpennis* (Miridae) is an important antagonist of the brown planthopper (*Nilaparvata lugens*) in rice production in Southeast Asia (cf. Fig. 3.2).

Dragonflies (Odonata)

Both damselflies (Zygoptera) and dragonflies (Anisoptera; Fig. 5.16) are predators. Their adults capture their prey mainly while flying. Because their larvae develop in water, the agroecological importance of dragonflies is largely limited to paddy rice production.

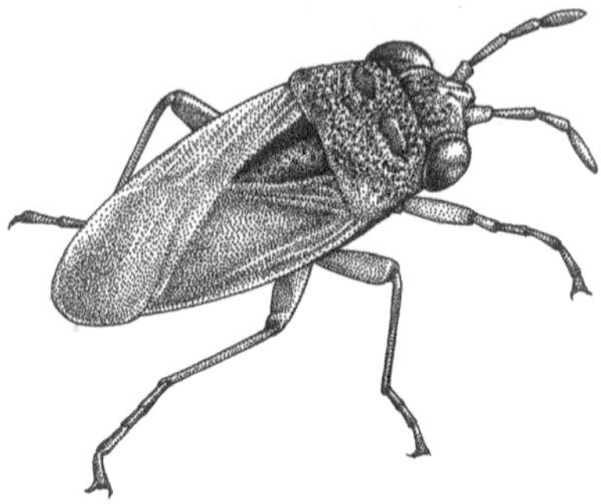

Fig. 5.15 A predatory true bug of the genus *Geocoris*

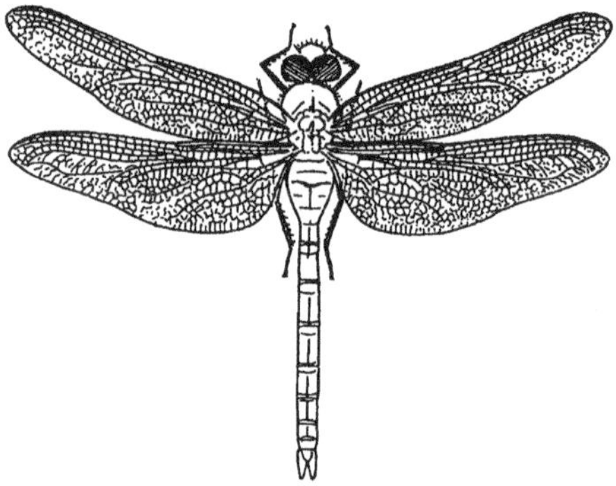

Fig. 5.16 A dragonfly (suborder Anisoptera of the order Odonata). Drawing by W. Lang

Spiders (Araneae)

All spiders, which are divided into approximately 60 families, are predators. Most species are generalists and feed on a broad range of arthropods (primarily insects). Measured by the number of individuals, they are the most common predators in many agroecosystems. Two main groups can be identified: species that construct webs to catch prey and species that hunt their prey.

The composition of the spider fauna in an agroecosystem is essentially affected by the geographic location, the crops that are produced, and by the type of management. In the production systems of Central and Northern Europe, primarily Linyphiidae (dwarf or money spiders) species are found. These species construct horizontal webs in which they catch mostly small insects (e.g. aphids). In most regions of the tropics and subtropics, a more diverse spider fauna can be found, e.g. in the production systems of southern North America. There, hunting species dominate; their range of prey includes caterpillars and phytophagous bugs. They are primarily represented by the lynx spiders (Oxyopidae), crab spiders (Thomisidae), and wolf spiders (Lycosidae; Fig. 5.17). The net spiders of this region are predominantly the long-jawed orb weavers or long jawed spiders (Tetragnathidae); the Linyphiidae are relatively rare (Nyffeler and Sunderland 2003).

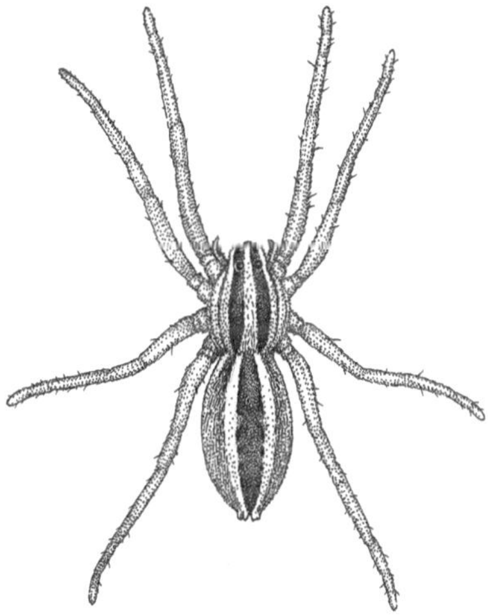

Fig. 5.17 A wolf spider (family Lycosidae)

The composition of spider communities can be markedly different in different crop stands. For example, in peanut, soybean, and alfalfa fields, ground-dwelling spiders (primarily Lycosidae) are the most common species, whereas plant-inhabiting spiders (both net-building and hunting species) predominate in stands of maize, cotton, sugarcane, and sorghum. Such differences are based primarily on factors such as structure of the crop stand, microclimate, prey availability, and disturbance frequency (Uetz et al. 1999).

Mites (Acari)

Several families of mites include predaceous species. These are more or less specialized and feed on small arthropods, for example phytophagous mites, thrips, whiteflies, and scale insects. As antagonists of phytophagous mites in agroecosystems and greenhouses, members of the family Phytoseiidae are important.

Roundworms (Nematoda)

Predatory nematodes generally live in the soil and prey on other nematodes, and/or insects. The latter primarily include the larvae and pupae of flies, beetles, and butterflies. Members of the genus *Steinernema* and *Heterorhabditis* enter the body openings of insects. Symbiotic bacteria live in the colon of these nematodes, and are released into the host organism. The bacteria fluidify the internal tissues of the insect and thus kill them. The fluidified tissue serves as food for a new generation of nematodes, which later live in the soil.

Pathogenic Microorganisms of Arthropods

Pathogens of arthropods may include bacteria, fungi, protozoa, and viruses. Some of these disease agents, especially viruses and protozoa, colonize the cells of their host and reproduce there. Most bacteria and fungi feed and reproduce extracellularly in or on the host organism. Among the bacteria, *Bacillus thuringiensis* has the greatest importance in agroecosystems. The effects of this pathogen occur as a result of release of toxic proteins in the digestive tract of insects (Sect. 5.2.1.2).

Many fungi that attack insects enter the host via the cuticula. The spores form a germ tube that develops into mycelia in the host organism. The insect dies as a consequence of the destruction of its tissues and/or the effect of toxins. Subsequently, the fungus grows out of the dead animal and forms new spores on its surface. An example of an entomopathogenic fungus is *Beauveria bassiana*, which attacks numerous insect species from various orders, as do most entomopathogenic fungi.

5.2.4.2 Relationships Between Antagonists and Pests

It is expected that antagonists in agroecosystems limit the number of pests and thus keep yield losses of crops to a minimum. This is only possible when they exert a control over the pest population, which means that the antagonists must be the factor that determines the upper limit of the growth of the pest population. The criterion for this is **predation rate**, in other words the portion of the pest population that is killed by the antagonists within a specific period of time. To limit the population density of the pests (in the ideal case before reaching the damage threshold), the predation rate must be at least as high as the growth rate of the pest population. When the number of pests increases either by reproduction or immigration of individuals, the number of individuals killed must also increase, which means that predation rate must increase appropriately. The precondition for this is an increase in the number of antagonists by reproduction or immigration.

The following sections examine more closely important factors, particularly the predation rate, that can affect the relationships between populations of antagonists and populations of prey species in agroecosystems.

Prey and Host Specificity of Antagonists

Similar to phytophages, there are specialists and generalists among the predators and parasitoids. The specialists feed on a limited range of animal species from specific taxonomic groups (usually genus, family, or order). Monophagous specialists, which only utilize a single species, are the exception among both predators and parasitoids. Generalists are species that feed on members of different orders. The factors that affect prey or host selection by antagonists are more complex than the factors involved in selection of host plants by phytophages, and are less well understood.

Prey selection by predators. An important criterion in food selection by predators is the body size of the potential prey. Most predators can only prey on animals that are smaller than themselves, but the prey must also be more than a minimum size to be attractive to the predator. Thus, the optimum body size of the prey of many spiders is approximately 50–80% of the spider's body size (Nentwig 1987). This is not true for ontogenesis, and there is a very wide range of sizes of prey organisms for individual spider species.

The range of prey utilized can also be related to the strategy of food acquisition. Most specialist predators are active hunters in their search for prey and may be aided in this by odorants or other characteristics of the prey or the host plant of the prey. Predators with a sit-and-wait strategy and predators that use traps (e.g. spider webs) are predominantly generalists.

Host selection by parasitoids. Most species of parasitoids are relatively specialized, both in terms of host species and the host's stage of development (egg, larva, pupa, adult). The host range of parasitoids is usually limited to members of a specific family and only in rare cases comprises different orders. In their search for hosts, parasitoids that attack phytophagous insects can be attracted by volatile compounds released by damaged plants (Sect. 5.2.4.3).

Defence Strategies of Prey Species

Animals have numerous options to defend themselves and/or to escape from predators, including the following strategies:

- Many arthropods have mechanisms of **chemical defence** against attackers. These are based on the release of secretions from glands or urticate hairs, for example those of some caterpillars. The defence substances utilized by phytophages are often secondary plant metabolites that they ingested with their food and that deter or harm the attacker.
- **Camouflage** can occur by way of blending in with their habitat and/or by imitation of objects that are of no interest to predators, for example twigs and leaves (Fig. 5.18).
- Discovery can also be hindered or prevented by finding hiding spots or by leading a furtive lifestyle, for example, in rolled-up leaves or in plant stems.
- Another way of preventing contact with feeding enemies is **flight**.

In individual species, defence is often based on a combination of different strategies. A specific reaction can also depend on the situation and change in relation to the predator species, which, in turn, affects predation success.

Example: The pea aphid (*Acyrthosiphon pisum*) is an important pest of alfalfa in North America. The most common antagonists of this species are the ladybird beetle *Coccinella septempunctata* and three predaceous bug species. The main

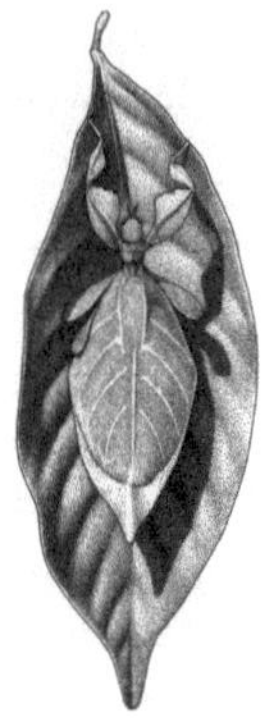

Fig. 5.18 The "walking leaves" (order Phasmida), found mainly in tropical Asia, which imitate living or dead leaves or twigs both in form and colour

strategy used by the aphids to escape from their predators is to drop from the plant. In the absence of predators, fewer than 7% of the aphids dropped. Dropping more than doubled (14%), when one of the predaceous bugs was present. Nearly 60% of the aphids dropped when the ladybird beetle was present.

The different reactions of the aphids are probably primarily because the ladybird is recognized as a predator from the vibrations it produces on the plant. Because the ladybird is significantly larger than the bug species, the disturbance caused by it is also greater. In reaction to this, the aphids release an alarm pheromone that elicits the dropping response in conspecifics (Losey and Denno 1998a).

Coincidence of Antagonists and Pests

The precondition for the occurrence of interactions between antagonists and pests is their temporal and spatial coincidence. The extent of the overlap of the respective populations also has substantial effect on predation success.

Temporal coincidence. This depends on the life cycles and phenology of the species. The presence and the abundance of pests and their antagonists in an agroecosystem is affected by a variety of factors. These include the duration of development and reproduction success, and environmental factors (especially climate and weather). The timing of the appearance of the species in the course of the year (phenology), the number of generations, and the temporal changes in abundance are also related to these factors.

Example: Development of the cockchafer (*Melolontha melolontha*), which is found in large parts of Europe, takes 4 years, on average. During this time, the larvae (grub) live in the soil and feed on roots. In warm regions, development is complete in 2–3 years whereas in colder regions it can take up to 5 years.

An example of seasonal and annual variations in the abundance of pest species is shown in Fig. 5.19.

Spatial coincidence. The environmental conditions that prevail in an agroecosystem and its surroundings can affect the colonization behaviour of

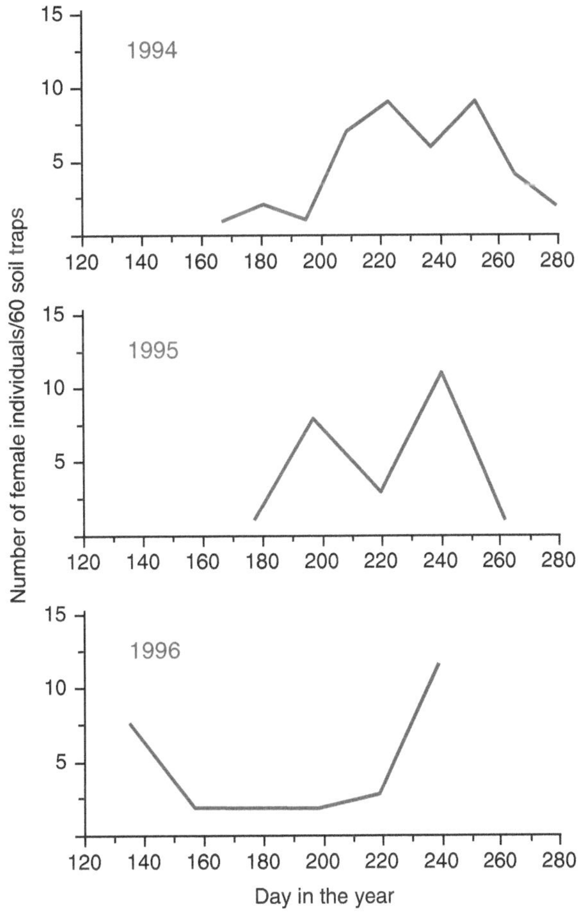

Fig. 5.19 Densities of female individuals of the wolf spider species *Hogna helluo* in a soybean field in Ohio (USA) in 3 consecutive years. The causes of the differences are unclear. It is possible that this species only produces one generation every 2 years (Based on Marshall et al. 2002)

antagonists and pests, and thus their population density. The factors of primary importance for this are:

(a) The **microclimatic conditions**, which are affected by plant density and the structure of the soil surface, are important to the presence and abundance of many species in a system. For soil-dwelling predators, which include most of the ground beetles, many rove beetles, and some spiders (e.g. wolf spiders), soil cover, either plants or mulch, is important (Sect. 5.2.4.5).
(b) The species and variety-dependent growth form and habitus of a crop, and the plant density, shape not only the microclimate but also the **spatial structure** of a plant stand. For example, these factors are of crucial importance in the colonization of agroecosystems by web-building spiders that require appropriate structures on which to attach their webs. The mobility of some species can also be impaired by spatial barriers that can result from dense vegetation.

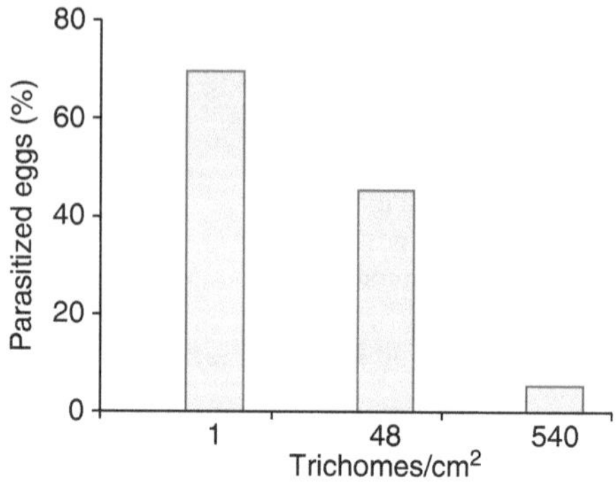

Fig. 5.20 Parasitization of eggs of the butterfly *Helicoverpa zea* by the ichneumon wasp *Trichogramma pretiosum*, on cotton, in relation to variety-specific trichome density on the leaves of the plants (Based on Treacy et al. 1986)

(c) The different animal species found in an agroecosystem colonize **different habitats**. These are essentially the soil, the soil surface, and the vegetation layer, but also specific parts of plants (e.g. the interior of the stalk, the flower, or the lower surface of leaves). Depending on where the pest and antagonist species primarily live, the probability of their encountering each other will vary.

Example: Pests with below ground stages in their life cycles, for example Diptera, whose larvae feed on plant roots, rarely come into contact with antagonists living on the soil surface, for example as ground beetles. The females of the cabbage fly (*Delia radicum*) lay their eggs on the root hairs of the host plant or in adjoining cracks in the soil. Most of the ground beetles are not able to find such deposits under the soil surface and their predation success is correspondingly low. This also applies to the eggs of the Western corn rootworm (*Diabrotica virgifera*), which are laid in the soil of maize stands and are not detected by ground beetles (Kirk 1982).

(d) **Morphological characteristics of the plant**, for example growth form and structure of their surfaces, can effect search behaviour and likelihood of the antagonist acquiring prey on the one hand and the distribution and chances of prey individuals taking flight or hiding on the other.

Example: The larvae of the parasitoid *Trichogramma pretiosum* develop in the eggs of the butterfly *Helicoverpa zea*, whose caterpillars, as polyphagous pests, are found on many crop species. In cotton, the success of predation is highly affected by the variety-specific trichome density on the leaves. The denser the hairs, the lower the number of eggs attacked (Fig. 5.20). The reason for this is primarily the limited mobility of ichneumon wasp females on hairy leaves.

(e) The prerequisite for colonization of agroecosystems by animal species that can only exist there during the cropping period is the presence of suitable **habitats in the surroundings**, which often serve for hibernation. The more suitable the living conditions and the higher the chance of survival outside the system, the greater is the potential of the respective populations to reach high densities in crop stands. This applies not only to many antagonists but also to some pest species. For example, the cereal aphid species

Metopolophium dirhodum has very specific requirements and uses roses (*Rosa* species) for overwintering.

Prey Availability

The prey available to antagonist species can be of importance from a variety of perspectives with regard to interactions between pests and predation success. The factors which primarily affect this are:

Pest density. For many antagonists, agroecosystems become attractive predation sites only when the pest density is high. However, at that point they no longer exert a regulatory effect on the development of the pest population, because they arrive at a time when the greatest crop damage has already occurred.

Example: Ladybird beetles (Coccinellidae) often act as antagonists of aphids but have little importance in agroecosystems (e.g. in cereal stands) in the control of populations of these pests. Most often, the number of ladybirds is too low at the beginning of the cropping period to be able to prevent an increase in the density of the aphids. Only with a relatively high prey availability does the number of ladybirds also increase (e.g. Dean 1974; Honěk 1982).

When the number of aphids in the crop stand reaches a specific threshold, the ladybirds also oviposit their eggs. The emerged larvae then encounter a high availability of prey. This relationship was shown in experiments by Hemptinne and Dixon (1997) in a wheat field in Belgium (Fig. 5.21). Both the larvae and the adults of the ladybirds profit from the large number of aphids, which are present near the end of the growing season in such crop stands. Even when predation success high, yield losses can no longer be prevented.

The dwarf or money spiders (Linyphiidae) commonly found in cereal fields of Europe also usually have no significant effect on limitation of the aphid population below the damage threshold. The number of spiders usually increases only in the course of the cropping season, which is the time the aphid population has already reached its highest density (Fig. 5.22). Thus, cereal fields also are attractive to dwarf spiders only when they provide sufficient prey (Sunderland et al. 1986; Harwood et al. 2001).

Availability of alternative prey species. Most predator species that appear as antagonists of pests in agroecosystems are not highly specialized and can, therefore, also feed on arthropods that are not pests. Such alternative prey species can determine the number of antagonists, and their influence on pest populations in various ways:

(a) In some agroecosystems, large populations of alternative prey species are present before an increase of pests occurs. Such prey availability can be attractive to predators and thus lead to an increase in their population density, even at an early stage of cropping. When the density of the pest increases, their antagonists are already present in large numbers.

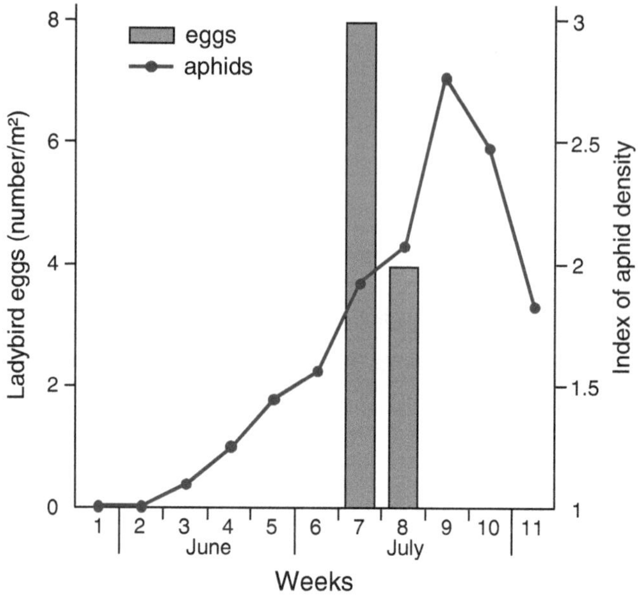

Fig. 5.21 Development of a population of the cereal aphid species *Sitobion avenae* and number of eggs of ladybirds (Coccinellidae) in a wheat field in Belgium (Based on Hemptinne and Dixon 1997)

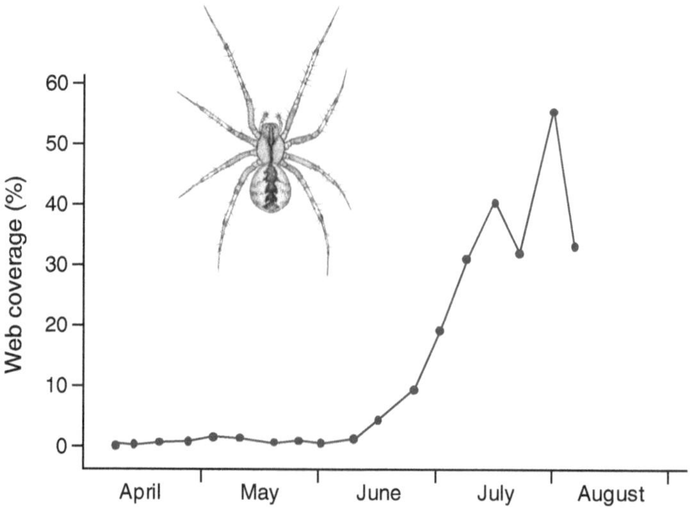

Fig. 5.22 Population development of dwarf or money spiders (Linyphiidae), measured by the degree of coverage of the horizontal webs in a wheat field in England (Based on Sunderland et al. 1986)

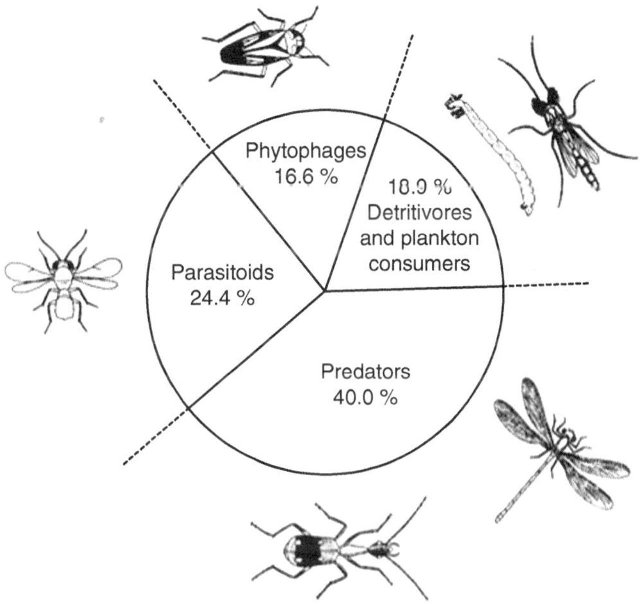

Fig. 5.23 Proportion of the different trophic groups of the arthropod species recorded in paddy rice fields of Java, Indonesia (Based on Settle et al. 1996; drawing by W. Lang)

Example: In the growing season, rice paddies are usually flooded with several centimetres of water. In this aquatic habitat, a variety of insects that feed on the detritus or plankton are present. Many of these species are larvae of Diptera, primarily non-biting midges (Chironomidae) and shore flies (Ephydridae). The adults of these species hatch at the water surface and appear on rice plants, but are not phytophagous. They become the prey of many predator species that are also antagonists of rice pests. In an experiment in Indonesia, it was found that the adults of these Diptera reached their highest abundance approximately 30 days after the rice was transplanted. Subsequently, the density of predators also increases. The number of rice pests only increases approximately 50–60 days after transplanting. At this time, a relatively high density of their antagonists is already present in the fields. Overall, approximately 19% of all arthropod species of the rice field community are detritus and plankton feeders. The proportion of the total community of these alternative prey species is thus even higher than that of the phytophagous pests (Settle et al. 1996; Fig. 5.23).

(b) Another situation occurs when antagonists colonize a crop stand on which high densities of pests and alternative prey species are present simultaneously. If the antagonist does not have a high preference for the pest, then the alternative prey species can make up a large portion of the antagonist's diet and, thus, the number of pests consumed is correspondingly lower than it would otherwise be.

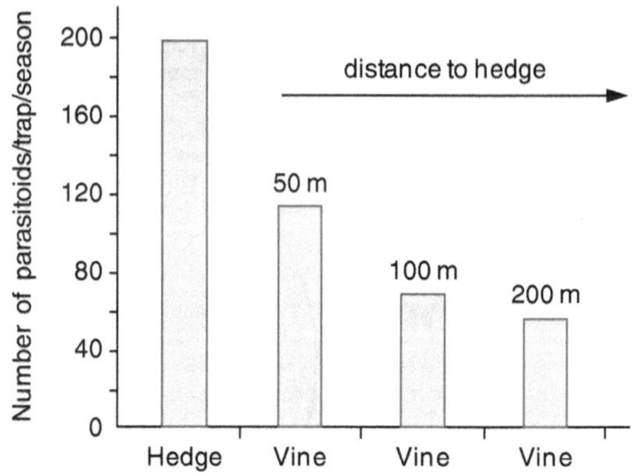

Fig. 5.24 Number of individuals of dwarf wasps of the Mymaridae family (egg parasitoids of the grape leafhopper) in the area of a hedge and in an adjoining parcel of vineyard in eastern Switzerland. The insects were collected by use of yellow traps over the course of one season (Based on Remund and Boller 1996)

(c) The presence of alternative prey or host species outside the agroecosystem can also be of importance to interactions between the antagonist and the pest:

- When crop stands have lower prey availability than habitats in the surroundings, they are less attractive to mobile predators.
- In phases or seasons in which there is no crop present and thus there are no pests, the antagonists require alternative sources of food, assuming this time is not a resting period of the antagonist. Species that act as alternative hosts during the absence of the utilized pests are important to parasitoids.
- **Example:** The grape leafhopper (*Empoasca vitis*) is a pest in European viticulture. Parasitoid wasps from the Mymaridae family act as antagonists of the grape leafhopper and develop in their eggs. The grape leafhopper hibernates outside the vineyards as an adult, which means that the eggs of this species are not available during this time. However, because the parasitoid also produces larvae during this time, it requires an alternative host. Some species of leafhoppers which overwinter in the egg stage on hedge plants are candidates for this. In a study in Switzerland, it was found that eggs of the leafhopper genus *Ribautiana* serve the parasitoids as a winter host and are frequently found on wild roses and on blackberry. In spring, when the grape leafhoppers return to the grape vines and deposit their eggs, these are again utilized by the parasitoids. At the same time, further generations of parasitoids develop in the hosts of the hedge plants, from where the adults also fly into the vineyards. There, the number decreases with increasing distance from the adjoining hedges (Fig. 5.24; Remund and Boller 1996).

Development stage of the prey. For many groups of insects (e.g. butterflies, saw fly, gall midges), only the larvae are pests; the adults usually only use the crop stands to deposit their eggs. Because of their short stay and high mobility, they are rarely found by predators in the crop stand. The predatory activity of antagonists in agroecosystems is therefore limited to specific stages of development of pest species, which are only a part of the total population of a species.

Whether a reduction in the number of eggs, larvae, or pupae in agroecosystems affects the development of such pest populations in the long term depends largely on the living conditions of the adults, i.e. the factors that cause their mortality. When the survival and reproductive success of the adults is high, a correspondingly high incidence of the larvae on the crop can result.

5.2.4.3 Relationships Between Antagonists and Crop Plants

Many predators and parasitoids have relationships, in various ways, with plants, which can also effect their interactions with their prey or host species.

Induced Indirect Plant Defence

In addition to the various mechanisms of direct defence against herbivores presented in Sect. 4.5.4.3, plants have mechanisms that attract natural enemies of herbivores after damage of plant tissues. This type of interaction between plants and antagonists of plant enemies is called **induced indirect plant defence**. Its underlying mechanism is release of volatile substances as reaction to an attack, leading to so-called herbivore-induced plant volatiles (HIPVs) in the air, which are then perceived by antagonists, for example parasitoids or predators. The HIPVs mainly comprise mixtures of terpenoids, but also include such other substances as green leaf volatiles (alcohols and aldehydes), which are responsible for the odour of damaged leaves (e.g., the odour of freshly mown grass). They are emitted either at the site of damage or systemically from undamaged parts of the affected plants. The blends can be complex, comprising hundreds of compounds, some of which are not produced by intact or artificially damaged plants and others of which are newly synthesized in response to herbivore attack (Dicke and Baldwin 2010).

The quantity and the composition of HIPVs released by the plants vary depending on the type of herbivore that causes the damage. As for direct induced defence, indirect defence is initiated by elicitors originating from the organism causing the damage. Overall, the blends of volatiles released from damaged plants are frequently specific, depending on the type of herbivore and its age, abundance, and feeding guild (specialist or generalist). Antagonists not only detect individual compounds of herbivore-induced blends, but also respond to the blend as a whole depending on the ratios of the components present (Clavijo McCormick et al. 2012).

It also has been shown that antagonists can differentiate between different HIPV mixtures. Thus, they are able to identify whether a plant has been attacked by a suitable host or by another species.

Example: In a field experiment, two groups of tobacco plants were established, one with caterpillars of the tobacco budworm (*Heliothis virescens*) and another with caterpillars of the cotton bollworm (*Helicoverpa zea*). Both species are members of the owlet moths (Noctuidae), and both are generalists. A third group

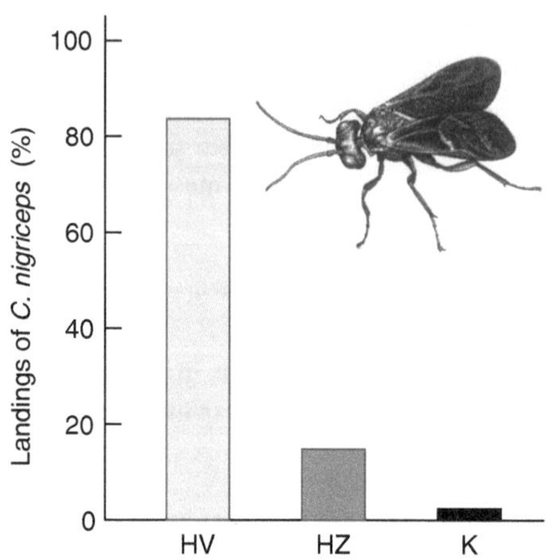

Fig. 5.25 Proportion of landings (in percent) of females of the ichneumon wasp *Cardiochiles nigriceps* on different tobacco plants in the field: plants attacked by *Heliothis virescens* (*HV*), plants attacked by *Helicoverpa zea* (*HZ*), and undamaged plants (*K*) without herbivores (Based on De Moraes et al. 1998)

of plants was kept free from herbivores and served as a control. The parasitoid ichneumon wasp *Cardiochiles nigriceps* (Braconidae) uses the caterpillars of *H. virescens* as hosts for the development of its larvae, but does not attack the caterpillars of the *H. zea*.

In field trials, *C. nigriceps* females distinguished between the odours emanating from tobacco plants infested by their host, *H.virescens*, and those from plants infested by *H. zea* or from undamaged plants. Females selected host over non-host infested plants in approximately 83% of plant visits (Fig. 5.25). Thus, this parasitoid is able to identify infestations of its host on the basis of chemical cues involving either quantitative or qualitative differences in the composition of volatile compounds produced and emitted by the plant.

Other studies have revealed that parasitoids are attracted by the volatile plant substances, and not by substances released by the caterpillars. When cotton was used instead of tobacco as the food plant for the two caterpillar species, *C. nigriceps* again showed a pronounced preference for plants with caterpillars of *H. virescens*. Thus, the parasitoids recognize their host even on different plant species, even though these plants release different mixtures of terpenoids in reaction to the same herbivore species (De Moraes et al. 1998).

In general, HIPVs have been interpreted as being indirect defences. However, given that it is so far unknown whether HIPVs benefit the fitness of a plant, their defensive function remains to be verified. Moreover, herbivores, pathogens, pollinators, and competitors also respond to HIPVs. In addition, neighbouring plants also emit volatiles that provide a background odour. These considerations enrich the evolutionary context of HIPVs and complicate predictions regarding their adaptive value. HIPVs have more functions than solely as carnivore attractants and should be seen in the context of an infochemical web that overlays the food webs of a community (Dicke and Baldwin 2010).

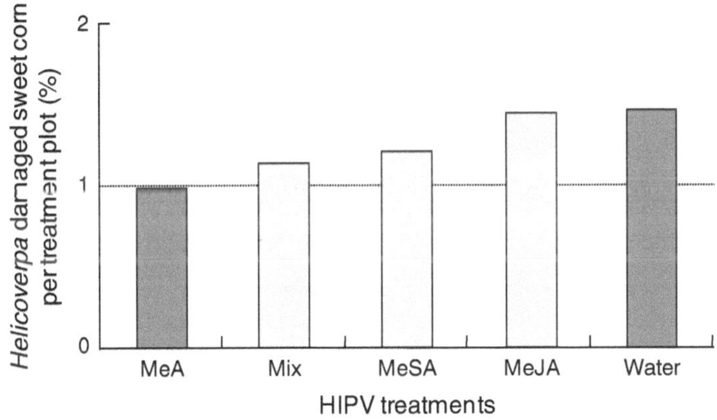

Fig. 5.26 Proportion of *Helicoverpa* spp. damage to sweet corn cobs for different HIPV treatments as attractants and for control (water) in plots with buckwheat (*Fagopyrum esculentum*) as reward plant (Based on Simpson et al. 2011)

Nevertheless, the principle of herbivore-induced plant defence can be of use in pest control, adopted to attract natural enemies into crops from surrounding habitats.

Example: Simpson et al. (2011) conducted field studies to test a novel biological control approach, the "attract and reward" method. It uses synthetic HIPVs to improve immigration of beneficial taxa into crops (attraction) and nectar plants to maintain their populations in the field (reward). This approach was tested for sweetcorn, broccoli, and wine-grapes with several HIPV formulations, using methyl anthranilate (MeA), methyl jasmonate (MeJA), methyl salicylate (MeSA), and a mixture of these. Buckwheat (*Fagopyrum esculentum*) was used as reward. Increased predator and parasitoid abundance (assessed by use of non-attractive sticky traps) were demonstrated for several of the HIPV treatments. Nectar plant reward increased catches of parasitoids in all three crop species tested and increased predators in sweet corn and broccoli. Increases in natural enemy numbers were correlated with effects at the first and second trophic levels. Significantly fewer larvae of pest caterpillars (*Helicoverpa* spp.) were found on sweetcorn plants from plots with reward. Furthermore, significantly less *Helicoverpa* damage was evident on cobs for one of the HIPV treatments (MeA, Fig. 5.26). The overall results of this study suggest that application of synthetic HIPVs can enhance recruitment of natural enemies and buckwheat was a suitable resource subsidy plant for increasing abundance and residency.

Indirect Effects of Plants on Antagonists

Plants can have effects on antagonists that occur indirectly via the herbivorous prey. Herbivores ingest secondary plant metabolites while feeding, which can

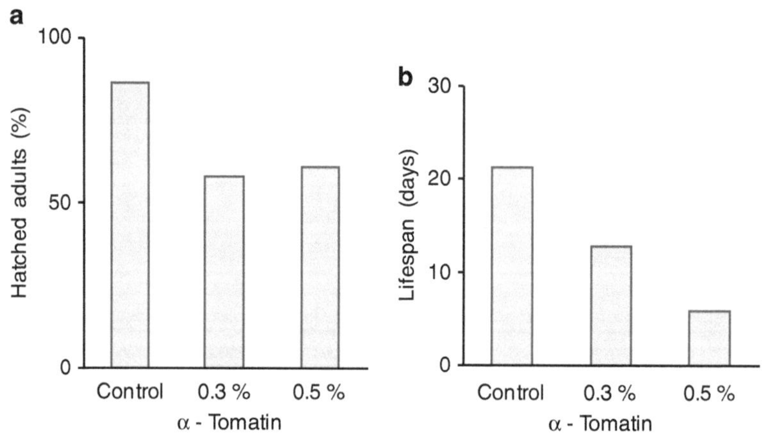

Fig. 5.27 Effects of the alkaloid α-tomatin on (**a**) the number of hatched individuals and (**b**) the lifespan of adults of the ichneumon wasp *Hyposoter exiguae*. The percentage values refer to the concentration of α-tomatin in the food of the caterpillars of *Heliothis zea*, which served the parasitoid larvae as hosts (Based on Campbell and Duffey 1979)

accumulate in their body tissues. In some cases, concentrations can be reached that are harmless to the herbivore, but that have negative effects on antagonists that feed on the herbivore species.

Examples: The cabbage aphid (*Brevicoryne brassicae*) is specialized on crucifers (Brassicaceae). The plants of this family contain glucosinolates (cf. Sect. 4.5.5.2), which are found in different quantities depending on the species and variety. Consumption of *B. brassicae* by the larvae of the ladybird beetle, *Adalia bipunctata*, increases mortality. Cabbage aphids that were captured on rapeseed (*Brassica napus*) had a lethal effect on 40% of the ladybird larvae. Of the larvae that captured this prey on yellow mustard (*Sinapsis alba*) or on black mustard (*Sinapsis nigra*), all died. The toxic effect is a result of isothiocyanate, which is formed by enzymatic breakdown of glucosinolate during food uptake by the aphids. Rapeseed contains lower quantities of glucosinolate than the two mustard species. Therefore, the concentrations of these compounds found in the aphids vary and finally result in different mortality among ladybird larvae. In contrast with *A. bipunctata*, the cabbage aphids are adapted to high concentrations of glucosinolate and isothiocyanate and are not harmed by these substances (Francis et al. 2001).

Such effects are also found among parasitoids. The caterpillars of the cotton bollworm (*Helicoverpa zea*) ingest the alkaloid α-tomatin when feeding on tomato plants. An antagonist of *H. zea* is the ichneumon wasp *Hyposoter exiguae* whose larvae develop in the caterpillars and are harmed by the alkaloid. The substance affects the number of adults that hatch (Fig. 5.27a) and their lifespan (Fig. 5.27b). The caterpillars of *H. zea* are not very sensitive to α-tomatin (Campbell and Duffey 1979).

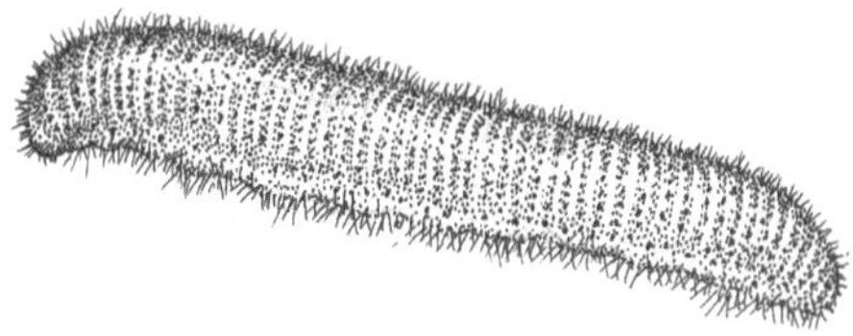

Fig. 5.28 Caterpillar of the small cabbage white (*Pieris rapae*)

Not only secondary plant metabolites but also other factors that affect the quality of food can have an effect on the relationship between antagonists and pests.

Examples: On nitrogen-poor cabbage plants, a larger proportion of the caterpillars of the small cabbage white (*Pieris rapae*; Fig. 5.28) was captured by predators than on plants fertilized with nitrogen. The explanation for this is that the time required for the development of caterpillars increases with decreasing nitrogen concentrations in the food and therefore, there is a higher probability that the larvae fall victim to a predator. Parasitoids prefer caterpillars on nitrogen-rich plants. A possible reason for this is the improved nutrient status for the parasitoid larvae of such host individuals (Loader and Damman 1991).

Seven-spotted ladybirds (*Coccinella septempunctata*) were fed with pea aphids (*Acyrthosiphon pisum*) which had been reared on two different food plants (alfalfa and faba bean). Faster larval development, greater survival, and larger size were observed for the offspring of beetles that consumed aphids from alfalfa than for those that received aphids from faba beans. This can be explained by the different nutritional value of the plant species. Aphids reared on alfalfa take up greater quantities of fatty acids than those reared on faba beans and thus provide a more energy-rich source of food for the ladybirds (Giles et al. 2002).

It has occasionally been shown that the δ-endotoxins produced by transgenic Bt crops (Sect. 5.2.2.1) have a negative affect on antagonists that capture their prey on such plants.

Example: The ichneumon wasp *Parallorhogas pyralophagus* is a parasitoid of the caterpillar of the butterfly *Eoreuma loftini*, which appear as stem borers in maize in America. Mortality was significantly higher for larvae of the ichneumon wasps that utilized hosts on Bt maize than for those that attacked caterpillars on conventional maize (Fig. 5.29a). The surviving larvae developed more slowly and had a shorter lifespan as adults (Bernal et al. 2002). Similar effects have also been observed for larvae of the green lacewing *Chrysoperia carnea* that were fed with the caterpillars of two different pest species. When the prey animals contained Bt toxins from transgenic maize plants, mortality of the green lacewing larvae was approximately one-third higher than for those that received their prey from maize plants without Bt genes (Fig. 5.29b; Hilbeck et al. 1998).

In contrast with these results, no negative effects on survival and reproduction success were observed for a ladybird species (*Hippodamia convergens*) that consumed aphids (*Myzus persicae*) fed on Bt potatoes (Dogan et al. 1996).

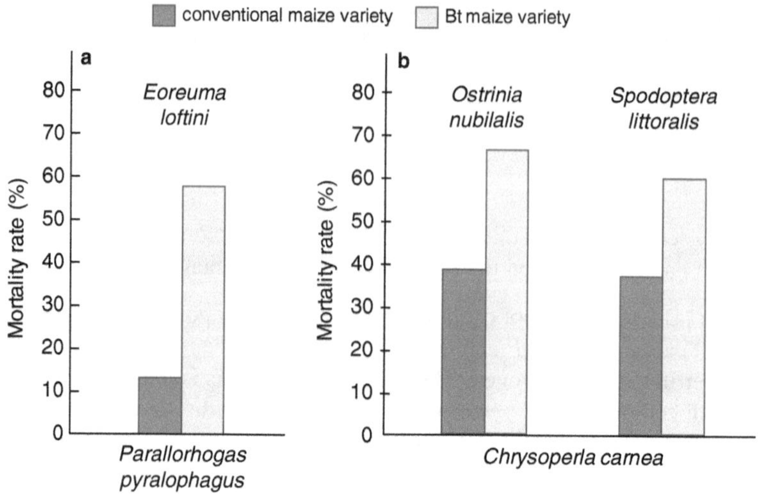

Fig. 5.29 Mortality of antagonists that fed on herbivores of transgenic Bt maize and on conventional maize varieties: (**a**) larvae of the ichneumon wasp *Parallorhogas pyralophagus* in the caterpillars of the stem borer *Eoreuma loftini* (Based on Bernal et al. 2002); (**b**) larvae of the green lacewing *Chrysoperla carnea* fed with caterpillars of *Ostrinia nubilialis* and *Spodoptera littoralis* (Based on Hilbeck et al. 1998)

5.2.4.4 Relationships Between Antagonist Species

In agroecosystems, there are generally several species of antagonist present at the same time and these can have direct or indirect relationships with each other. This variety of interactions can result in very different effects on the populations of pest species.

Feeding Relationships

Some predators and parasitoids are not, or not only, natural enemies of pests, but may also prey on other antagonists. Two situations can be identified in this context:

1. Two antagonist species (A and B) have a pest species as common prey, but antagonist B also feeds on antagonist A (Fig. 5.30a). Sometimes the two antagonist species feed on each other (Fig. 5.30b). Both situations can result in fewer pests being preyed upon in the presence of both antagonists than in the presence of one antagonist species alone.

   Example: In California, the larvae of green lacewings (*Chrysopidae*) are important consumers of pests of cotton, primarily aphids (*Aphis gossypii*). A variety of predaceous bug species also occur in the fields; these prey on both aphids and green lacewing larvae, and some also use plant sap or cotton pollen as a food source. Some development stages of the bugs also serve as prey for green

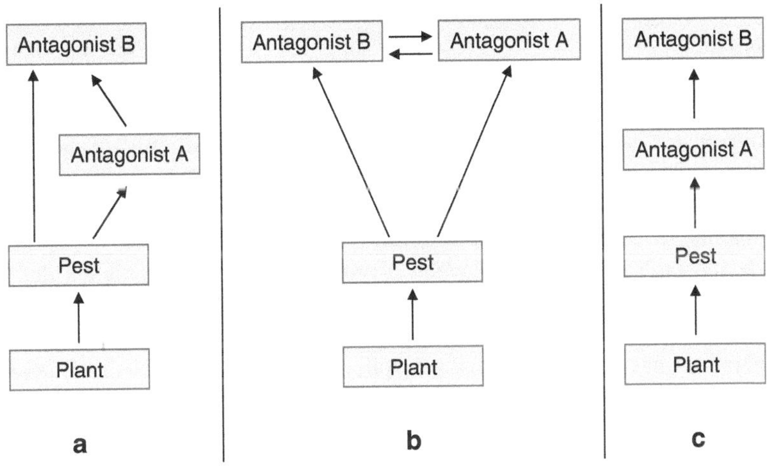

Fig. 5.30 Possible feeding relationships between two antagonist species and a pest species. (**a**) two antagonists (A and B) have a pest species as common prey, and antagonist B also feeds on antagonist A. (**b**) Both antagonist species have a common prey species, but also feed on each other. (**c**) Antagonist A has a pest species as prey, antagonist B only feeds on antagonist A and is thus a second order antagonist

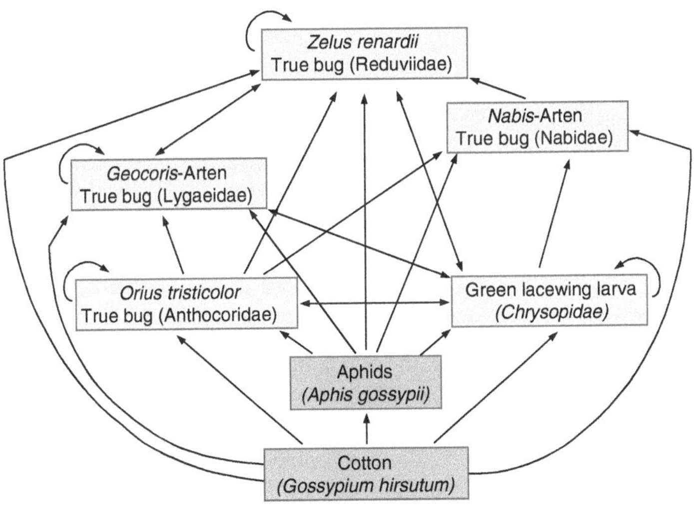

Fig. 5.31 Feeding relationships between the antagonists of aphids in cotton fields in California (Based on Rosenheim et al. 1999)

lacewing larvae, however. Overall, very complex feeding relationships exist between these species (Fig. 5.31). In comparative experiments it has been shown that green lacewing larvae can significantly reduce the number of aphids in the absence of other predators. In contrast, the total community of predators has little effect on the aphid density. The reason for this is the reduction in the number of green lacewing larvae by the bugs. At medium aphid density, approximately 94% of the green lacewing larvae were preyed upon by the bugs; at high aphid density, the proportion was approximately 60% (Rosenheim et al. 1999; Rosenheim 2001).

2. Antagonist A has a pest species as prey. Antagonist B feeds only on antagonist A and is thus a second-order antagonist (Fig. 5.30c). This type of interaction can occur both among predators and parasitoids. In the latter case, the second-order antagonist is termed a hyperparasitoid.

   A special situation occurs when predators feed on prey animals that have already been attacked by parasitoid larvae. Sometimes the predators then have a stronger effect on the population of parasitoids than on the prey species. In Australia it has been shown that caterpillars of the small cabbage white (*Pieris rapae*) that are parasitized by the ichneumon wasp *Apantele glomeratus* are more vulnerable to predation by ants of the genus *Iridomyrmex* than are unparasitized larvae. Presumably, the parasitized individuals are easier to capture because of their impaired condition (Jones 1987).

   Not only pests, but also predatory insects, for example ladybirds, serve as hosts for some parasitoid species. In northern Germany, parasitization of the adults of the ladybird *Coccinella septempunctata* by the ichneumon wasp species *Dinocampus coccinellae* in the hibernating sites in spring can reach 33%. In cereal fields, up to 12.5% of the individuals were attacked (Triltsch 1996).

Indirect Relationships

Different species of antagonist are interdependent not only because they feed on each other, but also because they feed on the same prey species. With regard to the effects on a pest in agroecosystems, they can either hinder each other in this activity or they can complement each other. Depending on the situation, this can have different effects on the total number of pests that are captured:

- In instances of interference, i.e. the encounter of individuals of different antagonist species in search of prey, the situation may often result in direct competitive interaction (e.g. local displacement of one species by another). As a consequence, predation success of individual antagonist species can be lower than it would be in the absence of the other species.
- An additive effect occurs when the total number of prey individuals captured by two or more antagonist species in a specific period of time is as high as the sum of the number of prey individuals that is captured by each of these antagonists alone.
- A synergistic effect exists when the total number of prey individuals captured by two or more antagonists is higher than the sum of the number of prey individuals that is captured by each of these antagonists alone.

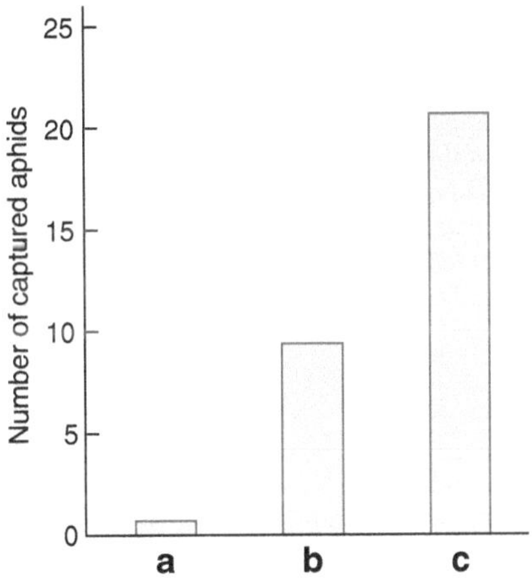

Fig. 5.32 Number of pea aphids (*Acyrthosiphum pisum*) captured by predators over the course of 24 h on alfalfa plants in different experimental setups: (**a**) by the ground beetle *Harpalus pennsylvanicus*, which captures its prey on the ground, (**b**) by the ladybird *Coccinella septempunctata*, which preys on the plants, and (**c**) by both predator species together (Based on Losey and Denno 1998b)

The occurrence of additive and synergistic effects is most likely when the individual antagonist species forage in different habitats within an agroecosystem, and/or have different forage strategies.

Example: In experiments, the effects of two predator species on the number of pea aphids (*Acyrthosiphum pisum*) on alfalfa plants were examined. The predators were the seven-spot ladybird (*Coccinella septempunctata*), which forages on the plants, and the ground beetle *Harpalus pennsylvanicus*, which captures its prey on the ground. Over time periods of 24 h each, the effects of each predator species on the aphids were first observed individually in systems that consisted of one plant, 30 aphids released on to the plant, and one individual of the predator species. The ground beetles alone did not have a significant effect on aphid density on the plants (Fig. 5.32a). The ladybird alone captured approximately one-third of the prey individuals present (Fig. 5.32b). When both predators were present together, the number of aphids was reduced by approximately two-thirds, i.e., combined predation by both predators was substantially higher than would have been expected on the basis of adding their individual consumption (Fig. 5.32c). This synergistic effect is explained by the dropping behaviour of the aphids. In the presence of the foliar-foraging ladybird beetle, aphids fall to the ground where they become vulnerable to attack by the ground-foraging ground beetle. Some aphid individuals are therefore killed before they are able to return to the plant (Losey and Denno 1998b).

5.2.4.5 Conservative Biological Pest Management

The fact that antagonists in an agroecosystem kill pests is, alone, not sufficient to draw any conclusions about the resulting effects. Predators and parasitoids only exert effective pest control when they limit the densities of pest populations and thus, in the end, reduce damage to the crop. Often, however they are unable to

Fig. 5.33 Percentage of cases in which parasitization of pest species in mixed cropping was higher than, lower than, or approximately the same as in sole cropping. The data are based on analysis of 26 individual studies (Based on Coll 1998)

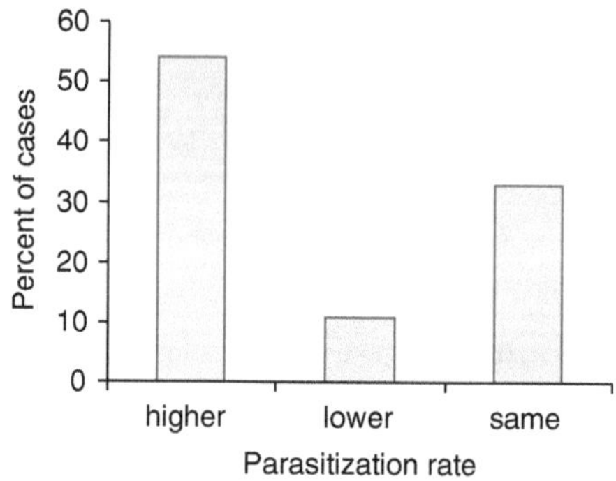

achieve this, because they occur in insufficient numbers or because they appear in the agroecosystem at the wrong time (Sect. 5.2.4.2).

With appropriate measures to protect or promote antagonists, attempts can be made to increase their numbers in the agroecosystem and thus to achieve more effective pest management. This approach is called **conservative biological pest management** and is essentially based on modification of the habitat structure of the cultivation system or its environment.

Mixed Cropping

Numerous experiments have dealt with the question of whether and how the numbers of species and individuals of antagonists in mixed cropping systems differ from those in one-crop systems. Analysis of more than 200 studies showed that for approximately half (53%) of all predator species and three-quarters (75%) of all parasitoid species numbers of individuals in mixed cropping were higher than in one-crop systems. Furthermore, approximately half (52%) of the phytophagous species in mixed crops were present in lower numbers than in sole crops (Andow 1991).

The results of this analysis show that establishment of mixed cropping can contribute to promotion of the populations of antagonists and to reduction of pest densities. However, the question of whether the lower individual densities of the phytophagous species in the mixed cropping systems can always be attributed to the effect of antagonists remains unanswered.

Similar analysis has been conducted for studies based solely on parasitoids (Coll 1998). Not only were the numbers of individuals in different production systems compared but also the success of parasitization. In more than half of the cases, the latter were higher in mixed crops than in sole crops (Fig. 5.33)

In most cases in which the number of antagonists in mixed crops was higher than the number in the one-crop system, this can be attributed to the greater structural diversity of the mixed cropping system (Langellotto and Denno 2004). How this factor improves the living conditions of the species cannot always be clearly stated.

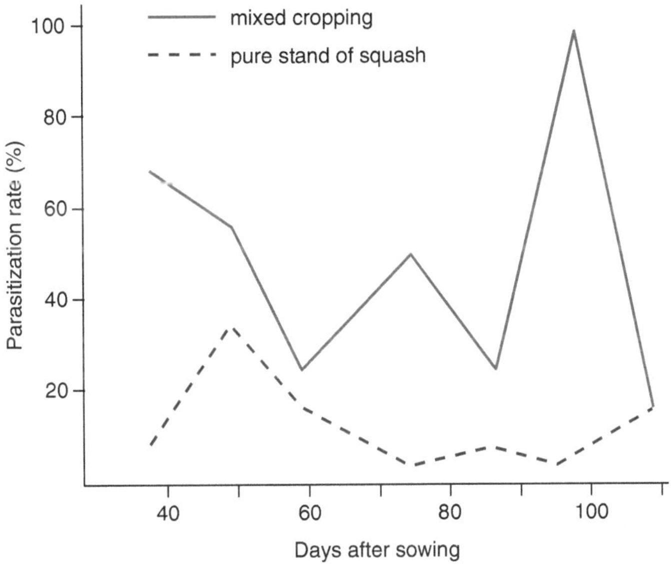

Fig. 5.34 Parasitization of the eggs of the butterfly *Diaphania hyalinata* in pure squash stands and in traditional mixed system of squash, maize, and cowpea in Mexico (based on Letourneau 1987)

It might, for example, be related to the greater availability of alternative prey species, to more suitable forage conditions, or to the availability of nectar and pollen as additional sources of food. Among the antagonists that utilize flowers as resources are the ladybirds, hoverflies, and some green lacewing species, in addition to the adults of parasitoids. Generally valid explanations of the different numbers of individuals of antagonists in mixed and one-crop systems cannot be made, because the species and varieties of the crops used, their patterns of cultivation, and the species of antagonist are very diverse. It is also often observed that different antagonist species do not react in the same way to modifications in the production system.

Examples: The caterpillars of the butterfly *Diaphania hyalinata* occur as pests in cucurbit crops in Central America. A study in Mexico examined whether antagonists in a traditional mixed cropping system of maize, cowpea, and squash had a greater effect on the density of the caterpillar population than in pure squash stands. Parasitization of the eggs and caterpillars of *D. hyalinata* (Fig. 5.34) by different species of ichneumon wasps was significantly higher in the mixed cropping system than in the pure stand of squash. However, the responses of natural enemies to vegetational diversity were not uniform. Most predators were not significantly affected by cropping patterns, and some were more abundant in pure squash stands than in the traditional system (Letourneau 1987).

A study in Uganda examined how the number of individuals of antagonists varies between one-crop of cowpea and mixed crops (cowpea with sorghum, and cowpea with mung bean). Although the densities of ladybirds and hover fly larvae

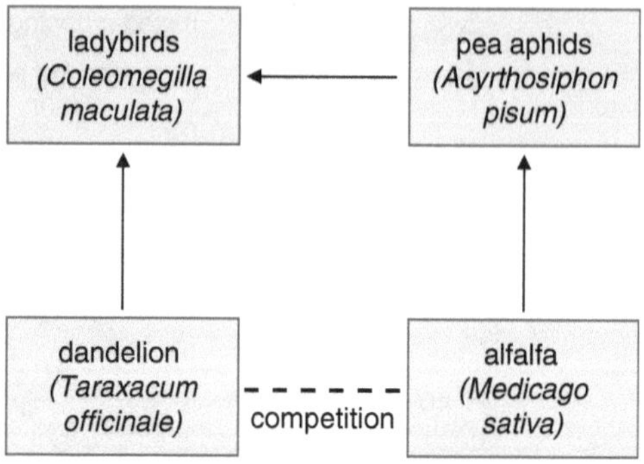

Fig. 5.35 Feeding and competition relationships between different species in an alfalfa field in Wisconsin, USA (Based on Harmon et al. 2000)

were not affected by the cultivation system, those of spiders and predaceous bugs of the genus *Orius*, were more abundant in the one-crop of cowpea than in both mixed crops (Nampala et al. 1999).

Establishment of mixed cropping systems often contributes to management of pests. However, it cannot be assumed that this method will always be successful, because not all antagonist species react to modifications of the production system in the same way.

Integration of Wild Plants

The natural growth or sowing of wild plant species, which do not serve as crops, can promote antagonist populations in a similar way to the establishment of mixed cropping systems.

Examples: In many alfalfa fields in the United States, dandelion (*Taraxacum officinale*) occurs a wild plant. A study showed that the density of these plants was positively correlated with the number of ladybirds (*Coleomegilla maculata*), which feed on the pollen of the dandelion flower as an additional source of food. At the same time, there was a negative correlation between the number of ladybirds and the number of pea aphids (*Acyrthosiphon pisum*). The latter are found as pests in alfalfa but are not found on dandelions. The stands of dandelion thus indirectly reduce pest density. However, in this study by Harmon et al. (2000) it remained unclear how aphid density on the one hand and competition between the dandelions and alfalfa on the other affects yields. The relationships between the species are shown in Fig. 5.35.

In an apple orchard in Switzerland, strips of wild plants were established between the rows of trees. In the flowering phase of the wild plant species, different predators (primarily spiders, bugs, ladybirds, and green lacewings) appeared in greater numbers on the apple trees than in control areas without wild plant strips. The rosy apple aphid (*Dysaphis plantaginea*) and the apple aphid (*Aphis pomi*) were found in lower densities in the area with wild plants than in the control areas. This can be attributed to the activities of the antagonists. The positive effect of the

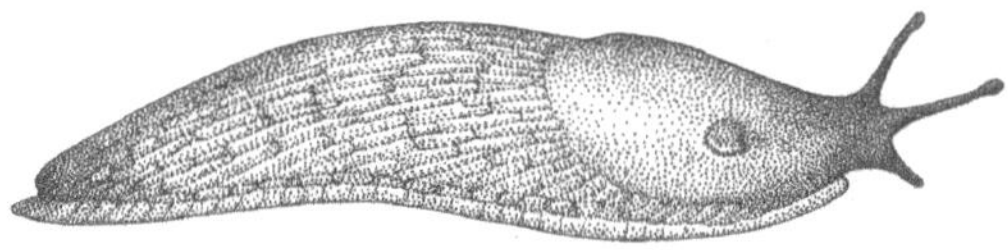

Fig. 5.36 Spanish slug (*Arion lusitanicus*). This species probably originates from the Iberian Peninsula and is today found in large parts of Central Europe

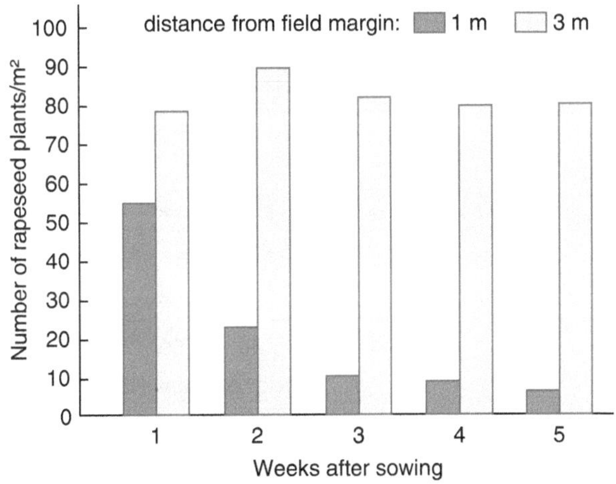

Fig. 5.37 Number of rapeseed plants per square meter at different distances from the field margin, from which different slug species (primarily *Arion lusitanicus*) invade the field (Based on Frank 1998)

wild plant stands on these species is based essentially on the additional availability of food (nectar, pollen, and aphids) found on the wild plants (Wyss 1995).

Natural or sown stands of wild plants in or around agroecosystems may be attractive not only to antagonists but also to pests that attack crop plants.

Example: Field margins are an ideal habitat for slugs because of the shady and moist microclimate of the soil surface. A study in Switzerland showed that the Spanish slug (*Arion lusitanicus*; Fig. 5.36) and other snail species invade adjoining rapeseed fields from these edge habitats, predominantly at night. Up to approximately 1 m inside the field, young rapeseed plants were often completely eaten by the snails whereas no damage occurred 3 m from the field edge (Fig. 5.37; Frank 1998).

Soil Cover

The extent of coverage of the soil surface by stands of wild plants or undersown crops is primarily of importance for soil-dwelling antagonists, for example ground beetles and some spider species. It has often been shown that the number of individuals of such antagonists is higher on soil with dense plant cover than on bare surfaces (Fig. 5.38). However, not all species react similarly to the plant cover

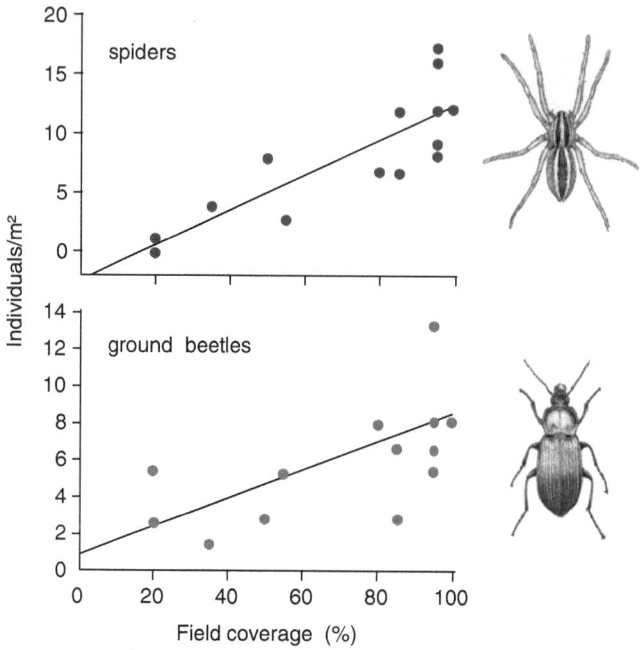

Fig. 5.38 Relationships between soil coverage by wild plants and the density of individuals of spiders and ground beetles in a soybean field in Ohio, USA (Based on Halaj et al. 2000)

density. Ground beetles include species that are mainly found in shaded areas and others that prefer bare ground.

An increase in the density of antagonists may also be achieved by covering the soil with mulch (harvest residues or other organic material).

Example: By covering the soil with straw after planting of potatoes in a field in North America, yield losses caused by Colorado potato beetles were significantly reduced. This pest produces two generations per season. The most important antagonists of the first generation were ground beetle species that climb up the potato plants and primarily prey on the second and third development stages of the potato beetle larvae. With the emergence of the second generation of potato beetles, primarily ladybirds, the larvae of the green lacewing *Chrysoperla carnea*, and the predatory bug *Perillus bioculatus* were present instead of the ground beetles. These species primarily feed on the eggs and on the first larval stage of the pest. The number of different antagonists on the potato plants was higher on straw covered plots than on uncovered plots during the entire growing season (Fig. 5.39). The higher antagonist density resulted in reduction of potato plant leaf area loss by a factor of 2.5 and in potato yields approximately 30% higher than for plots without straw mulch (Brust 1994).

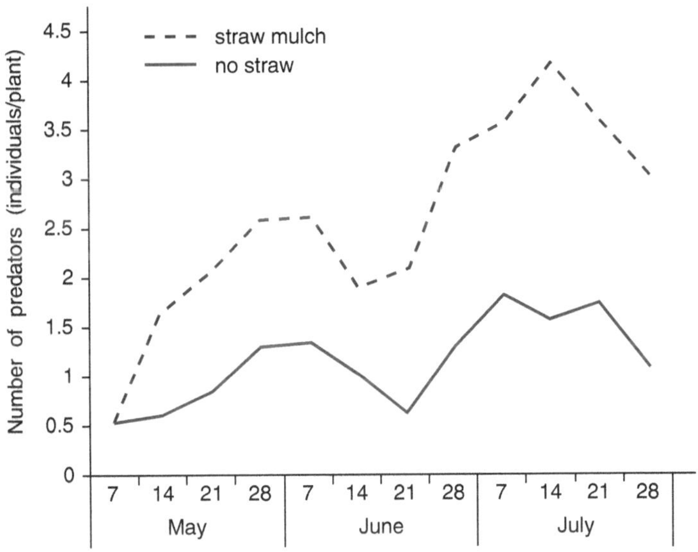

Fig. 5.39 Numbers of individuals of predatory arthropods on potato plants in parcels with straw mulch and in parcels with bare soil in a field in Indiana, USA (Based on Brust 1994)

5.2.4.6 Inundative Biological Pest Control

Antagonists and pests often have different development cycles or phenology; this can result in temporal differences in the appearance and abundance of the respective species (Sect. 5.2.4.2). Even with directed measures to protect or promote antagonists, it is often not possible to increase their numbers at the right point in time, as is necessary for effective control of pests. One solution to this problem is the rearing of specific antagonists under artificial conditions and then releasing them in large numbers into fields or greenhouses. This approach is called **inundative biological pest control**. The principle of this method is to increase the number of antagonists in relation to the pest population to the extent that they are able to reduce the density of a pest significantly within a short period of time. Thus, with this type of management, a similarly strong effect is expected as would result from the application of insecticides. Generally, it is not expected that the respective antagonist species will become established permanently. Depending on the need, they must be released every season or several times during a season. Successful application of the inundative method is based on two necessary preconditions:

- The antagonists selected must, at least, have a preference for the target pest species, to limit their use of other prey or host species, especially other antagonists.
- Both the timing and the number of individuals to be released must be carefully determined. In fields, it must be expected that a small or large portion of the

Fig. 5.40 Ichneumon wasp of the genus *Trichogramma* on the egg of an insect

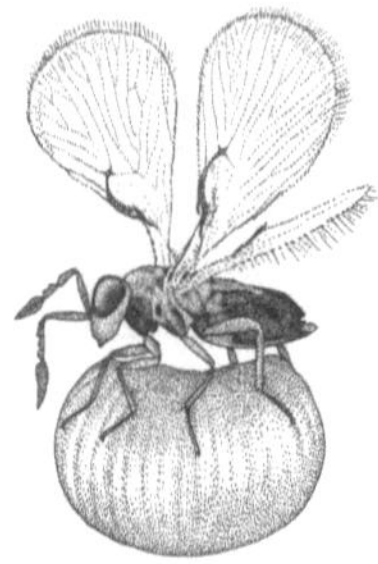

individuals will either actively or passively (e.g. by wind) disappear from the target area.

Overall, representatives from almost all the antagonist groups described in Sect. 5.2.4.1 are used in the inundative method. Many species are commercially produced and are available for purchase.

Examples: Parasitoid wasps of the genus *Trichogramma* (Fig. 5.40) attack the eggs of different orders of insects (e.g. butterflies, true bugs, and beetles) and are successfully used in inundative pest management worldwide (primarily against caterpillars). For crops such as sugar cane, wheat, maize, or Brassica species, parasitization by *Trichogramma* of 60–80% can be achieved, and thus the yield losses caused by their hosts are substantially reduced (Li 1994).

One of the most important maize pests in Europe is the European corn borer (*Ostrinia nubialis*), whose caterpillars feed inside the maize-cobs and the stems of the plant. From the Mediterranean region, this butterfly species expands to the north and is currently a problem primarily in the south of Germany. *Trichogramma* species (especially *T. evanescens* and *T. brassicae*) are important in corn borer management. They are applied by way of parasitized host eggs that are exposed in the fields on cartons or in capsules, from which the parasitoids hatch. In most years, two applications of approximately 100,000 individuals per hectare are sufficient to reduce pest density so far that the yield losses remain low or are completely prevented.

The predatory bug *Perillus bioculatus* is an important antagonist of the Colorado potato beetle in North America and feeds on the eggs and young larvae of this pest. By directed release of *P. bioculatus*, the number of eggs of the potato beetle in the field can be significantly reduced, as has been shown in experiments by Hough-Goldstein and Whalen (1993; Fig. 5.41).

5.2.4.7 Classical Biological Pest Control

Biological control of pests in a region to which they were introduced from another region is called **classical biological pest control**. The specific characteristic of this method is that management of such pest species is conducted with antagonists that also originate from another region, generally from same region of origin as the pest.

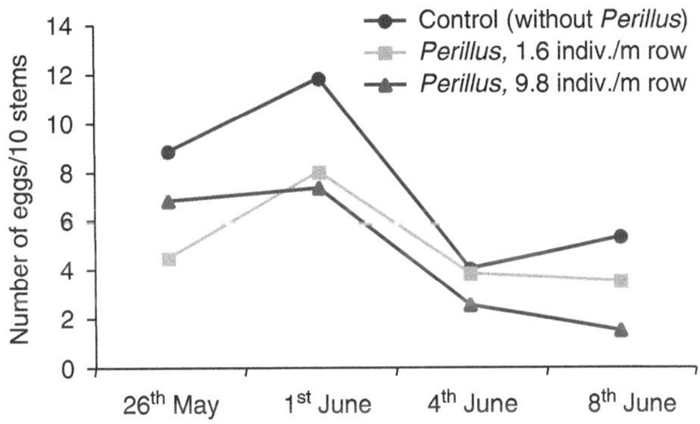

Fig. 5.41 Number of eggs of the Colorado potato beetle (*Leptinotarsa decemlineata*) on potato plants for different densities of the predatory bug *Perillus bioculatus*, which was released in fields in Delaware, USA (Based on Hough-Goldstein and Whalen 1993)

Examples: In 1887, the Californian citrus industry was facing ruin because of a massive attack on the citrus plantations by the cottony cushion scale *Icerya purchasi*. This species, which sucks plant sap, probably originates from Australia, where the search for natural enemies of the pest was subsequently undertaken. These include the larvae and adults of the ladybird *Rodolia cardinalis*. Their release in the citrus plantations led to control of the pest within 1 year. *I. purchasi* was not entirely eliminated, but survived in small and dispersed populations. These were not important to citrus production, but were sufficient to maintain populations of *R. cardinalis*. Problems only emerged in the 1940s, when DDT was applied to the citrus crops and killed not only pests but also the ladybird beetles. For biological control of *I. purchasi*, it was necessary to repeat introduction and rearing of *R. cardinalis*.

St. John's wort (*Hypericum perforatum*) is an herbaceous plant native to the Mediterranean region and was introduced to California in approximately 1900. There, it developed into an invasive weed (cf. Sect. 3.3.3) and, after four decades, covered large areas of pasture land. Not only did it greatly reduce the grazing area for cattle, sheep, and horses, St John's wort also contains the alkaloid hypericin, which is toxic to mammals. Ingestion makes the skin sensitive to sunlight and affects the performance and welfare of the animals. In the 1940s, two leaf beetle species, *Chrysolina hyperici* and *C. quadrigemina* (Chrysomelidae), were introduced from Europe and released in California as biological control agents. Both species are specialist feeders on *H. perforatum*, feeding on shoot tips and leaves (the larvae) and on flower buds (the adults). The species became successfully established and within a few years were able to reduce the area infested by *H. perforatum* to less than 1% of its former extent.

The overwhelming success of such projects is mainly because the introduced antagonist species do not have important natural enemies or competitors in their new habitats, and also because they are highly specialized on the introduced pest species.

There are many examples of classical biological pest control projects in which the desired effects have been achieved; most such approaches have failed. In only approximately 16% of all cases has relevant reduction of the respective pest species

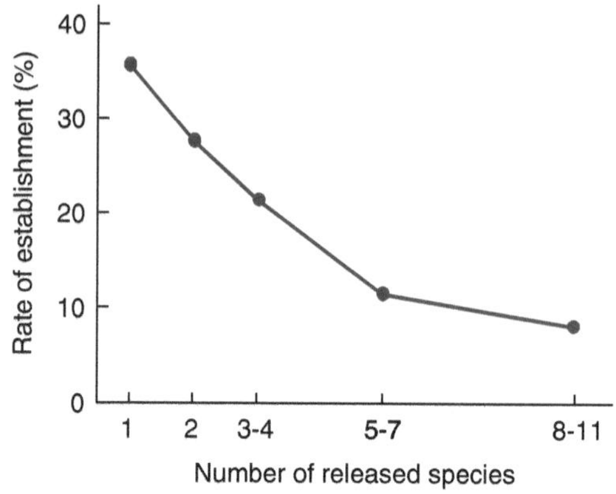

Fig. 5.42 Relationship between rate of establishment and number of simultaneously released antagonist species in classical biological control projects (Based on Ehler and Hall 1982)

been achieved by release of introduced predators or parasitoids (Myers et al. 1989; Waage and Mills 1992). Factors that may explain such failures are primarily:

- Natural enemies of the antagonist: Predators, parasitoids, or animal pathogens can reduce the population density of introduced species and, in extreme cases, lead to extinction of a population.
- The prey or host range of the antagonist: When introduced antagonist species are not highly specialized on the host species, it is possible they also feed on other prey or host species and thus threaten the success of the project.
- The abiotic environment: The climatic conditions that exist in the new habitat of the antagonists are usually not exactly identical with those in their area of their origin. The effects of the climate may, therefore, also impair or prevent establishment or reproduction of the introduced antagonist species whereas the same conditions are tolerated by the introduced pest species.
- Competitors: In many classical biological pest-control projects, not only one but several antagonist species were released in the target area. It has been shown that the rate of establishment of introduced natural enemies is inversely related to the number of species released simultaneously (Ehler and Hall 1982; Fig. 5.42). In many cases, this can be explained by interspecific competition for the common prey or host species, which leads to extinction of the inferior species. When the remaining antagonist species do not have a large effect on the pest population, because of low specialization or other factors, effective control can no longer be achieved.

Overall, the successes and failures of classical biological control projects must be seen against the background of the special situations created by humans (i.e., the accidental introduction of a pest and the intended introduction of antagonist species). It is, therefore, hardly possible to predict the outcome of predator–prey interactions in the area of introduction.

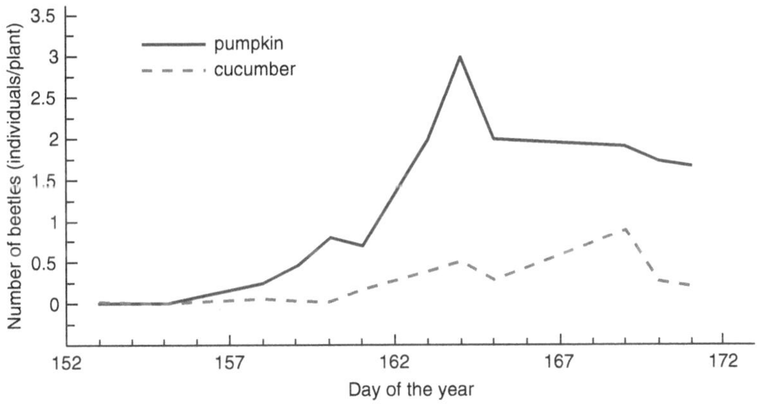

Fig. 5.43 Number of individuals of the striped cucumber beetle (*Acalymma vittata*) per plant in a field (in Maine, USA) which was planted with cucumbers and pumpkin in equal ratio (Based on Radin and Drummond 1994)

5.2.4.8 Biological Pest Management with Plants

Not only antagonists, but also plants can be used for biological pest management. Although conservative biological pest management is intended to promote populations of antagonists, the density of pests on crops can also be directly reduced by specific modification of the plants in agroecosystems. Such measures include the establishment of plant species that are preferred by pests over other food sources (the crop plant) and which therefore serve as **trap crops**. Trap crops have been defined as plant stands grown to attract insects or other organisms, for example nematodes, to protect target crops from pest attack, preventing the pests from reaching the crop or concentrating them in a specific part of the field where they can be economically destroyed (Hokkanen 1991).

Examples: The striped cucumber beetle (*Acalymma vittata*) feeds primarily on cucurbits (Cucurbitaceae) but has different preferences for individual species within this plant family and mainly feeds on pumpkin (*Cucurbita maxima*). This attribute can be used to protect cucumbers (*Cucumis sativus*) from attack by this pest. When half of a crop stand is composed of cucumbers and the other half of pumpkin, at least 70% of the beetles colonize the pumpkin plants (Radin and Drummond 1994; Fig. 5.43).

Lygus bugs (genus *Lygus*) are polyphagous plant sap suckers that occur as pests on many agricultural crops. Because of their wide range of host plants, varieties of wild plant species which are preferred to the crops can be used as trap crops. In the production of lettuce (*Lactuca sativa*) in Sweden, legume species (clover species, alfalfa, common vetch) and mugwort (*Artemisia vulgaris*) have been used for this. Five to thirty times more lygus bugs became established on the legume plants rather than on the lettuce plants. For mugwort the density of the pest was 100 times higher than on the lettuce (Rämert et al. 2001).

Meloidogyne incognita is a root gall nematode that is important as a pest primarily in the tropics and has a wide range of host plants. In experiments in Cuba, lettuce and radish (*Raphanus sativus*) were planted as nematode trap crops and then completely removed from the field after 1 month of growth. By this means it was possible to reduce nematode density in the soil by 50%. For subsequently grown crops, for example tomato, pumpkin, or cowpea, 10–20% higher yield was achieved on the sanitized fields than on plots without previous sanitation with trap crops (Cuadra et al. 2000).

In different situations it has been shown that expectations of the trap crop method were not, or were only partially, fulfilled; insufficient differences between the food preferences of the pests are primarily responsible for this. These are not only related the respective plant species but are also affected by such factors as fertilization or the nitrogen content of the plant (Sect. 4.5.3.2) and pest density.

Example: Use of mustard and rapeseed as trap crops to protect broccoli (*Brassica oleracea*) from attack by the harlequin bug *Murgantia histrionia* in North America was only successful when the pest species appeared in small numbers. At high population density, individuals of the pest population began to move to the broccoli plants (Ludwig and Kok 1998).

5.3 Management of Phytopathogens

In comparison with phytophagous pests, the life strategies of agents of plant diseases are more diverse, which results not only in differences, but also limitations, in the possibilities of control.

5.3.1 Viruses

To reproduce, viruses depend on the host cells they infect. Direct chemical control of phytopathogenic viruses can hardly be achieved, because the agents attack the plant cells and interfere with nucleic and amino acid metabolism. Repression of viral activity by use of chemical inhibitors would also cause significant disorder in the metabolism of the plant. It is, however, possible to limit the spread of viruses, in particular by control of their respective vectors (Sect. 4.6.1) and use of virus-free seeds and plant material. Infected seeds and seedlings are often the primary source of virus diseases.

5.3.2 Bacteria

The most effective way of controlling bacteria is by use of antibiotics. These are substances that are usually produced by bacteria and fungi and have negative effects on the metabolism and the reproduction of other microorganisms (primarily bacteria) and/or their ability to survive. Today, antibiotics also include numerous synthetic compounds. Antibiotics affect either specific groups of microorganisms or a more or less broad range of species.

However, use of such compounds as plant-protection products is problematic for several reasons. First, use of antibiotics can have negative effects on other species in the environment (e.g. soil microorganisms). Second, there is the threat of development of resistance to antibiotics. This is not necessarily limited to the target species, but can also occur in bacteria species that are human and animal pathogens. These can then no longer be controlled by use of antibiotics with a similar mode of action. Nevertheless, antibiotics are sometimes used for plant protection. An example is the use of streptomycin against fire blight (Table 4.9) in pome fruit production. Because of potential ecological and health risks, this antibiotic, which is also used to treat tuberculosis, lost its registration as a plant-protection product in the European Union in 2004. In some countries, for example Germany, streptomycin continues to be used, at least in exceptional cases, against fire blight. Chemical control of phytopathogenic bacteria by use of other substances (e.g. copper compounds) is successful in a few cases only.

5.3.3 Fungi

For control of phytopathogenic fungi, a broad range of fungicides is available. Amongst the longest used of these substances are inorganic copper and sulfur compounds, which were used as long ago as the nineteenth century, primarily in vineyards and orchards. Almost all modern formulations are based on organic substances that interfere with the metabolism of fungi in different ways. Depending on their effect, fungicides can be classified into two groups:

1. **Protective (preventive) substances**, which also include copper and sulfur compounds, are applied to the surfaces of plants or seeds and thus prevent fungal attack. Such substances prevent the germination of spores and/or formation of the germ tubes, for example by impairment of enzyme activity or the mitochondrial respiration chain.
2. **Curative (healing) substances** are usually taken up systemically by the plant and serve to kill the agent of the disease after infection has occurred. This works by impairment of different metabolic processes of the fungus.

Similarly to pest management, there is also a variety of methods used for protection of crops from pathogens (primarily from fungi) that avoid the application of chemical substances. These include the breeding of disease-resistant varieties, which is conducted with the same methods and objectives as in breeding for pest resistance (Sect. 5.2.2). No great progress has yet been made in the development of transgenic varieties with resistance to pathogens. Biological management is also far less important in the treatment of plant diseases than it is for pests. Success has, however, often been achieved by application of specific microorganisms (bacteria and fungi), which have negative effects on some phytopathogenic fungi in the soil. They can harm the disease agent by:

- parasitization of the fungal hyphae,
- release of enzymes which dissolve the cell wall of the fungal pathogen, or
- suppression of the growth of its population because of competition for food in the rhizosphere.

Different methods of cultivation can also contribute to limiting the development of populations of phytopathogens. These include crop rotation with species that are not attacked by the agent, and incorporation of organic materials into the soil to promote naturally occurring antagonists. In addition, plant spacing is also important, because the spread of fungal spores is primarily promoted by high plant density. For this reason, the establishment of mixed cropping can also make an important contribution to reduction of the incidence of specific phytopathogens. This is particularly important in the humid tropics, where conditions are ideal for reproduction of phytopathogenic fungi. Not only integration of different species, but also of different varieties of a crop species, can be successful.

Example: In China, the incidence of rice blast fungus (*Magnaporthe grisea*) in a rice variety can be significantly reduced by growing this variety (in a 4:1 ratio) with another variety that is not susceptible. Yield was 89% greater for disease-susceptible rice varieties planted in mixtures with resistant varieties, and blast was 94% less severe than when the rice was grown in a monoculture (Zhu et al. 2000).

References

Andow DA (1991) Vegetational diversity and arthropod population response. Annu Rev Entomol 36:561–586

Bedos C, Cellier P, Calvet R, Barriuso E (2002) Occurrence of pesticides in the atmosphere in France. Agronomie 22:35–49

Bernal JS, Griset JG, Gillogly PO (2002) Impacts of developing on *Bt* maize-intoxicated hosts on fitness parameters of a stemborer parasitoid. J Entomol Sci 37:27–40

Brust GE (1994) Natural enemies in straw-mulch reduce Colorado potato beetle populations and damage in potato. Biol Control 4:163–169

Campbell BC, Duffey SS (1979) Tomatine and parasitic wasps: potential incompatibility of plant antibiosis with biological control. Science 205:700–702

Charlet LD, Knodel JJ (2003) Impact of planting date on sunflower beetle (Coleoptera: Chrysomelidae) infestation, damage, and parasitism in cultivated sunflower. J Econ Entomol 96:706–713

Clavijo McCormick A, Unsicker SB, Gershenzon J (2012) The specificity of herbivore-induced plant volatiles in attracting herbivore enemies. Trends Plant Sci 17:303–310

Coll M (1998) Parasitoid activity and plant species composition in intercropped systems. In: Pickett CH, Bugg RL (eds) Enhancing biological control. University of California Press, Berkeley, pp 85–119

Cuadra R, Cruz X, Fajardo JL (2000) The use of short cycle crops as trap crops for the control of root-knot nematodes. Nematropica 30:241–246
De Moraes CM, Lewis WJ, Paré PW, Alborn HA, Tumlinson JH (1998) Herbivore-infested plants selectively attract parasitoids. Nature 393:570–573
Dean GJ (1974) Effects of parasites and predators on the cereal aphids *Metopolophium dirhodum* (Wlk.) and *Macrosiphum avenae* (F.) (Hem., Aphididae). Bull Entomol Res 63:411–422
Delobel AG (1982) Effect of sorghum density on oviposition and survival of the sorghum shoot fly, *Atherigona soccata*. Entomol Exp Appl 31:170–174
Dicke M, Baldwin IT (2010) The evolutionary context for herbivore-induced plant volatiles: beyond the 'cry for help'. Trends Plant Sci 15:167–175
Dogan EB, Berry RE, Reed GL, Rossignol PA (1996) Biological parameters of convergent lady beetle (Coleoptera: Coccinellidae) feeding on aphids (Homoptera: Aphididae) on transgenic potato. J Econ Entomol 89:1105–1108
Dosdall LM, Florence LZ, Conway PM, Cowle NT (1998) Tillage regine, row spacing, and seedling rate influence infestations of root maggots (*Delia* spp.) (Diptera: Anthomyiidae) in canola. Can J Plant Sci 78:671–681
Ehler LE, Hall RW (1982) Evidence for competitive exclusion of introduced natural enemies in biological control. Environ Entomol 11:1–4
Finch S, Skinner G, Freeman GH (1976) The effect of plant density on populations of the cabbage root fly on four cruciferous crops. Ann Appl Biol 83:191–197
Francis F, Lognay G, Wathelet JP, Haubruge E (2001) Effects of allelochemicals from first (Brassicaceae) and second (*Myzus persicae* and *Brevicoryne brassicae*) trophic levels on *Adalia bipunctata*. J Chem Ecol 27:243–256
Frank T (1998) The role of different slug species in damage to oilseed rape bordering on sown wildflower strips. Ann Appl Biol 133:483–493
Giles KL, Madden RD, Stockland RE, Payton ME, Dillwith JW (2002) Host plants affect predator fitness via the nutritional value of herbivore prey: investigation of a plant-aphid-ladybeetle system. BioControl 47:1–21
Halaj J, Cady AB, Uetz GW (2000) Modular habitat refugia enhance generalist predators and lower plant damage in soybeans. Environ Entomol 29:383–396
Hardin MR, Benrey B, Coll M, Lamp WO, Roderick GK, Barbosa P (1995) Arthropod pest resurgence: an overview of potential mechanisms. Crop Prot 14:3–18
Harmon JP, Ives AR, Losey JE, Olson AC, Rauwald KS (2000) *Coleomegilla maculata* (Coleoptera: Coccinellidae) predation on pea aphids promoted by proximity to dandelions. Oecologia 125:543–548
Harwood JD, Sunderland KD, Symondson WO (2001) Living where the food is: web location by linyphiid spiders in relation to prey availability in winter wheat. J Appl Ecol 38:88–99
Heinrichs EA, Aquino GB, Chelliah S, Valencia SL, Reissig WH (1982) Resurgence of *Nilaparvata lugens* (Stål) populations as influenced by method and timing of insecticide applications in lowland rice. Environ Entomol 11:78–84
Hemptinne JL, Dixon AF (1997) Are aphidophagous ladybirds (Coccinellidae) prudent predators? Biol Agric Hortic 15:151–159
Hilbeck A, Baumgartner M, Fried PM, Bigler F (1998) Effects on transgenic *Bacillus thuringiensis* corn-fed prey on mortality and development time of immature *Chrysoperla carnea* (Neuroptera: Chrysopidae). Environ Entomol 27:480–487
Hokkanen HM (1991) Trap cropping in pest management. Annu Rev Entomol 36:119–138
Holland JM, Reynolds CJ (2003) The impact of soil cultivation on arthropod (Coleoptera and Araneae) emergence on arable land. Pedobiologia 47:181–191
Honěk A (1982) Factors which determine the composition of field communities of adult aphidophagous Coccinellidae (Coleoptera). Zeitschrift für angewandte Entomologie 94:157–168
Hough-Goldstein J, Whalen J (1993) Inundative release of predatory stink bugs for control of Colorado potato beetle. Biol Control 3:343–347

James DG, Stevens MM, O'Malley KJ, Faulder RJ (1999) Ant foraging reduces the abundance of beneficial and incidental arthropods in citrus canopies. Biol Control 14:121–126

Jones RE (1987) Ants, parasitoids, and the cabbage butterfly *Pieris rapae*. J Anim Ecol 56:739–749

Kirk VM (1982) Carabids: minimal role in pest management of corn rootworms. Environ Entomol 11:5–8

Langellotto GA, Denno RF (2004) Responses of invertebrate natural enemies to complex structured habitats: a meta-analytical synthesis. Oecologia 139:1–10

Letourneau DK (1987) The enemies hypothesis. tritrophic interactions and vegetational diversity in tropical agroecosystems. Ecology 68:1616–1622

Li L-Y (1994) Worldwide use of Trichogramma for biological control of different crops: a survey. In: Wajnberg E, Hassan SA (eds) Biological control with egg parasitoids. CAB International, Wallingford, pp 37–53

Loader C, Damman H (1991) Nitrogen content of food plants and vulnerability of *Pieris rapae* to natural enemies. Ecology 72:1586–1590

Losey JE, Denno RF (1998a) The escape response of pea aphids to foliar-foraging predators: factors affecting dropping behaviour. Ecol Entomol 23:53–61

Losey JE, Denno RF (1998b) Positive predator-predator interactions: enhanced predation rates and synergistic suppression of aphid populations. Ecology 79:2143–2152

Ludwig SW, Kok LT (1998) Evaluation of trap crops to manage harlequin bugs, *Murgantia histrionica* (Hahn) (Hemiptera: Pentatomidae) on broccoli. Crop Prot 17:123–128

Marshall SD, Pavuk MP, Rypstra AL (2002) A comparative study of phenology and daily activity patterns in the wolf spiders *Pardosa milvina* and *Hogna helluo* in soybean agroecosystems in southwestern Ohio (Araneae, Lycosidae). J Arachnol 30:503–510

Milbrath LR, Weiss JJ, Schatz BG (1995) Influence of tillage system, planting date, and oilseed crucifers on flea beetle populations (Coleoptera, Chrysomelidae). Can Entomol 127:289–293

Mochizuki A, Nishizawa Y, Onodera H, Tabel Y, Tokl S, Habu Y, Ugaki M, Ohashi Y (1999) Transgenic rice plants expressing a trypsin inhibitor are resistant against rice stem borers, *Chilo suppressalis*. Entomol Exp Appl 93:173–178

Morrill WL, Kushnak GD (1999) Planting date influence on the wheat stem sawfly (Hymenoptera: Cephidae) in spring wheat. J Agric Urban Entomol 16:123–128

Myers JH, Higgins C, Kovacs E (1989) How many insect species are necessary for the biological control of insects? Environ Entomol 18:541–547

Nampala P, Adipara E, Ogenga-Latigo MW, Kyamanywa S, Obuo JE (1999) Effect of cowpea monocultures and polycultures with sorghum and greengram on predatory arthropods. Ann Appl Biol 135:457–461

Nentwig W (1987) The prey of spiders. In: Nentwig W (ed) Ecophysiology of spiders. Springer, Berlin, pp 249–263

Nyffeler M, Sunderland KD (2003) Composition, abundance and pest control potential of spider communities in agroecosystems: a comparison of European and US studies. Agric Ecosyst Environ 95:579–612

O'Neal ME, Difonzo CD, Landis DA (2002) Western corn rootworm (Coleoptera: Chrysomelidae) feeding on corn and soybean leaves affected by corn phenology. Environ Entomol 31:285–292

Radin AM, Drummond FA (1994) An evaluation of the potential for the use of trap cropping for control of the striped cucumber beetle, *Acalymma vittata* (F.) (Coleoptera: Chrysomelidae). J Agric Entomol 11:95–113

Rämert B, Hellqvist S, Ekbom B, Banks JE (2001) Assessment of trap crops for *Lygus* spp. in lettuce. Int J Pest Manag 47:273–276

Reimer NJ, Cope M-L, Yasuda G (1993) Interference of *Pheidole megacephala* (Hymenoptera: Formicidae) with biological control of *Coccus viridis* (Homoptera: Coccidae) in coffee. Environ Entomol 22:483–488

Remund U, Boller E (1996) Bedeutung von Heckenpflanzen für die Eiparasitoide der Grünen Rebzikade in der Ostschweiz. Schweizerische Zeitschrift für Obst- und Weinbau 132:238–241

Rosenheim JA (2001) Source-sink dynamics for a generalist insect predator in habitats with strong higher-order predation. Ecol Monogr 71:93–116

Rosenheim JA, Limburg DD, Colfer RG (1999) Impact of generalist predators on a biological control agent, *Chrysoperla carnea*: direct observations. Ecol Appl 9:409–417

Rothamsted Research (2012) www.rothamsted.ac.uk/index.php

Sauerborn J (1989) The influence of temperature on germination and attachment of the parasitic weed *Orobanche* spp. on lentil and sunflower. Angewandte Botanik 63:543–550

Settle WH, Ariawan H, Austuti ET, Cahyana W, Hakim AL, Hindayana D, Sri LA, Sartanto P (1996) Managing tropical rice pests through conservation of generalist natural enemies and alternative prey. Ecology 77:1975–1988

Simpson M, Geoff M, Gurr GM, Simmons AT, Wratten SD, James DG, Leeson G, Helen I, Nicol HI, Orre-Gordon GU (2011) Attract and reward: combining chemical ecology and habitat manipulation to enhance biological control in field crops. J Appl Ecol 48:580–590

Sunderland KD, Fraser AM, Dixon AF (1986) Field and laboratory studies on money spiders (Linyphiidae) as predators of cereal aphids. J Appl Ecol 23:433–447

Thacker JR (2002) An introduction to arthropod pest control. Cambridge University Press, Cambridge

Treacy MF, Benedict JH, Segers JC, Morrison RK, Lopez JD (1986) Role of cotton trichome density in bollworm (Lepidoptera: Noctuidae) egg parasitism. Environ Entomol 15:365–368

Triltsch H (1996) On the parasitization of the ladybird *Coccinella septempunctata* L. (Col., Coccinellidae). J Appl Entomol 120:375–378

Uetz GW, Halaj J, Cady AG (1999) Guild structure of spiders in major crops. J Arachnol 27:270–280

Waage JK, Mills NJ (1992) Biological control. In: Crawley JJ (ed) Natural enemies: the population biology of predators, parasites and diseases. Blackwell, Oxford, pp 412–430

Wolfenbarger LL, Phifer PR (2000) The ecological risks and benefits of genetically engineered plants. Science 290:2088–2093

Wyss E (1995) The effects of weed strips on aphids and aphidophagous predators in an apple orchard. Entomol Exp Appl 75:43–49

Zhu Y, Chen H, Fan J, Wang Y, Li Y, Chen J, Fan JX, Yang S, Hu L, Leung H, Mew TW, Teng PS, Wang Z, Mundt CC (2000) Genetic diversity and disease control in rice. Nature 406:718–722

Chapter 6
Production and Management of Livestock Resources

To improve food supply for the animals and to increase livestock productivity, different strategies and production systems have been developed, ranging from traditional pastoral grazing to factory farming using industrially produced feedstuff (Fig. 6.1). The different systems are characterized by different levels of intensification and management, and are also accompanied by different consequences for the sustainability of the systems and the environment.

6.1 Grassland-Based Production Systems

For millions of years, interrelationships between large wild ungulates and natural grazing ecosystems have existed in different regions of the world, including tropical savannas, temperate steppe, prairie grasslands, and the arctic tundra. Grazing ecosystems support more herbivore biomass than any other terrestrial habitat. The long co-evolutionary history of grasslands and ungulate grazers has led to high sustainability of the systems. This is mainly explained by the seasonal and short-term migratory behaviour of ungulates, the stimulation of compensatory growth of plant shoots after defoliation, the removal of phenologically older, less productive tissue, and enhancement of mineral availability by increasing nutrient cycling within patches of their waste. Furthermore, ungulates in grazing ecosystems modify vegetation structure, with the result that herbivores increase their own foraging efficiency (Frank et al. 1998).

However, domestic livestock grazing differs in two main aspects from wild ungulate grazing:

- domestic livestock biomass on pasture and rangeland tends to be higher than wild ungulate biomass in grazing ecosystems, and
- herding of domesticated ungulates by humans is different from the spatial and temporal patterns of wild ungulate grazing.

K. Martin and J. Sauerborn, *Agroecology*, DOI 10.1007/978-94-007-5917-6_6,

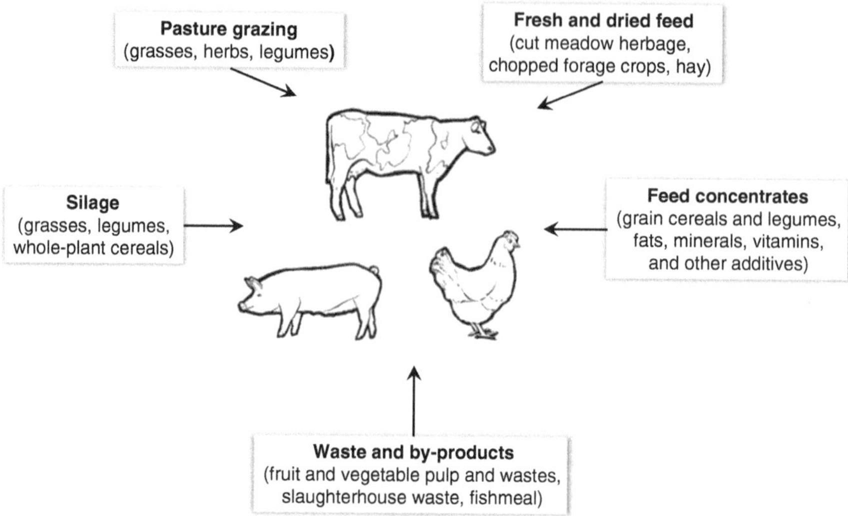

Fig. 6.1 Categories of livestock feed (Drawing by Ch Allgaier)

Therefore, stocking density (animals per hectare) and stocking rate (animals per hectare per unit time) are important to the sustainability of the entire system. To maintain the balance between forage demand and recovery of the vegetation, several factors must be managed across the landscape—the number of individuals, their spatial and temporal distribution, and species, size, and age of the animals used in a grazing system.

6.1.1 Natural Grasslands

These systems are characterized by direct consumption of forage produced with a minimum of energy and nutrient input. The forage in these systems comprises native or naturalized grass and forb species. Among agricultural systems, grazing systems are probably the closest to natural ecosystems (Laca and Demment 2002).

Extensive livestock grazing in rangelands (pastoralism) is most common in semi-arid regions, where it is traditionally based on sheep and camels. It can also be found in other regions, for example reindeer pastoralism under subarctic conditions, but was absent from America and Australia before the arrival of Europeans. In its original form, pastoralism is highly nomadic (Sect. 2.5.1.2) with free selection of forage on suitable land and no management of stocking numbers. Movement is restricted by spatial patchiness, seasonal availability, the quality of feeding resources, and access to water.

Another type of extensive livestock production is conducted in controlled systems in which the land is individually owned and usually fenced (ranching). The United States is an example of the gradual transition from common ownership of cattle ranching systems prevalent in the nineteenth century to enclosed systems today. Ranching is the dominant system in North America, Australia, and parts of South America.

The use of natural grasslands on a limited land area requires control over land use to maintain or improve forage production for livestock. Fencing not only indicates ownership but also provides opportunities for manipulating livestock grazing patterns, to improve uniformity of grazing. Grazing systems management may include:

- **Seasonal grazing:** Use of a particular area for only a part of the year, enabling land that is not being grazed to regenerate and produce new forage.
- **Rotational grazing:** Division of rangeland into several grazing areas, each of which is then grazed in sequence throughout the grazing period. Rotations may include the burning of patches to stimulate the growth of fresh grasses.
- **Herding:** Control of livestock movement under the direction of a person (sometimes assisted by dogs) to manipulate grazing distribution of the herded animals.
- **Water availability:** Water is often limited in semi-arid regions, and areas far from sources may not be grazed as heavily as areas close to water. More evenly distributed water resources can increase forage use nearby and improve the overall uniformity of grazing. New water sources are usually created by drilling wells, constructing ponds, building pipelines, or developing springs.

6.1.2 Secondary Grasslands

Different types of grassland became established after the human removal of forests by logging and burning. In the tropics, especially South and Central America, large areas of rainforest have been converted into pasture for cattle ranching. However, one of the oldest and most important centres of meadow and pasture farming is Europe. The history of European grassland in regions with cool–temperate climate dates back approximately 4,000 years. Originally, natural grasslands in temperate Europe were limited to very few sites where forest failed to grow, for example high mountain elevations or coastal marshlands.

6.1.2.1 Semi-natural Pastures and Meadows

In Europe, the first pastures developed gradually as a result of grazing and browsing of cattle in woodlands. Later, the evolution of meadows was much affected by regular cutting of grass for haymaking; this has been practised in temperate Europe for more than 2,000 years to maintain feed supply during winter. The agricultural

Fig. 6.2 The annual meadow grass (*Poa annua*) is one of the most common grass species and has worldwide distribution (Drawing by Ch. Allgaier)

grasslands of Europe are dominated by approximately 20 widely distributed major grass species, including bent grasses (*Agrostis* spp.), rye grasses (*Lolium* spp.), and meadow grasses (*Poa* spp.; Fig. 6.2).

According to Scholz (1975), it is reasonable to assume that most of the valuable meadow and pasture grass species of temperate Europe originated and evolved in historic times as a result of the intervention of man. In addition to the native species (members of the natural vegetation cover), several species or varieties have migrated, with differentiation from their wild ancestors (the minor part), or evolved in situ by hybridization and introgression (the major part). Therefore, European meadow and pasture grasses are often regarded as semi-domesticated plants.

Such semi-natural grasslands consist of unsown vegetation that is maintained by livestock grazing and/or mowing and has not been substantially modified by intensive fertilization, drainage, or herbicide use. They are characterized by high richness of herbaceous plants, supporting a high diversity of species at higher levels of the food web, for example insects, spiders, and birds. In the current agricultural landscape of Europe, these types of habitat are rare, and are restricted to marginal areas, which are typically used for sheep grazing (Sect. 3.3).

6.1.2.2 Intensively Managed Grassland and Fodder Production

In temperate regions, grassland systems are usually based on permanent pasture and meadows or are part of a mixed grassland–arable rotation (Sect. 2.4.1.1).

At the beginning of the twentieth century, most traditional grasslands were agriculturally improved and their management was intensified. pH and soil nutrient content of acidic soils of low fertility were increased by liming and regular fertilization. Artificial drainage was introduced to remove surplus water from wet soils. These measures were accompanied by, and followed by, higher stocking and frequency of defoliation. Thus, diverse traditional grasslands became replaced by intensified pastures and meadows, characterized by a few productive forage species and low overall biodiversity (Isselstein et al. 2005).

Grassland forage of higher quality and quantity became particularly necessary in connection with developments in dairy farming, mainly in the industrialized countries:

- Development of large, specialized dairy units with large numbers of cows is usually accompanied by reduced access to pasture grazing. In Denmark, for example, grazing is only used by approximately half of the herds that comprise more than 90 cows (Kristensen et al. 2005).
- In the European Union, milk production per cow has more than doubled in the past 40 years, because of improvements in animal genetics and breeding, and has increased from around 4,000 l per year in the 1960s to around 8,000 l per year today. There are even high-producing breeds that produce twice this amount.

Profitable dairy and cattle production therefore requires various strategies for fodder production, land management, and forage conservation.

6.1.2.3 Permanent Grassland Production

Despite intensification measures, for example fertilization, liming, and water management, swards of permanent secondary grasslands may lose their quality because of the disappearance of valuable components, for example tall grasses and leguminous plants. The main reason for the decline in yield and quality includes the introgression of inferior grasses and herbs that gain competitive advantage within the swards. Less palatable plant species may lock up nutrients or immobilize nutrients, because of slow decomposition, and thereby limit opportunities for re-establishment of palatable species with greater nutrient requirements These processes are often promoted by inaccurate fertilizer application, soil compaction by passage of heavy machinery or animals, excessive cutting, and selective grazing. Oversowing of seeds of grasses and legumes in the old sward, with herbicide application to suppress unwanted plant species, is common but not always successful practice. In easily convertible soils in a moderate climate, for example that in North-West Europe, it is common practice for farmers to renovate their grassland regularly by destroying the old sward with herbicides and ploughing, followed by a new establishment.

6.1.2.4 Temporary Grassland Production

A common method of fodder production, especially in mixed crop-livestock farming systems, includes the establishment of perennial grasslands with a production phase of up to 5 years as a component of arable rotation (Sect. 2.4.1). These temporary systems typically consist of mixtures of grasses and legumes used for cutting or grazing. Animal production is usually higher in mixed swards than in pure grass swards, and with increasing proportions of legumes in the sward. White clover

Fig. 6.3 White clover (*Trifolium repens*) (Drawing by Ch. Allgaier)

(*Trifolium repens;* Fig. 6.3) is the most important legume species of such systems in temperate Europe, followed by red clover (*T. pratense*), mixed such with small-size grasses as *Lolium perenne*, *Poa pratensis*, or *Festuca rubra*.

In warmer and dryer regions of Europe, alfalfa (*Medicago sativa*) is widely sown in pure stands and used for both grazing and as forage over several years. Nitrogen fertilization is not recommended for legume–grass mixtures because it reduces the proportion of legumes and is unnecessary for adequate plant growth in these systems if they are seeded with approximately 50% legumes. However, legumes may fail to persist beyond 3 years, especially under intensive grazing with high nutrient inputs in the form of manure.

The benefits of legumes in livestock-production systems include symbiotic fixation of nitrogen (Sect. 3.7.4.1) and high nutritive value because of higher crude protein content. Higher voluntary food intake and higher milk yield per cow have both been demonstrated for cattle feeding on pure legume or grass–clover stands rather than fertilized pure grass.

Establishment of temporary grass–legume mixtures before subsequent arable crops increases soil fertility in terms of soil organic matter accumulation and reduces losses of soil nitrate by leaching. Ploughing the grassland leads to the mineralization of the organic matter in the upper soil layer. This results in higher yields, and lower nitrogen fertilization requirements to achieve optimum yields.

The benefits of grassland–arable systems and use of manure to maintain soil fertility are especially important in organic and other low-input mixed farming systems with the objective of efficient crop and livestock production with minimum input and minimum losses to the environment.

6.1.3 Annual Fodder Crop Production

Grazing systems offer only limited potential for intensification, and livestock production therefore is becoming increasingly crop-based. Most of the increase in meat and milk production during recent decades has been achieved by increasing the production in mixed and industrial production systems, and much less so in pastoral systems. Non-pasture forage that was originally used to cover seasonal feed gaps is now used with different levels of supplemental feed throughout the year. Annual forage and fodder crops are used worldwide as important sources of feed for dairy systems. Depending on the crop, they are established in different seasons and various ways, for example:

- single main crop,
- undersown or intercropped with other crops or forage,
- break crop or stubble seed between the growing seasons of field crops, and
- in rotation with other forage crops.

Potential use of different annual forage include grazing, fresh feeding, conservation as silage (see below), or grain. The following are important groups and species (after Havilah 2002):

Maize is an important component of dairy cattle feed worldwide. It is high in energy but has low protein content. Usually the entire plants are harvested, cut, and stored as silage.

Sorghum and millet are suitable for production in areas that are relatively dry during summer; they are used for grazing, fresh food, or silage. In many subtropical regions, such crops as maize, sorghum, and millet are dual purpose—their grain provides food for humans and their residues are used as feed for livestock. Farmers in mixed crop-livestock systems sometimes value the crop residues as much as the grain, owing to their importance as a feed for livestock, particularly in the dry season (Herrero et al. 2010).

Small-grain cereals, for example wheat, oats, barley, and rye can also contribute to livestock feeding and can be grazed as early feed in autumn or used for production of hay, silage, or grain.

Fodder beet or mangelwurzel (*Beta vulgaris*) is primarily used as feed for cattle and sheep. It is typically grown in cool-temperate climates as a single crop during summer, but in warm-temperate to sub-tropical climates it can be sown in autumn as a winter crop. The roots are stored and used as feed after chopping or used for silage.

Grain legumes, notably soybean and cowpea, are used as fodder crops and are produced during summer in tropical and subtropical regions. They can be grazed, used for silagen or harvested as grain. Other annual legumes, especially clover species, are used as a break crop in dryland cropping rotation.

Brassicas (species from the family Brassicaceae) provide high quality and highly digestible feed. Brassicas are useful for extending the summer feed supply or

Fig. 6.4 Rapeseed or oilseed rape (*Brassica napus*). Special cultivars are also called canola (Drawing by Ch. Allgaier)

producing autumn and winter feed. They are usually fed fresh after green chopping or are grazed. Important species include rapeseed (*Brassica napus*, Fig. 6.4), turnip (*Brassica napa*), and leafy kale (*Brassica oleracea* var. *medullosa* and *viridis*).

6.1.4 Fodder and Feed Concentrates

In general, livestock feed must provide the basic nutrients required for animal production, including carbohydrates, proteins, fats, minerals, vitamins, and other micro-nutrients. In addition, or as an alternative to the natural food requirements of livestock species, feed that contains a high density of energy and is rich in total digestible nutrients is provided to the animals to fulfil the demands of meat, milk, or egg production. Depending on the species, the age of the animal, and production purposes, such feed concentrates are blended or formulated differently.

Dairy cow and cattle raising is principally based on roughage, including fresh feed from grazing, hay, and/or silage of meadow grasses, including legumes, maize, and other crops.

Haymaking, the traditional method of fodder conservation for centuries, has been gradually replaced by **silage** production in many farming systems. Silage is a high-moisture fodder produced by controlled fermentation of sugars in plant material by lactic acid bacteria. It is made from the entire green plant either by placing the cut fresh material in a silo or by wrapping it in plastic sheets.

In intensive production systems, however, up to a third of total feed is provided as **feed concentrates**. These are produced from cereal grain (for example corn, barley, sorghum, and wheat) and provide animals with energy because of their starch, oilseed meal (especially soybean and rapeseed), protein, and fat content. In addition, concentrates may include citrus pulp, sugarbeet pulp, brewer's, and distiller's residues, molasses (by-products of sugar production), minerals, and vitamins. The overall feed is often provided as compound of basic roughage mixed with the concentrates. However, because the digestive systems of ruminants are not designed for grain, cattle can develop severe health problems, including ruminal acidosis, bovine mastitis, liver abscesses, and sudden death syndrome.

Pigs and poultry produced in landless factory farms are solely fed with concentrates. Usually more than half of these compounds consist of grain cereals, mainly corn (for pigs) and wheat (for chicken). The most important source of protein both for pigs and poultry is soybean meal with proportions of up to one fourth of the concentrates. Soybean meal is a by-product of soybean oil production. More than 97% of globally produced soybean meal is fed to livestock whereas only 3% of the harvested beans is directly fed to animals (FAO 2006a).

Additional ingredients include vegetable or animal fats including fish, meat, or bone meal, minerals and vitamins, flavour enhancers, and, in egg production, synthetic dyes. On industrial poultry farms, especially, the feed may contain a range of additives which can cause a variety of health problems in humans. These include antiparasitic and antifungal compounds, antibiotics, and arsenic to promote growth and/or prevent disease. Use of antibiotics has been shown to select for antibiotic resistance in both commensal and pathogenic bacteria. Furthermore, fats from a variety of sources may contain contaminants, for example dioxins, polychlorobiphenyls (PCBs), and metals.

6.2 Environmental Effects of Livestock Production

Livestock production is the major sector of agricultural land use and is responsible for a variety of environmental problems around the globe. Major concern are outlined below:

- land degradation in relation to pasture land use and effects on biodiversity,
- contamination and pollution of the environment, especially resulting from industrial livestock production systems, and
- indirect effects related to the production of and the global trade in feed crops and livestock products.

Further concerns regarding global livestock production and global change are discussed in Chap. 8.

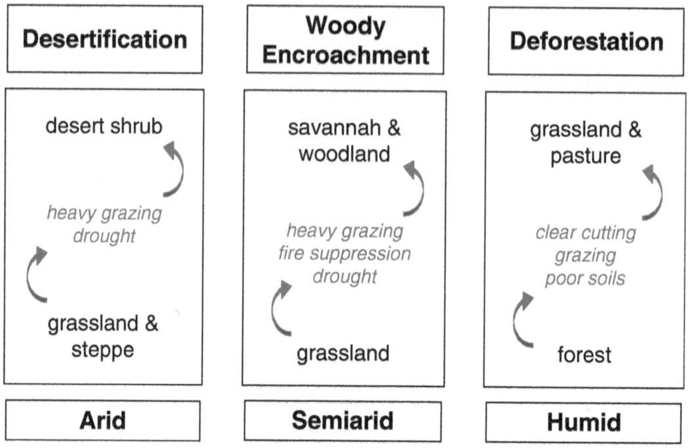

Fig. 6.5 Three regional syndromes resulting from managed grazing practices under different climatic conditions (Based on Asner et al. 2004)

6.2.1 Land Degradation

Degradation of pasture land results from different processes which lead to reduction of the resource potential and productivity of ecosystems, with the different effects depending on climatic region (Fig. 6.5). Approximately 20% of the world's pastures and rangeland have been degraded to some extent, and the proportion may be as high as 73% in dry areas (UNEP 2004).

6.2.1.1 Desertification

The development or expansion of desert-like conditions in regions where deserts would not naturally develop under the existing climatic conditions, is called **desertification**. This process is a major problem in arid regions (with a vegetation period of less than 90 days; Sect. 7.2.1.2), and the rate of desertification has been estimated to be greater for grazing land than for other land use. Of the total land area classified as being affected by desertification, 93% is located in rangelands (UNEP 1992). Desertification of pasture land results in replacement of the herbaceous cover by shrub cover and bare soil (Asner et al. 2004). The single greatest cause of grassland degradation is overgrazing by livestock, because of excessive stocking, resulting in several interlinked effects on vegetation, soil, and hydrology. Reduction or removal of the vegetation cover exposes the bare soil and reduces the supply of dead organic material, which leads to reduction of soil organic matter. Trampling by animals causes compaction of the soil with reduction of pore space, followed by a reduction of infiltration capacity and an increase in surface runoff, which in turn enhances overall susceptibility to soil erosion.

6.2.1.2 Woody Encroachment

Increasing woody plant cover on pasture land is an important aspect of land degradation in the semi-arid zones (with a vegetation period of approximately 90–180 days and higher annual precipitation than in arid regions). Under these conditions, overgrazing of the herbaceous cover creates competitive advantages for woody seedlings. Woody and herbaceous life-forms utilize soil moisture from different depths in the soil profile. Woody plants take advantage of deeper soil moisture, and grasses and herbaceous plants access moisture mainly in the upper soil layers. Because of the relatively high precipitation during the vegetation period, rainwater infiltrates deep soil layers and thereby creates soil moisture conditions that favour woody species. Surface soil compaction by trampling animals increases evaporative losses from bare surface soils and further changes the conditions for grasses and herbs for the worse. Increased surface litter cover, carbon storage, and changes in soil nutrient distribution during and subsequent to the establishment of woody plants in grazed systems may constitute positive feedback and increase the likelihood of additional woody plant encroachment. As a result, woody plant encroachment may continue to occur even after livestock grazing pressure is reduced (Asner et al. 2004; Archer et al. 1995).

6.2.1.3 Deforestation

Whereas major conversion of the temperate European landscape dates back hundreds to thousands of years, replacement of forest by cattle pastures is still proceeding in the humid tropics. Ranching-induced deforestation is common in Central and South America, and, to a very limited extent, in Central Africa and South-East Asia. Management of such pastures is typically extensive, with low levels of external inputs and no fertilizer use. Repeated burning is often used to remove woody regrowth and weeds and to renew nutrient availability in pastures. However, because of the low nutritional quality of the grasses, soil degradation, and the invasion of unpalatable species, such pastures are usually abandoned within a period of less than 10 years.

6.2.2 Environmental Effects of Industrial Livestock Production

During the past four decades, global per capita meat production has increased more than 60%. In recent years, livestock production from industrial enterprises has grown twice as fast as that from more traditional mixed farming systems and more than six times faster than from grazing systems. At the turn of the century, industrial enterprises accounted for 74% of the world's total poultry production and 40% of its pig meat (FAO 2003).

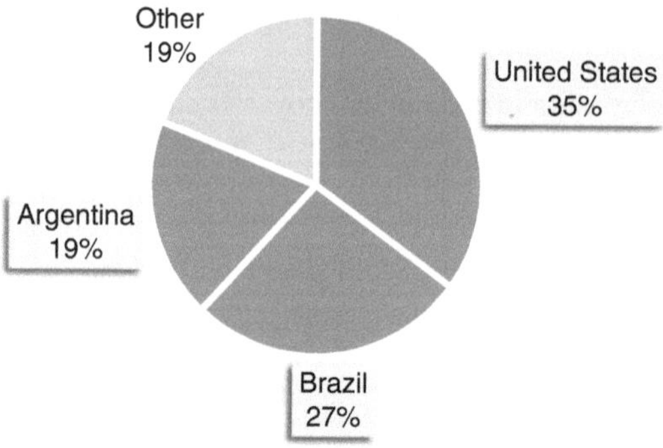

Fig. 6.6 Countries with the main contributions to global soybean production in 2011 (Based on FAOSTAT data)

More than one third of global cereal production is currently used to meet the feed requirements of industrial livestock animals, mostly as feed concentrates. This is equivalent to food for three billion people.

For production of meat, substantially more energy is required than for the production of plant foodstuffs. To produce 1 kg of beef, approximately 7 kg of grain are needed. For pork the relationship is approximately 1:4 and for chickens 1:2. Hence, livestock production is much less energy-efficient than food crop production. This includes such accelerated inputs to agriculture as machinery, fuel, fertilizers, pesticides, and water for feed crop production, processing of feed concentrates, and global trade in feed and meat. This is related to the fact that feed production and use are largely unconnected in industrial systems. In addition, feed production increases the pressure on natural ecosystems. This becomes especially evident for global soybean and soybean meal production, which is mostly used for animal feed. In 2011, the United States, Brazil, and Argentina together accounted for more than 80% of global soybean production (Fig. 6.6), mainly genetically modified varieties. These nations are also the main soybean exporters. Worldwide, the top consumer of soybean and soybean meal is China, and is followed by the European Union and the United States.

World soybean production has increased substantially in recent decades, with an increase of 27% between 2000 and 2010 alone (Fig. 6.7). The largest share of this increase was contributed by South American countries, where mainly pasture land and natural vegetation have been transformed into soybean fields.

Overall, production and trading of livestock feedstuff is associated with massive worldwide transfer of energy and nutrients, depleting resources in the regions of production and causing nutrient overload and pollution in the regions of final use. This is especially true for nitrogen from animal waste products. Livestock excreta are the main source of ammonia (NH_3) in the global nitrogen cycle. Ammonia is an important atmospheric pollutant with a wide variety of effects. Most agricultural

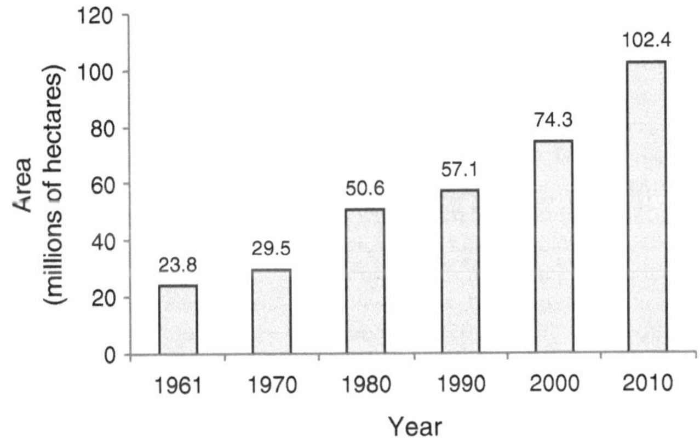

Fig. 6.7 Increase in world soybean production area between 1961 and 2010 (FAOSTAT data)

ammonia emissions are released from livestock production (Sect. 3.7.4.2), including animal housing and slurry storage systems (approx. 45%) and spreading of manure on land (approx. 30%). Grazing animals contribute approximately 25% to ammonia volatilization (Beusen et al. 2008). Essentially, all emitted NH_3 is returned to the surface by atmospheric deposition, thereby contributing to soil acidification via various processes (Sect. 3.7.4), eutrophication, and losses of biodiversity in natural ecosystems.

In addition to fertilizers, animal excreta are also a large source of nitrate (NO_3^-), which reaches surface water systems (rivers, lakes, reservoirs, and coastal marine ecosystems) and groundwater, affecting ecosystem function including fish production and drinking water quality. Approximately 30–40% of the world's lakes and reservoirs have been affected, to a greater or lesser extent, by water eutrophication.

Livestock production also contributes to the anthropogenic greenhouse effect (Sect. 8.1.3) via the emission of carbon dioxide (CO_2), methane (CH_4), and nitrous oxide (N_2O). The last of these is also involved in depletion of the atmospheric ozone layer.

References

Archer S, Schimel DS, Holland EA (1995) Mechanisms of shrubland expansion: land use, climate or CO_2? Clim Chang 29:91–99

Asner GP, Andrew J, Elmore AJ, Olander LP, Martin RE, Harris AT (2004) Grazing systems, ecosystem responses, and global change. Ann Rev Environ Resour 29:261–299

Beusen AHW, Bouwman AF, Heuberger PSC, Van Drecht G, Van Der Hoek KW (2008) Bottom-up uncertainty estimates of global emissions from global agricultural production systems. Atmos Environ 42:6067–6077

FAO (2003) World agriculture: towards 2015/2030 – An FAO perspective. Food and Agriculture Organization of the United Nations, Rome. Earthscan Publications, London, and http://www.fao.org/docrep/005/y4252e/y4252e00.htm

FAO (2006) Livestock's long shadow. Environmental issues and options. Food and Agriculture Organisation of the United Nations, Rome, http://www.fao.org/docrep/010/a0701e/a0701e00.HTM

FAOSTAT. The statistical database of the Food and Agriculture Organization (FAO), Rome. http://www.faostat.fao.org/

Frank DA, McNaughton SJ, Tracy BF (1998) The ecology of the Earth's grazing ecosystems. Bioscience 48:513–521

Havilah EJ (2002) Annual forage and pasture crops – species and varieties. In: Roginski H, Fuquay JW, Fox PF (eds) Encyclopaedia of dairy science. Academic, London, pp 1131–1142

Herrero M, Thornton PK, Notenbaert AM, Wood S et al (2010) Smart investments in sustainable food production: revisiting mixed crop-livestock systems. Science 327:822–825

Isselstein J, Jeangros B, Pavlu V (2005) Agronomic aspects of biodiversity targeted management of temperate grasslands in Europe – a review. Agron Res 3:139–151

Kristensen T, Søegaard K, Kristensen IS (2005) Management of grasslands in intensive dairy livestock farming. Livest Prod Sci 96:61–73

Laca EA, Demment MW (2002) Livestock production systems. In: Hudson RJ (ed) Management of agricultural, forestry and fisheries enterprises Encyclopedia of Life Support Systems (EOLSS). Eolss Publishers, Oxford, http://www.eolss.net

Scholz H (1975) Grassland evolution in Europe. Taxon 24:81–90

UNEP (1992) World Atlas of desertification. United Nations Environment Programme, Edward Arnold, London

UNEP (2004) Land degradation in drylands (LADA): GEF grant request. United Nations Environment Programme, Nairobi

Chapter 7
Climate Zones and Land Use

7.1 Global Atmospheric Circulation

Because of the relationship between latitude and global radiation (Sect. 4.1), the radiation balances of the regions near the equator and the poles are very different. The area between the equator and approximately 40° north and south receives more energy than it returns to space, which means that the radiation balance of these regions is positive. Between the 40th latitude and the poles, the balance is negative. However, these conditions do not lead to a steady increase in temperatures around the equator while the polar regions become progressively cooler, because a transfer of energy occurs between the lower and higher latitudes via air and ocean currents.

The global air exchange, initiated by the unequal radiation balance, makes up the **global atmospheric circulation**. As a result of the Earth's rotation and other factors, global atmospheric transport is not a consistent flow between the equator and the poles, but rather made up of different systems, which are referred to as zonal cells (Fig. 7.1).

The starting point of atmospheric transfer of energy is the Earth's surface near the equator, where warm moist air rises into the upper atmosphere as a result of heating. This air masses cools with increasing altitude, which leads to the formation of clouds. Because of the high water vapour content of the rising air, the resulting precipitation is heavy and can exceed 100 mm per day, producing the high rainfall of the wet–tropical rainforest areas. As a result of the rise of the warm air, low pressure areas develop the vicinity of the equator between latitudes of approximately 5° north and south, that are referred to as the equatorial low-pressure trough or simply the equatorial trough. At the borders of the troposphere, the rising air masses begin to flow towards the north and the south. The troposphere is the lowest layer of the atmosphere in which weather (formation of clouds and precipitation) occurs. It extends approximately 8 km at the poles and 16 km at the equator.

In the region of the so-called horse latitudes (about 30° north and south), the air masses gradually sink, and heat the air by compression. This process leads to the formation of the subtropical high pressure belt. In these areas, cloudless and dry

K. Martin and J. Sauerborn, *Agroecology*, DOI 10.1007/978-94-007-5917-6_7,

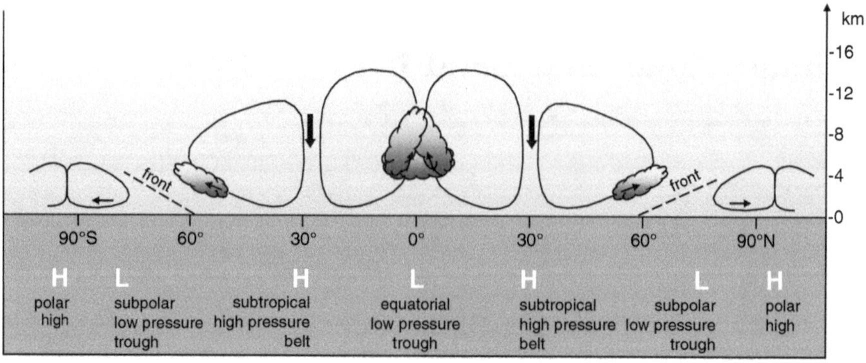

Fig. 7.1 Schematic representation of the global atmospheric transport system

conditions dominate and also characterize the hot deserts of the earth. From this zone, a portion of the surface air moves back toward the equator completing the circulation system known as the **Hadley cell**.

This moving air is also deflected by the Coriolis effect, creating the northeast and the southeast trade winds which converge in the equatorial area, and is therefore also known as the **intertropical convergence zone (ITCZ)**. When it lies near the equator, it corresponds with the equatorial trough. However, the location of the ITCZ varies with the seasons of the year as the subsolar point (the point where the sun is perceived to be directly overhead, i.e., in zenith) moves between the tropic of Cancer (23.5° N) and Capricorn (23.5° S), but with a temporal delay of 1–2 months. The shift of the subsolar point during the course of the year within the tropics of Cancer and Capricorn is because of the tilt of the Earth's axis of 23.5° and the subsequent angle of the sun.

The shift of the ITCZ is responsible for the occurrence of wet and dry periods within the tropics. The highest precipitation occurs where the ITCZ is located at a specific time of the year. The areas close to the equator are constantly affected by the ITCZ throughout the year, which is why it usually rains every month in these regions. Because the ITCZ crosses the equator twice per year, two peak rain periods occur there (in April and in October). The latitudes within the tropics but further from the equator are affected by the ITCZ once per year only. The duration of the rainy season decreases in the direction of the tropics of Cancer and Capricorn.

These conditions are simplified, because the ITCZ is not actually perfectly parallel with the equator. Instead, its course is affected by the distribution of the land masses and oceans in the respective areas (Fig. 7.2).

These factors determine the formation of **monsoon winds** in South and Southeast Asia. These are wind systems that reverse their direction twice a year, because of the seasonal change from high pressure to low pressure over large continental areas. In the summer of the northern hemisphere, the large Asian land mass is subject to intense heating and a low pressure system develops above the area. As a result, the ITCZ expands beyond the tropic of Cancer. The southeast trade winds,

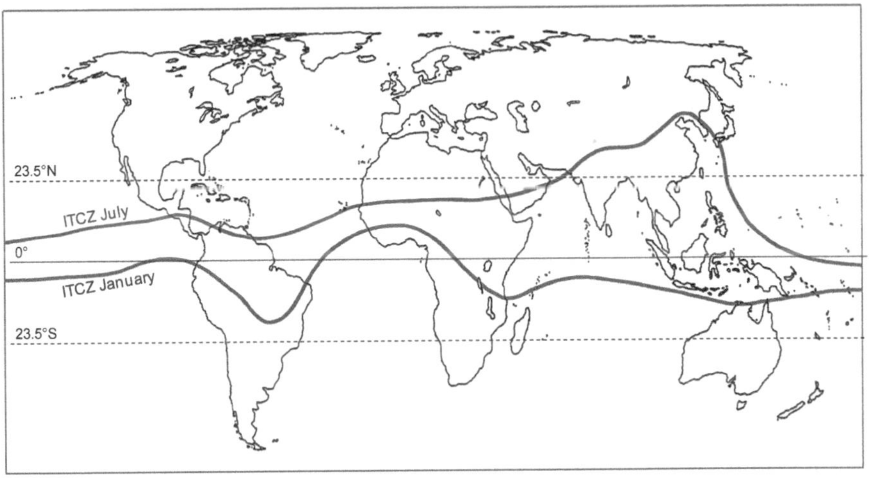

Fig. 7.2 The maximum extension of the Intertropical Convergence Zone (ITCZ) to the north (in July) and to the south (in January)

which reach the ITCZ from the southern high-pressure belt and form the **summer monsoon**, take up moist air over the Indian Ocean, which produces rain in southern Asia. The summer monsoon occurs in South Asia primarily from June to October. During this time, approximately 80% of the average annual rainfall occurs, and the agriculture of the region is highly dependent on this. In the northern winter, the opposite conditions exist. In other words, a high-pressure system forms over Asia and the ITCZ is located south of the equator. The northern trade winds, which correspond to the **winter monsoon**, originate over the Asian continent. Because the winds take up only little moisture there, South Asia experiences low precipitation during this period.

North and south of the Hadley Cell, the so-called planetary frontal zones are found and extend from approximately 30–60° latitude. These essentially make up the **Ferrel cells**, in which the outer-tropical westerlies predominate. At 60° north and south latitude, the westerlies collide with the polar easterlies, i.e., cold air travelling from the poles. This collision results in frontal uplift and the creation of the subpolar lows or mid-latitude cyclones. Most of this lifted air is directed to the polar vortex where it moves downward to create the polar high and the **polar cells**. Overall, the air currents in the temperate latitudes between 30 and 60° are very inconstant. As a result of disturbances which originate in the polar zone, the irregular formation of smaller low and high-pressure areas occurs, after which they are transported, and finally disintegrate.

Regional deviations from the overall conditions of global atmospheric transport are the result of the relative proportions of land masses and oceans, the differential heating of oceans and land masses in the course of the year, and the topography of land surfaces (mountains, plains). The global distribution of energy between the different latitudes depends not only on atmospheric circulation but is also strongly affected by oceanic currents. The oceans take up solar energy in the equatorial regions. Surface currents flow in the direction of the poles, where they cool and sink into deeper layers and then flow back to the equator. This is, for example, the cause of the Gulf Stream, which results in higher water and air temperatures in coastal areas of western and northern Europe, primarily in the winter months, than are found at the same latitudes on the east coast of North America.

7.1.1 Climate

The meteorological phenomena that make up the average conditions in the atmosphere at a particular location on the Earth's surface are referred to as the **climate**. This is determined by two components:

1. **Climate elements** are the measurable, physical properties of the atmosphere. These include temperature, radiation, precipitation, and wind, humidity, clouds, and air pressure, which can also affect each other.
2. **Climate factors** are geographic conditions that effect the climate elements and thus also determine the climate of a region. The climate factors mainly include geographical site characteristics, for example latitude, distance to oceans, altitude, and slope aspect, which primarily affect temperature and precipitation conditions.

The climate elements are measurable variables whose average values can be used to describe the climate of a locality. The respective values are based on time periods of several years, with 30 years regarded as representative. The difference between climate and weather is given by the time aspect, i.e. **weather** refers to the current condition of the atmospheric elements and their variations over short periods (hours or days) whereas climate can be defined as the weather averaged over a long period. Weather data collected at specific points in time are used to calculate the climate.

The climate that dominates a location can be presented graphically as a **climograph** by using the monthly average temperature and precipitation at a specific location. The climograph presentation based on Walter and Lieth (1967) shows the average monthly precipitation and average monthly temperature as lines (Fig. 7.3).

The climate of a location is the result of the interaction of the global atmospheric transport system and the local climate factors.

7.2 Climate Zones

Large areas of the Earth characterized by a relatively uniform climate are known as **climate zones**. Such zones can be classified on the basis of different criteria. The most common classification systems use measured climate element data, of which

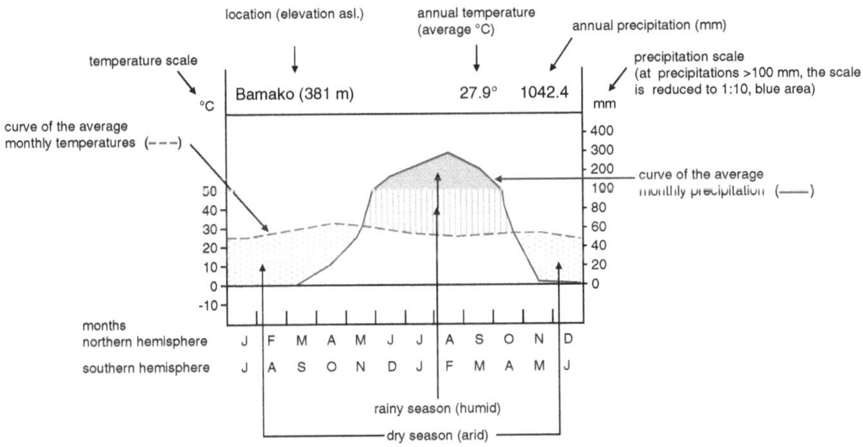

Fig. 7.3 Explanation of the climograph developed by Walter and Lieth (1967)

temperature and precipitation are of most importance ecologically. The climate zones are generally connected with each other by more or less wide transition zones in which a gradual change in the climate occurs. There are, therefore, no distinct boundaries between climate zones; instead they are defined by threshold values of particular climate elements. Another problem in the determination of climate zones is the scale of differentiation. Areas characterized on a large scale are not completely homogeneous units, but rather include smaller or larger regions with climate conditions that deviate from the average. Depending on interpretation, such areas can be regarded as subregions, can be treated as separate climate zones, or can be disregarded. Overall, no common system exists for classification of climate zones.

In addition to temperature and precipitation conditions, vegetation may also be included as a criterion in the differentiation of climate zones. The plant cover of the Earth consists of different types of vegetation, which can be characterized by their dominant growth and life forms and by their structural characteristics, and can be termed "formations". Examples of such formations are tropical forest, savanna, grassland, tundra, and taiga. Regions of the Earth that are characterized by specific dominant formations are referred to as **vegetation zones** or **biomes**. The latter include not only plants but also all the organisms of the respective ecosystems. The global distribution of vegetation zones is largely determined by climate, which means there is a close relationship between the different climates and the vegetation zones or biomes (Table 7.1).

The feasibility of agriculture, and types of land use, are also closely related to the climate conditions in different regions of the Earth. Arable cropping, livestock grazing, and forestry are the predominant types of global land use. A factor of major importance to the suitability of arable cropping in many of the climate zones is the length of the **vegetation period**. This period covers the time span within a year in

Table 7.1 Climate zones of the Earth and their corresponding vegetation zones/biomes

Climate zone	Vegetation zone/biome
Tropics	
Humid tropics	Evergreen tropical rainforest
Wet–dry tropics	Deciduous tropical forests and savanna
Subtropics	
Arid subtropics	Deserts and semi-deserts
Winter humid subtropics (Mediterranean climate)	Sclerophyll forest
Humid subtropics	Evergreen mixed forests
Temperate latitudes	
Humid–temperate latitudes	Summer–green deciduous forests
Dry–temperate latitudes	Steppes, deserts
Boreal climate zone	
Boreal	Boreal forest (Taiga)
Subpolar and polar climate zone	
Tundra	Tundra
Ice sheet	Ice deserts

which plant growth is possible, and is characterized by two threshold values (Schultz 2002):

1. The **thermic threshold** is given by an average monthly temperature of 5°C, which means that productive plant growth can occur only when this value has been reached or exceeded. For many plant species, this threshold is much higher. Most tropical species require an average monthly temperature of at least 10°C. Maize requires approximately 12°C and rice 18°C. However, there are also variety-specific differences.
2. The **hygric threshold** is given by the amount of precipitation (in mm) that exceeds twice the value of the average monthly temperature (in °C). Based on the definition of Walter and Lieth (1967), which is apparent in the climographs (cf. Fig. 7.3), these conditions are found in the humid months. They correspond to those periods in which the precipitation curve is above the temperature curve. Under the opposite conditions, arid conditions prevail.

The time span during which the given climatic conditions are suitable for plant growth can be expressed by consideration of both threshold values. The time span represents the **hygrothermic vegetation period**, which can be identified for any location with help of the climograph.

The hygrothermic vegetation period covers the number of humid months in a year that have an average temperature above 5°C.

In the humid tropics (Sect. 7.2.1.1), the vegetation period covers all months of the year. For this reason, for example, three rice harvests per year can be achieved in the tropics whereas in many rice-production areas outside this climate zone the vegetation period is shorter and enables one harvest only (e.g. in the Po River Valley of northern Italy, the northernmost rice production area). The climatic limit of arable cropping is reached at a vegetation period of 3–4 months (Sect. 7.2.4).

However, the feasibility of arable cropping is not only determined by the length of the vegetation period. During the vegetation period other conditions must meet the specific requirements of the crop, especially for crops cultivated outside their region of origin. With the exception of nutrient availability, the primary factor is appropriate solar radiation. This includes the length of the day, which important to short and long day plants (Sect. 4.1.1.4), the radiation intensity, and the temperatures reached, which, for example, affect temperature sums (Sect. 4.1.2.2). To summarise, the duration of the vegetation period combined with the prevailing conditions of radiation determines the character of the agroecosystems and crop species typical of the different climate zones (Fig. 7.4).

7.2.1 Tropics

The tropics are mathematically defined as the area between the tropics of Cancer and Capricorn. However, when precipitation and temperature conditions are used as the criteria for the definition, significant deviations from the latitudes become apparent. Overall, the tropics extend from the equator to the subtropical desert zones, and a distinction is made between the humid or permanently wet tropics and the wet–dry or summer humid tropics.

7.2.1.1 Humid Tropics

The largest areas covered by the humid tropics are found within 10° latitude in South America (primarily the Amazon Basin), Africa (primarily the Congo Basin), and on the islands of South Asia (Fig. 7.5).

Climate and Vegetation

The humid tropics are characterized by the absence of seasons and relatively constant average monthly temperatures all year round in the range 26–28°C in the lowlands. However, the temperature differences that occur within 24 h are significantly greater with deviations of 6–12°C from the monthly average. Because of these conditions, the climate in the humid tropics is termed **daytime climate**. Annual precipitation is usually in the range 2,000–3,000 mm and monthly averages are usually above 100 mm. A typical climograph is shown in Fig. 7.6.

The natural vegetation of the humid tropics is **evergreen tropical rainforest**. The distinguishing characteristic of the trees, irrespective of their species, is that the

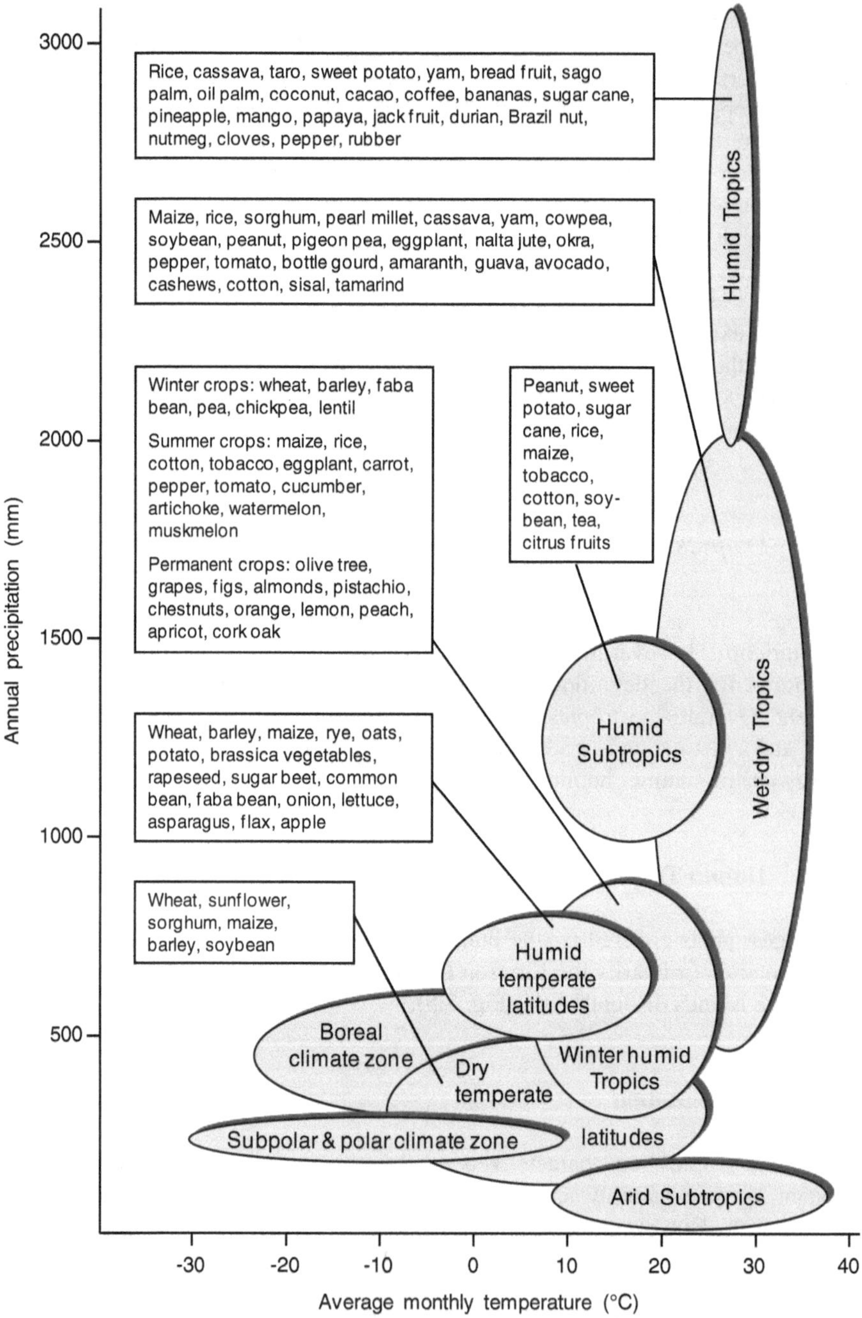

Fig. 7.4 Temperature/precipitation ecogram of the climate zones with some of the characteristic crop species

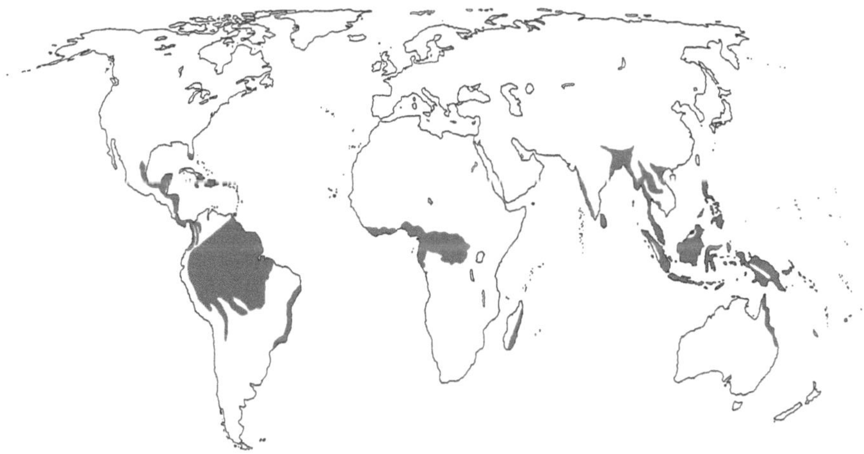

Fig. 7.5 Distribution of the humid tropics

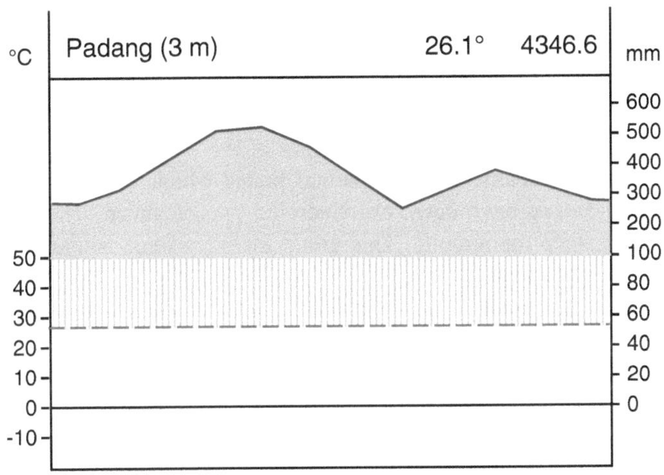

Fig. 7.6 Climograph from the humid tropics (Padang on Sumatra, Indonesia)

dropping of leaves occurs throughout the year. The forests usually have two or three stories, the uppermost layer of which attains a height of 50–70 m. This layer is formed by individual trees that extend beyond the tree layer below. Together, the understories of the forest form a closed leaf canopy. There is, therefore, a lack of light on the forest floor, which is why shrub and herbaceous layers are very scarce or entirely absent. In addition to trees, which make up 70% of all plant species in these forests, typical life forms are epiphytes and lianas (climbing plants). Epiphytes root not in the soil but instead colonize the branches or stems of trees, where light conditions are suitable. They are not parasites, but instead are dependent on nutrients that are delivered to them from the surroundings. Typical representatives of the epiphytes include the orchids (primarily in Southeast Asia), Bromeliaceae (primarily in South America), and ferns. Lianas, of which approximately 90% of all species are found in the tropics, root in the soil and form sprouts that climb up trees. With this strategy they reach light, without the need to develop a supportive stem. High species diversity is also typical of the humid tropical rainforest (cf. Sect. 3.3.1).

Land Forms and Soils

The occurrence of tropical rainforests is not dependent on particular land forms, parent materials, or soils. They are found on very old landforms and on young, volcanic parent material. The different land forms in the tropics essentially comprise four groups:

1. Old eroded areas, which are primarily found in West Africa and South America and have granite or gneiss as parent materials.
2. Shallow depressions, in which different types of sediment (limestone, sandstone, and claystone) have been deposited and which have existed at least since the Tertiary. Typical regions include the basins of the Amazon and Congo rivers.
3. Edge regions of oceanic and continental plates which were folded up in the Tertiary and are, or have been, characterized by volcanism. These landscapes usually have steep topography. This group includes some islands in Southeast Asia and the Caribbean.
4. Zones of current deposition of loose sediments that form plains along rivers and coastal regions.

The parent materials of the humid tropics were and are subject to rapid and intense chemical weathering because of the constantly high temperatures and high precipitation. The corresponding soils are very deep, usually reaching a depth of several meters. In many regions of the humid tropics, the dominant soil types are the **ferralsols** (especially in South America and Africa). These soils are characteristic of old landforms and are an advanced stage of soil development. Their distinguishing components are iron and aluminium, which remain after the leaching of soluble compounds in the course of soil development and are present as sesquioxides (Fe_2O_3 and Al_2O_3). The weathering is also associated with transformation of the materials. This affects, among others, the clay minerals, whose composition and structure change in the course of these processes. In the tropics, mainly clay minerals with a low cation-exchange capacity (CEC; Sect. 4.3.3.1) are generated. On the other hand, the anion-exchange capacity of ferralsols and other tropical soils rich in Fe and Al oxide is comparatively high, and high and often irreversible phosphate fixation commonly occurs (cf. Sect. 3.6.5). Overall, the strongly weathered soils of the humid tropics and of the wet–dry tropics are characterized by low nutrient availability and are essentially the result of three factors:

- high precipitation, which promotes leaching,
- low CEC, which results in a soil with low capacity to hold and exchange nutrients, and
- no mineral weathering, which means there is no supply of nutrients.

The ferralsols are also characterized by low pH (often <4), which further limits plant growth.

An important component of soil affecting fertility is humus (Sect. 4.3.5.3), which is relevant to plant growth both as a nutrient supplier and as a nutrient exchanger. Under the climatic conditions of the humid tropics, decomposition of organic material proceeds approximately five to ten times as fast as it does in the humid temperate latitudes. Because of the absence of seasons, plant litter is continuously dropped in the evergreen rainforests. The nutrients released from this material are particularly important to the nutrient supply of plants and must be used efficiently, because of the unfavourable conditions in the mineral soil. This is achieved primarily by fine roots, which are often found near the soil surface and are associated with fungi for most of the tree species of these ecosystems (cf. Mycorrhiza; Box 3.2).

In addition to the ferralsols, a variety of other soil types are found in the humid tropics, some of which are found primarily in other climate zones. Worth mentioning are the **andosols**. Although they make up only approximately 2% of the land area of the humid and wet–dry tropics, they have, in contrast with the ferralsols, high potential for agricultural use. They develop on recent volcanic ash and have, in young stages of soil development, slightly acidic pH range and a relatively high CEC. As weathering proceeds, nutrients are released, pH decrease, and fixation of phosphate increases. These soils are found, for example, on the Indonesian Islands of Java and Bali.

Land Use and Crops

The traditional type of land use in the humid tropics is arable cropping in the form of **slash-and-burn agriculture** (Sect. 2.4.1.1); this is still of great importance in many regions today (especially in South America and Africa). In this form of agriculture small forests areas are cleared with subsequent burning of the residual biomass. The resulting ash provides nutrients and increases the pH of the upper soil, such that temporarily suitable conditions are created for production of crops. In the humid tropics, however, such areas can rarely be used for more than 1–2 years, because the fertility of the soil declines rapidly. This is primarily the consequence of nutrient leaching and erosion during heavy rainfall. Subsequent agricultural use of the land requires a fallow period of approximately 20 years, during which secondary forest becomes established. This period does not primarily serve to

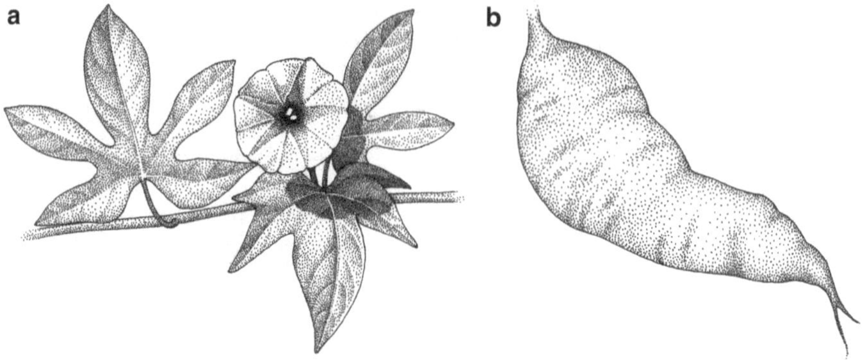

Fig. 7.7 Sweet potato (*Ipomoea batatas*). (**a**) shoot with leaves and flower, (**b**) root tuber

regenerate soil fertility, because this is rarely occurs under these ecological conditions. Rather, the main purpose of this forest phase is production of biomass, which must be available for production of ash in the next slash-and-burn activity. The diversity of crops produced in the slash-and-burn system is relatively low. In lowland areas, the crops include mainly upland rice and tuber crops, for example cassava and yam. Production of sweet potato (*Ipomoea batatas*; Fig. 7.7) and such vegetables as cabbage, beans, onion, pepper, and cucurbits (e.g. chayote, *Sechium edule*) is most often limited to higher elevations.

Other important types of land use in the humid tropics include:

Permanent arable cropping. In rainfed agriculture, this type of land use can be successfully undertaken only under specific conditions. These include, especially, fertile soils, for example the andosols, which are found in a few regions of this climate zone only (compare above). Apart from that, permanent arable cropping is usually possible only with irrigation, which is essentially limited to paddy rice production in the humid tropics (cf. Sect. 4.2.2).

Perennial crops and plantations. Production of tree and shrub crops (cf. Sect. 2.4.2.2) is more suited than slash-and-burn agriculture to the climate and site conditions of the humid tropics. The structure and function of perennial crop species are more similar to those of natural forest vegetation more than are those of annual crops; they also limit the extent of erosion and nutrient losses. This applies, particularly, to forest gardens and, to some extent, other agroforestry systems in which fruit trees, for example the jackfruit (*Artocarpus heterophyllus*; Fig. 7.8), perennial plants, for example bananas (Fig. 7.9), and climbers, for example pepper (*Piper nigrum*; Fig. 7.10), are grown. The greatest diversity of cultivated fruit trees on Earth is found in the humid tropics. On the islands of Indonesia, alone, at least 120 fruit tree species are cultivated; of these, however, only approximately a quarter is native to Southeast Asia (Scholz 1998).

Some of the plantation crops produced in the tropics have already been mentioned in Sect. 2.4.2.2. In the humid tropics, agriculture is primarily practised by

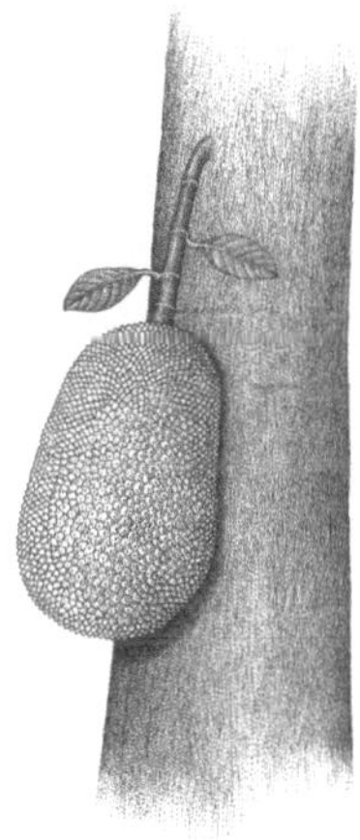

Fig. 7.8 The jackfruit tree (*Artocarpus heterophyllus*) is, like cacao, a cauliflorous species (cf. Fig. 2.25)

Fig. 7.9 Inflorescence of the banana (*Musa*×*paradisiaca*)

smallholders and produces a wide range of crop plants by traditional slash-and-burn agriculture on the one hand and commercial plantation production on the other (Table 7.2, Fig. 7.11). These products are partly used for subsistence and partly sold to traders (e.g. rice and coffee) or in local markets (e.g. vegetables).

Grazing management. Traditionally, this type of land use does not occur in the humid tropics. In recent decades, however, in regions of the South American humid tropics (primarily in the Amazon basin), large pasture areas have been created on which extensive cattle grazing is conducted.

Commercial timber harvesting. Much of the global demand for timber is met by the tropical forests. For many countries of this climate zone, export of tropical wood was, or is, economically important. In many regions, however, the forests have been

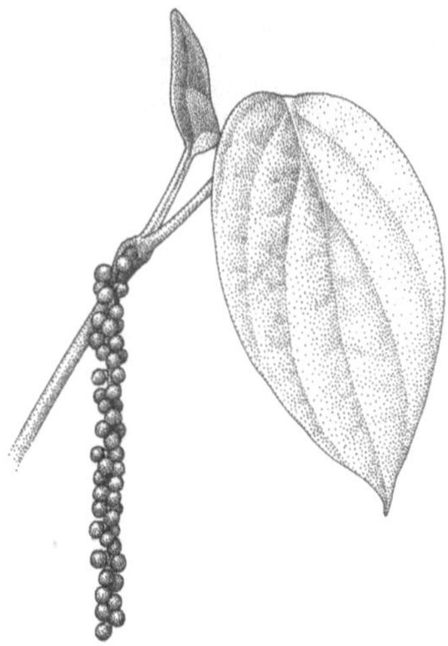

Fig. 7.10 Pepper (*Piper nigrum*), twig with leaves and fruits

decimated to such an extent that the international lumber trade is no longer important to the respective countries. In the Philippines, for example, less than 10% of the original forest cover remains.

The types of land use described have been mainly responsible for the progressive decline of tropical forests. Overall, however, the causes of forest destruction are various and can often be traced back to a combination of factors:

- Commercial logging often serves not only for timber extraction but also for creation of more agricultural land. This occurs, for example, in Indonesia and Malaysia, where deforested areas are used for production of palm oil (Fig. 7.12). These two countries currently account for more than 80% of global palm oil production. In Brazil, deforested areas are primarily used for cattle grazing.
- With commercial exploitation of forests, clearings and roads are created, which also offer peasant settlers easy access to the forested areas and facilitates their use for slash-and-burn agriculture. Because, with this method of production, new crop land is needed after only a few years, transformation of the land continues. In addition to this uncontrolled peasant colonization, governmental settlement or transmigration projects (primarily in South America and Indonesia) contribute to this process.

7.2.1.2 Wet–Dry Tropics

The wet–dry tropics are the regions adjoining the humid tropics and extend to approximately 15° latitude north and south (Fig. 7.13).

Table 7.2 List of crops grown in the fields and gardens of a small village in the Philippine lowlands (Cienda on Leyte Island)

Plant species	F	G	S
Starch plants			
Rice (*Oryza sativa*)	x		
Maize (*Zea mays*)	x	x	
Plantain (*Musa* sp.)		x	
Ube (*Dioscorea alata*)	x	x	
Taro (*Colocasia esculenta*)	x	x	
Cassava (*Manihot esculenta*)	x	x	
Arrowleaf elephant ear (*Xanthosoma sagittifolium*)			x
Sweet potato (*Ipomoea batatas*)	x	x	
Breadfruit (*Artocarpus alitilis*)		x	
Vegetable plants			
Okra (*Abelmoschus esculentus*)	x	x	
Winged bean (*Psophocarpus tetragonolobus*)		x	
Water spinach (*Ipomoea aquatica*)			x
Eggplant (*Solanum melongena*)		x	
Gourd (*Lagenaria siceraria*)		x	
Squash (*Cucurbita maxima*)		x	
Bitter melon (*Momordica charantia*)		x	
Onion (*Allium cepa*)	x	x	
Moringa (*Moringa oleifera*)		x	
Yardlong bean (*Vigna unguiculata sesquipedalis*)		x	
Fruit plants			
Lemoncito (*Citrus microcarpa*)		x	
Pomelo (*Citrus maxima*)		x	
Banana (*Musa x paradisiaca*)		x	x
Mango (*Mangifera indica*)		x	
Pineapple (*Ananas comosus*)		x	
Papaya (*Carica papaya*)		x	
Rambutan (*Nephelium lappaceum*)		x	
Jackfruit (*Artocarpus heterophyllus*)		x	
Spices and sugar plants			
Pepper (*Capsicum annum*)		x	
Ginger (*Zingiber officinale*)		x	
Lemon grass (*Cymbopogon citratus*)		x	
Cacao (*Theobroma cacao*)		x	
Sugar cane (*Saccharum officinarum*)		x	
Oil plants			
Coconut (*Cocos nucifera*)	x		x
Fibre plants			
Abaca (*Musa textilis*)	x		
Pandanus (*Pandanus* spp.)			x

F = fields or plantations, **G** = garden, **S** = sporadic at other sites (Based on Szinicz et al. 2002)

Fig. 7.11 Peasant vegetable production in Sri Lanka

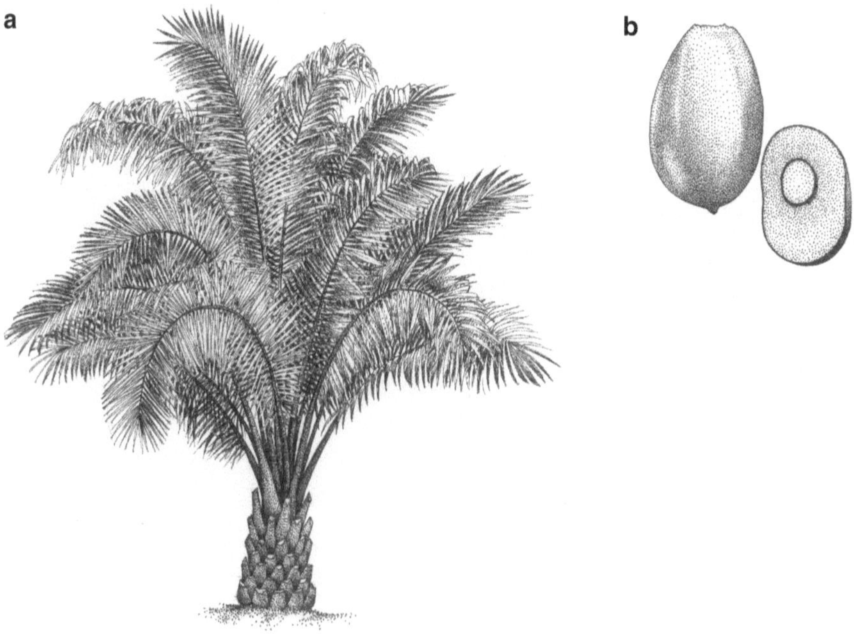

Fig. 7.12 Oil palm (*Elaeis guineensis*); (a) habitus and (b) fruit from which oil is extracted

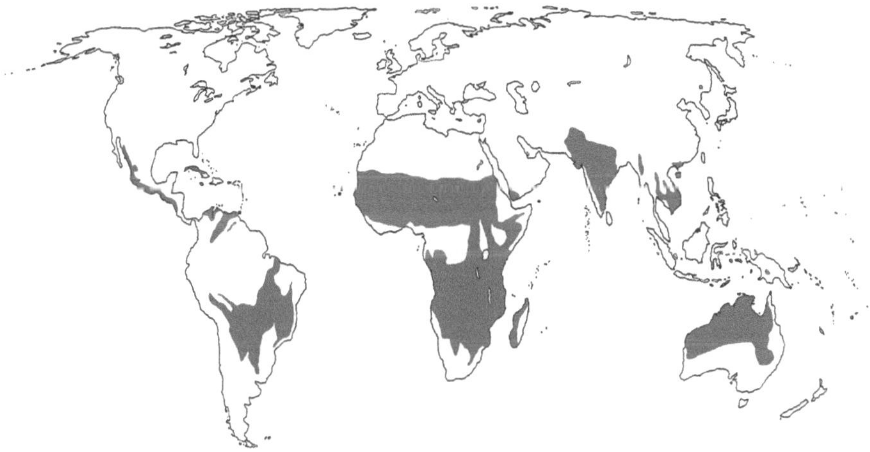

Fig. 7.13 Distribution of the wet–dry tropics

Climate and Vegetation

In the wet–dry tropics, average annual temperatures are between 24 and 30°C and average monthly temperatures do not fall below 18°C. Another characteristic of this climate zone is the occurrence of pronounced rainy and dry seasons. The duration of the respective seasons vary with increasing distance from the equator, which is why the wet–dry tropics cannot be characterized by a uniform climate but rather by a climatic gradient (Fig. 7.14).

This variation in the duration of wet and dry seasons also determines the type of vegetation—forests and different types of savanna. In detail, the following zones can be distinguished:

1. In areas where the annual precipitation is between approximately 1,500 and 2,000 mm and the duration of the dry season is 2–5 months, there is a transition from evergreen tropical lowland rainforest to **deciduous tropical forest**. Leaf fall in the dry season is an adaptation of the trees to the temporarily unfavourable water supply. In South and Southeast Asia, this type of forest is also referred to as **monsoon forest**. It comprises, primarily, economically valuable species, for example teak (*Tectona grandis*) and sal (*Shorea robusta*), the natural stands of which have been extensively logged.
2. Because of annual precipitation between 1,000 and 1,500 mm and a dry period of approximately 3–6 months, the stand density of the trees is reduced. Light conditions on the forest floor therefore improve and substantial grassland vegetation becomes established. These conditions are typical of the **tree or moist savanna**.
3. When the annual precipitation is approximately 500–1,000 mm and the duration of the dry season is 6–9 months, shrubs are the only woody plants that can survive alongside the grasses. This type of vegetation is called **shrub or dry savanna**.

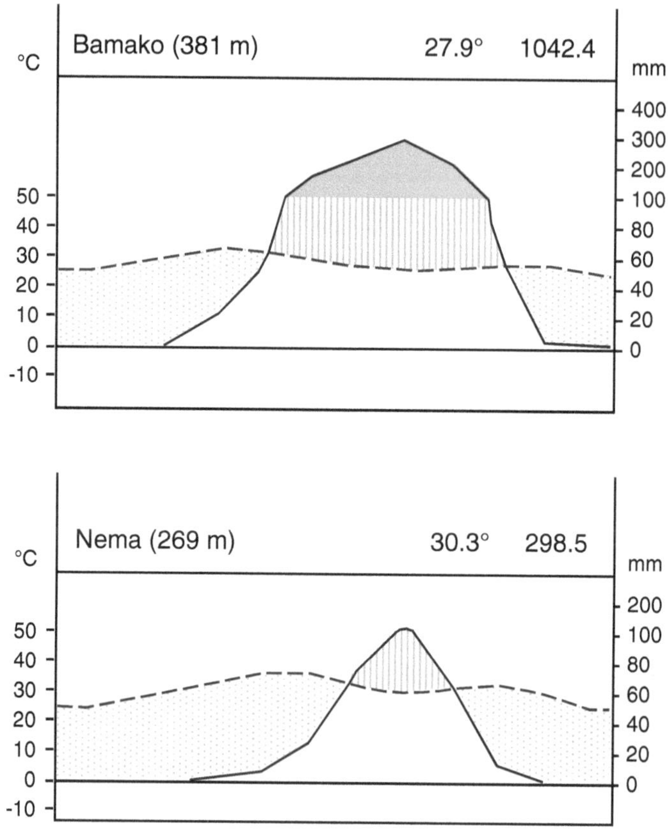

Fig. 7.14 Climographs from the wet–dry tropics: Bamako in Mali and Nema in Mauretania

4. The transition area between the savannas and the deserts is made up of the **thorn savanna**. This type of vegetation is found in zones with 200–500 mm annual precipitation within a period of less than 3 months.

Savannas are homogeneous plant communities made up of grasses and herbs with scattered woody plants, which can include trees, shrubs and bushes.

Material cycling, vegetation structure, and species composition of savanna ecosystems are determined not only by the climatic conditions but also by other factors. These include the grazing of wild animals and grass fires, which are caused naturally by lightning but more often by humans. Grazing animals are important consumers of the biomass produced, especially in the savanna ecosystems of Africa.

In many regions of the tropics, different types of savanna are also found in places with higher annual precipitation than would be expected. Such savannas are usually a result either of human activity (especially deforestation) or soil factors.

Large areas of the moist savannas of Africa are probably the result of human pressure on deciduous forests. Such areas developed from the forest clearings which resulted from logging, by establishment of grassland. In addition, frequent fires, caused by humans, result in the survival of few tree seedlings, preventing reforestation of such areas.

In dry savannas, excessive livestock density can lead to overgrazing, as a result of which grasses are suppressed and conditions suitable for woody plants are created (Sect. 6.2.1). This results in scrub encroachment and the creation of a type of savanna in which shrubs and bushes are more frequent than they would be on a natural site with similar precipitation. After the removal of these plants for firewood, such areas remain almost free of vegetation.

The processes that contribute to this degradation and, finally, desertification are initiated by human activity and strengthened by climatic effects (primarily naturally occurring periods of drought). The most important causes and effects of desertification include:

- **destruction of vegetation**, which occurs primarily as a result of overgrazing, expansion of agricultural areas, and the removal of wood, all of which increase the land's susceptibility to erosion;
- **alteration of the water balance** of the sites as a result of lowering of the water table and use of inappropriate irrigation techniques that lead to soil salinization (Sect. 4.2.2); and
- **degradation of soil**, which includes the loss of soil from vegetation-poor or agriculturally utilized areas by erosion, and loss of nutrients (e.g. as a result of shortened fallow periods).

Desertification not only affects areas adjoining deserts, but also areas in the wet–dry tropics, the dry temperate latitudes, and the winter humid subtropics of all continents.

Land Forms and Soils

Low hills and hollows which develop as a result of erosion and sedimentation during the rainy season are typical of savanna areas. Differences in the relief, which usually amount to a few metres only, result in small-scale variations in soil moisture. From such landscapes, so-called island hills or mountain ranges arise, which consist of parent rocks that are very resistant to weathering and therefore are eroded only slowly.

In the wet–dry tropics of Asia and South America, the predominant soil types are the **acrisols**, which may also be found in the humid tropics and in other climate zones. The properties of Acrisols resemble those of Ferralsols

(Sect. 6.2.1.1), but they are younger, less weathered and have a relatively higher yield potential. In the savanna regions of Africa, primarily **lixisols** are found; these, in turn, are similar to acrisols, but are at a less advanced stage of soil development. The most favourable soils for production of crops in the wet–dry tropics are the **nitisols**; these, however, make up only a small portion of all soils there. They develop from neutral or alkaline parent material (e.g. basalt, schist), still contain weatherable minerals, and are characterized by good water-retention capacity and a porous soil structure.

Both natural vegetation and the feasibility of agricultural land use are often affected by **laterite crusts** in savannas. In essence, these are hard layers of sesquioxides, the development of which requires alternation between wet and dry seasons. When the layer is located in close proximity to the soil surface it forms a water-impermeable layer on which water accumulates during the rainy season. Such sites are soil-induced (edaphic) savanna types and are unfavourable for trees, which is why grasses and palms usually dominate.

Land Use and Crops

In the moist savannas, a peasant, labour-intensive agriculture dominates, which is primarily based on annual crops and is practised as rainfed agriculture. The most important cereal species are maize, sorghum, pearl millet (the latter primarily in Africa), other millet species, and rice, which is the most important in the monsoon areas of Asia. A variety of bean species, for example cowpea (*Vigna unguiculata*) and soybean (*Glycine max*), are also of major importance. The latter is grown as a feed crop in intensive, commercial production, on large areas, primarily in the wet–dry tropics of Brazil (Cerrado region). Other important legume species are peanut (*Arachis hypogaea*) and pigeon pea (*Cajanus cajan*). The dominant root crops include cassava and yam species. Examples of typical vegetable species are eggplant (*Solanum melongena*), nalta jute (*Corchorus olitorius*), okra (*Hibiscus esculentus*), bell pepper (*Capsicum* species), tomato, and amaranth (*Amaranthus* species). Cotton and sisal are grown as fibre plants. Planned crop rotation is not widespread in peasant agriculture. Production of crops is traditionally based on mixed cropping, for which primarily cereal and legumes crops are used.

With the shortening of the rainy season and declining precipitation, the conditions for crop production in dry savanna areas become progressively less favourable. When precipitation is less than 500 mm per year, the agronomic production limit is reached in the summer rainy areas. This means that unrestricted rainfed agriculture is no longer possible. In Africa, sorghum and pearl millet are still produced in these areas, because these species have a higher drought tolerance than maize. Another disadvantage of dry savanna areas for plant production is

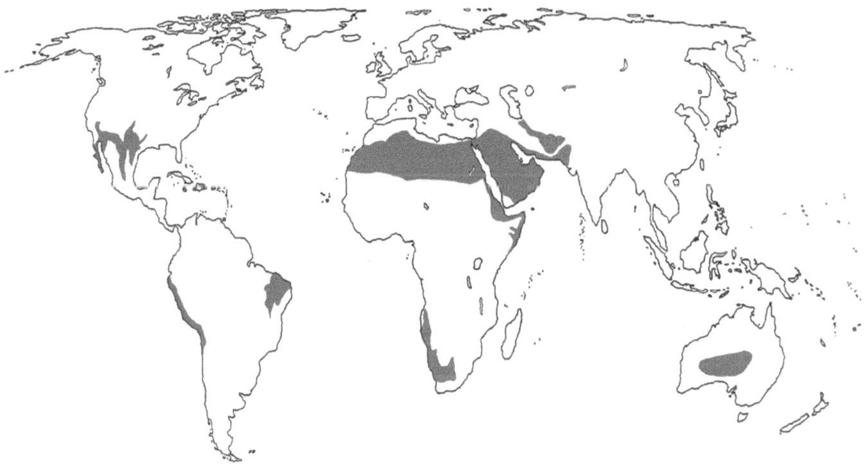

Fig. 7.15 Distribution of the subtropical deserts and semi-deserts

the pronounced variability in annual precipitation, which is why yields are not secured every year.

Under these conditions, livestock grazing becomes increasingly important. In many regions of the Sahel zone of West Africa, for example in Niger, agro-pastoral systems, in which **rotational fallow management** is practised, are traditionally found in the agronomic dry limit areas. Originally, this type of land use included mixed production of cereals (pearl millet, sorghum) and legume species (cowpea), interrupted by 15 years of grazed fallow. Today, the fallow period has been reduced to less than 5 years, mainly because of the increased human population. Even relatively unproductive areas that were previously used only as pasture land are increasingly used for arable cropping (Wezel and Boecker 1998). As a result of expansion of cropped areas on the one hand and the increase in livestock numbers (primarily cattle) on the other, competition for resources has increased in such areas. Conflicts especially occur in those areas where nomads and their cattle herds encounter sedentary farmers.

7.2.2 Subtropics

The term "subtropics" includes three climatically distinct zones that are found between the wet–dry tropics and the temperate latitudes.

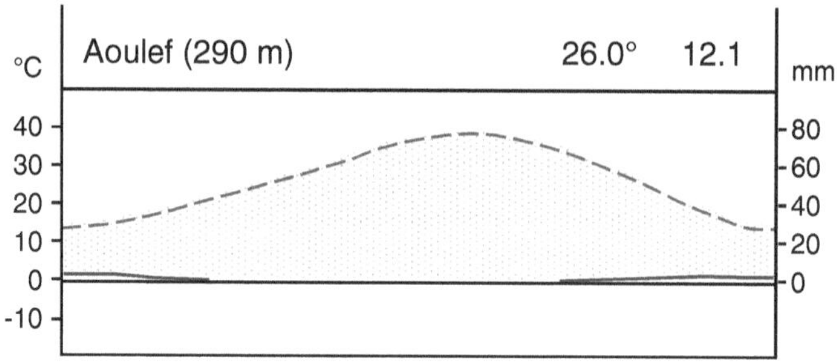

Fig. 7.16 Climograph from the Sahara (Aoulef, Algeria)

7.2.2.1 Arid Subtropics

The areas of the arid subtropics are found north and south of the equator in the range of the subtropical high pressure belt between approximately 15 and 35° latitude (Fig. 7.15).

Average annual temperatures usually range between 20 and 25°C. In the hottest months, average temperatures reach 30–35°C. In the other months, temperatures rarely drop below 10°C. Annual precipitation is usually less than 200 mm, whereas potential evaporation is more than 2,000 mm (Fig. 7.16). Areas with such conditions are defined as **hot deserts** (Walter and Breckle 1999). Precipitation can vary substantially from one year to the next and is not predictable, falling at different times of the year, depending on the region. For example, in the northern Sahara rain falls in the winter whereas in the south it falls in the summer; in the core area rain never falls.

Depending on precipitation conditions, vegetation structure and composition also differ within deserts. With relatively high rainfall of approximately 100–200 mm, the vegetation is usually diffuse, which means that plants (primarily low growing shrubs and grasses) are relatively evenly distributed over the landscape, but do not form a closed plant stand. Coverage ranges from 10 to 50%. Areas with such plant stands are often called **semi-deserts**. In desert areas with less precipitation, a contracted vegetation is found, which means that plants are only found in those few locations in which a sufficient supply of water is available to meet their needs.

Plant species with morphological and/or ecophysiological adaptations to long periods of drought, for example limitation of transpiration, storage of water in tissues (succulence), and CAM photosynthesis (Sect. 4.1.1.3), are **xerophytes**. However, in deserts, plants are also found that are not adapted to drought conditions. Such species only form aboveground live stages after rainfall, and endure the rest of the time as seeds or belowground organs (bulbs, rhizomes, or tubers). Such species make up the so-called ephemeral vegetation, which only appears after effective rains and is, therefore, in contrast with xerophytic permanent vegetation, only temporarily present.

Fig. 7.17 Distribution of the winter–humid subtropics (Mediterranean climate zone)

Depending on the surface material, deserts may also be classified into rock or boulder deserts, gravel deserts, and sand deserts with dunes. There are, in addition, troughs and hollows that are characterized by clay-rich or salt-rich deposits, termed pans. The soils of deserts are only slightly developed or do not exist, i.e. only mineral parent material is present.

Without the possibility of irrigation, which is usually only possible in river oases or groundwater oases, agriculture is not important in deserts and semi-deserts. The dominant type of land use is **extensive nomadic grazing**, which has a long tradition, especially in northern Africa. Although cattle grazing is still possible in the thorn savannas, natural grazing land in deserts and semi-deserts is primarily used for sheep, goats, and camels. Because of the low availability of fodder, the land required area per animal is very large. With excessively large herds, overgrazing is therefore a threat.

7.2.2.2 Winter–Humid Subtropics (Mediterranean Climate Zone)

Worldwide, the winter–humid subtropics are represented by five relatively small regions. They are found between 30 and 40° latitude north and south in the western parts of continents and lie in the Mediterranean area, in North America (California), in South America (part of Chile), on the southern tip of Africa, and in the southwest of Australia (Fig. 7.17). Because their climate conditions are similar to those found in the Mediterranean region, winter–humid subtropics are also called **Mediterranean climate zones**.

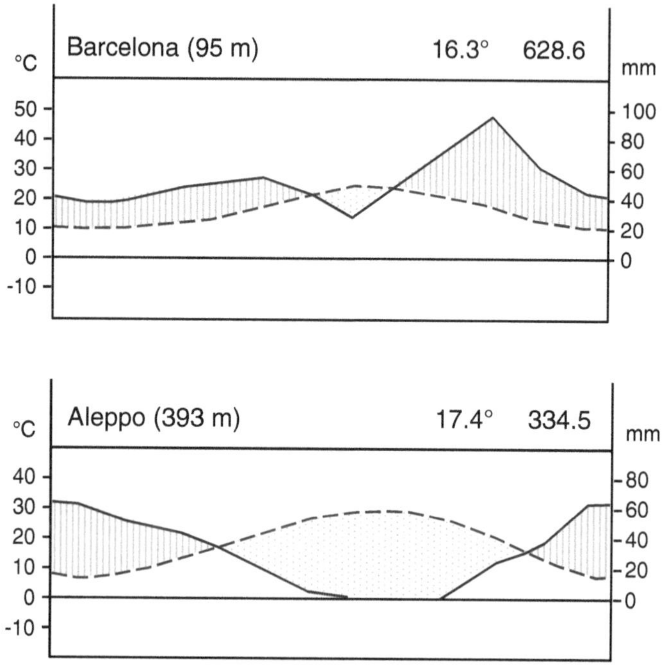

Fig. 7.18 Climographs from the winter–humid subtropics: Barcelona in Spain and Aleppo in Syria

Climate and Vegetation

Characteristics of this climate zone are hot, dry summers and wet, mild winters. In summer, these areas are usually affected by the subtropical high-pressure belt whereas in winter they are exposed to the westerlies. The average annual temperature usually reaches 15–18°C. In the summer months, the average ranges between 24 and 27°C, and in winter between 7 and 10°C (with occasional frost). Annual precipitation is within the range 300–900 mm (Fig. 7.18).

Characteristic of the natural vegetation of the winter humid subtropics are the **evergreen sclerophylls** with their relatively thick, stiff, and hard leaves that are an adaptation to the summer dry period and solar radiation. Sclerophylly has developed in many species, irrespective of taxonomy, in the different regions of the winter–humid subtropics.

In the Mediterranean region, oak forests (different *Quercus* species) were originally dominant, but these were almost entirely cleared thousands of years ago (Box 7.1). The emerged secondary vegetation consists of low-growing shrubs (maquis or chaparral) which, in turn, as a result of intense human activity (logging, fires, grazing), was often degraded to open heaths and grass communities (garrigue).

Box 7.1 "On the Hills of Attica[1]"

Contemporary Attica may accurately be described as a mere relic of the original land. There has been a constant movement of soil away from the high altitudes and what remains of her substance is like the skeleton of a body emaciated by disease. All the rich soil has melted away, leaving a country of skin and bone. When Attica was intact, her mountains were heavily forested and the country produced boundless pasture for cattle. The annual supply of rainfall was not lost as at present through being allowed to flow over the denuded surface into the sea, and so was received by the country in all its abundance into her bosom where she stored it in impervious potters' earth and so was able to discharge the drainage of the heights into the hollows in the form of springs and rivers with an abundant volume and a wide territorial distribution.

(Plato, 427–347 B.C.E.)

[1] Peninsula in southeast Greece.

Soils

Overall, a large diversity of soils is found in the winter–humid subtropics. Of these, the **chromic luvisols** are particularly typical, because they are found in all regions of this climate zone and make up between 2 and 29% of the area. The specification "chromic" (from gr. *chroma* = colour) refers to the intense red to brown colour. In the Mediterranean region, they develop primarily on carbonate rich parent material and are counted among the soils termed "*terra rossa*".

Approximately half of the area of the winter–humid subtropics is made up of soils with limited potential uses, because they are shallow, relatively impermeable to roots, and/or located on steep slopes. This soil's susceptibility to wind and water erosion is also correspondingly high, and large parts of the Mediterranean region are subject to these types of erosion. Nutrient-poor soils are primarily found in South Africa and Australia. Approximately 40% of the area of the winter–humid subtropics is made up of soils with moderate to high nutrient availability, although this proportion is substantially higher in Chile and substantially lower in South Africa (Jahn 2000).

Land Use and Crops

Because of the climatic conditions and the long history of land use, a diverse agricultural landscape associated with a wide range of crops developed in the Mediterranean region (Fig. 7.19). In the mild and rainy winters conditions are favourable for rainfed agriculture. For this, crops that are also widespread in

Fig. 7.19 Cultivated Mediterranean landscape in Morocco

the temperate latitudes are primarily grown. The high temperatures in summer enable the production of many species of tropical origin, if precipitation is sufficient or irrigation is possible.

In the southern and eastern peripheral areas of the Mediterranean region, the amount of precipitation is limited and very variable, which is why there is a high risk of drought-induced yield losses. For example, average annual precipitation in Tel Hadya (northwest Syria) is 343 mm, but in individual years deviations of more than 150 mm can occur (Fig. 7.20).

In areas with such conditions, barley (Fig. 7.21) can be grown, but the dominant type of land use is grazing (mainly with sheep and goats). Under favourable precipitation conditions, wheat and legumes (e.g. faba bean, pea, chickpea, lentil, and fodder legumes) are grown as **winter crops** in addition to barley, which is possible without the risk of drought in areas with 400–500 mm precipitation.

The **summer crops**, of which many have different regional importance, include maize, cotton, rice, sugarbeet, tobacco, and many vegetable and fruit species, for example eggplants, peppers, cucumbers, carrots (*Daucus carota*), tomatoes, watermelon (*Citrullus lanatus*), and muskmelon (*Cucumis melo*). Sunflower and soybean are of increasing importance. Some of these species can be grown without irrigation in areas with more than 700 mm annual precipitation and relatively short dry summers (usually in coastal regions).

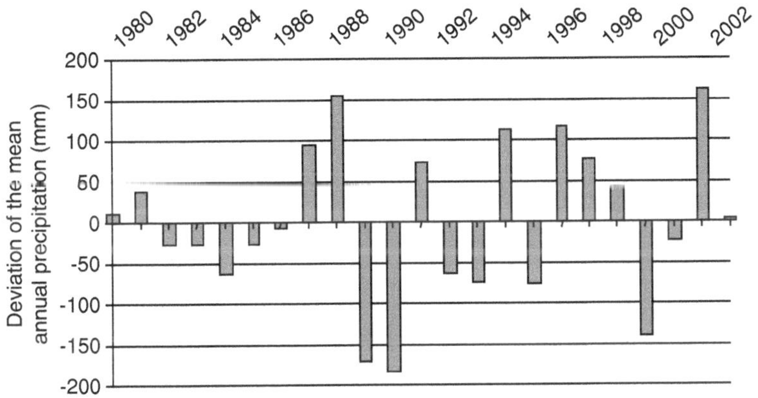

Fig. 7.20 Deviations of the annual precipitation from the long-term average (343 mm) between 1980 and 2002 in Tel Hadya, northwest Syria

Fig. 7.21 Barley (*Hordeum vulgare*)

In many parts of the Mediterranean, **perennial crops** are also of traditional importance. These primarily include grapes, olives, figs (*Ficus carica*), cork oak (*Quercus suber*), pistachio (*Pistacia vera*), chestnut (*Castanea sativa*), citrus species, for example lemon and orange, and a variety of stone fruits, for example peach, apricot, and almond (cf. Sect. 2.2.3). In addition, many fruit trees, which are also typical of the temperate latitudes, for example apples (*Malus domestica*) and sweet cherry (*Prunus avium*), are also found in the Mediterranean region. Many of the crops used in the Mediterranean region are also produced in other areas of the winter humid subtropics. For example, wine production is important in all the regions.

In some regions of North Africa, Western Asia, and Southern Europe with relatively high precipitation, annuals (e.g. cereals and grain legumes) are grown between the perennial tree cultures (e.g. olive, fig, and almond trees). Further types of land use include the production of wheat, with subsequent grazing on the stubble, and the combination of wheat undersown with legumes, for example subterranean clover (*Trifolium subterraneum*) and *Medicago* species. This system of ley-farming is widespread in the Mediterranean climate zone of Australia, where large areas are used both for cereal and animal production.

Fig. 7.22 Distribution of the humid subtropics

Because of favourable market conditions for fruit and early vegetables, production of such crops is widespread under soil-covering foil and plastic tunnels or in greenhouses. Primarily in coastal regions with mild winters, such species as eggplant, strawberry (*Fragaria* × *ananassa*), cucumber (*Cucumis sativus*), watermelon and muskmelon, peppers, and tomatoes are produced outside the respective season for the international market.

7.2.2.3 Humid Subtropics

The humid subtropics include five subregions which are, in contrast with the winter humid subtropics, located on the eastern sides of continents between 25 and 35° latitude of the northern and southern hemisphere (Fig. 7.22). The largest areas include the southern states of the USA and southeastern China.

The humid subtropics can be described as transition zones between the tropics and the temperate latitudes in which precipitation falls in all months of the year. The annual amount is usually in the range 1,000–1,500 mm. The average annual temperature is approximately 15–20°C. In the warmest months, averages of 25°C are reached; in the coldest months the average falls to 10°C (Fig. 7.23). These conditions are mainly found in coastal regions. In more continental regions, average temperatures in the coldest months can be less than 5°C. Similar to the winter humid subtropics, frost may sometimes occur during the cool season.

The natural vegetation consists of evergreen forests, which are composed of deciduous and coniferous species. Soils are predominantly Acrisols (Sect. 6.2.1.2).

Because of the high precipitation, rainfed agriculture can be practised in the humid subtropics and many tropical crops can be grown during the warmer months. These include peanut, sweet potato, sugarcane, rice, tobacco, maize, cotton, and soybean. The most important perennial crops include *Citrus* species and tea.

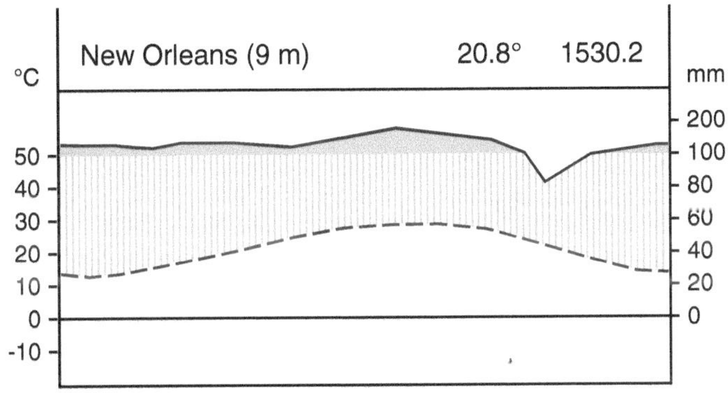

Fig. 7.23 Climograph from the humid tropics (New Orleans, Louisiana, USA)

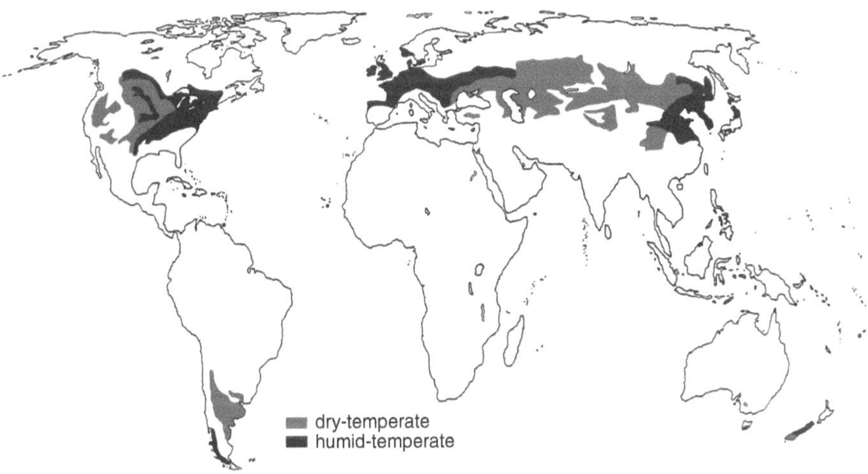

Fig. 7.24 Distribution of the humid and the dry temperate latitudes

7.2.3 Temperate Latitudes

The temperate latitudes, located in the area of the westerlies, extend from approximately 35–60° latitude and are primarily found in the northern hemisphere. They can be subdivided into humid and dry temperate climate zones (Fig. 7.24).

7.2.3.1 Humid–Temperate Latitudes

The distinct seasonal variations in average monthly temperatures and the relatively evenly distributed annual precipitation, which usually amounts to 500–800 mm, are characteristic of the humid–temperate latitudes. There are regional variations in the climatic conditions within the temperate latitudes, which are primarily the result of

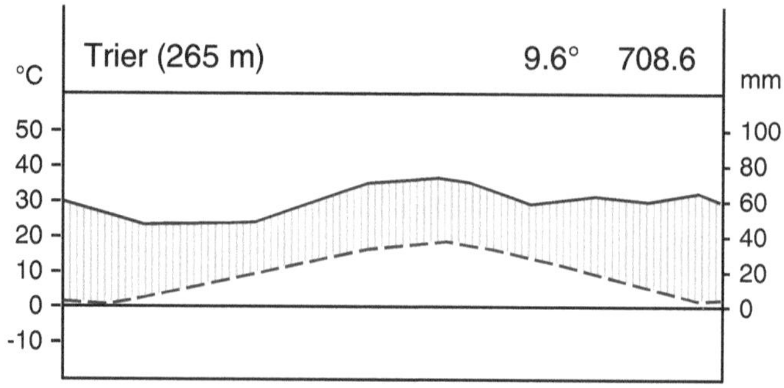

Fig. 7.25 Climograph from the humid–temperate latitudes (Trier, Germany)

the proximity of the regions to the ocean. For example, in different regions of Europe the prevailing conditions are:

- Eire is characterized by an oceanic climate with cool summers and mild winters in which frost rarely occurs. The lowest average monthly temperatures are usually between 5 and 6°C (January), the highest near 14–15°C (July). The average annual temperature is 9–10°C. Precipitation often exceeds 1,000 mm.
- In Germany, the lowest average monthly temperatures are usually −1 to +2°C, the highest between 17 and 19°C. Annual averages are approximately 8–11°C. The highest values within this range are found in the Rhine Valley. Annual precipitation is usually in the range 650–850 mm (Fig. 7.25).
- In Poland, a continental effect is observed. The lowest average monthly temperatures are between approximately −2 and −4°C, the highest approximately 16–18°C, and the annual average is 7–8°C. Annual precipitation is approximately 500–700 mm.

The natural vegetation of the humid temperate latitudes is summer–green **deciduous forest**. In Europe, the forests of this climate zone were originally dominated by beech (*Fagus sylvatica*) and oak (primarily *Quercus robur*), the stands of which had been entirely cleared by the Middle Ages and partially replanted subsequently. The woodlands found in the humid temperate latitudes of Europe today were all established by humans. In many regions, such coniferous species as Norway spruce (*Picea abies*) and Scots' pine (*Pinus silvestrus*) predominate.

Soils

Most of the soils in the humid–temperate latitudes developed after the last ice age and are therefore very young. As a result, they are usually weakly to moderately developed, relatively shallow, and usually have high mineral nutrient content and good CEC. The predominant soil types are Luvisols and Cambisols. The development of the **Luvisols** occurs by lessivation, i.e. translocation of clay from upper soil layers to lower soil layers. These soils primarily form on loamy parent material, are

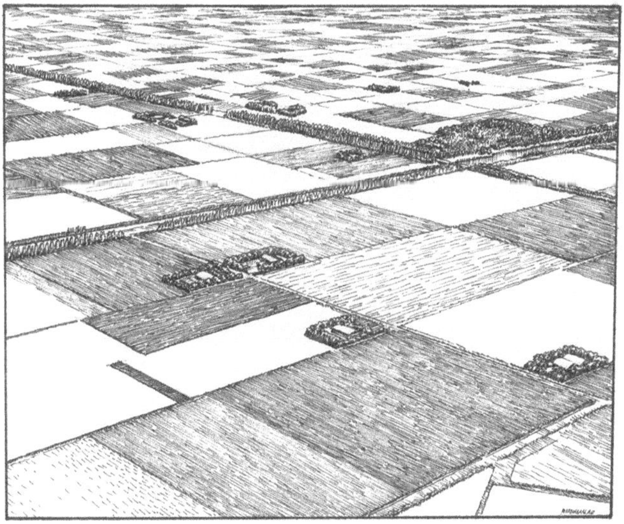

Fig. 7.26 Intensively managed agricultural landscape in the Netherlands

slightly acidic, and are, overall, favourable for agriculture. In the **Cambisols**, the horizon differentiation is weak. The properties of these soils are highly affected by the parent material and nutrient content, pH, and other characteristics can vary substantially. Cambisols with particularly unfavourable conditions for agriculture (e.g. on granite or sandstone) are mostly used for forestry.

Land Use and Crops

Most areas of the humid temperate latitudes have high potential for agricultural utilization, primarily because of favourable soil properties, relatively high and consistent water supply, which enables rainfed agriculture, and long vegetation periods of more than six months. This climate zone largely covers regions of high population density in the most developed regions of the world, both industrially and agriculturally (eastern parts of the United States, Central Europe, Japan, South Korea). Because of the favourable natural and economic conditions, highly industrialized intensive agriculture has developed in these regions (Fig. 7.26).

The most important crop plants of the humid–temperate latitudes include the cereal species wheat, barley, rye, (Fig. 7.27) and maize (also rice in eastern Asia), the root crops potato (Fig. 7.28) and sugarbeet (Fig. 7.29), and rapeseed as oil crop. In many regions, soybean is grown (especially in the United States), but apart from that, grain legumes are of only limited importance. With the exception of apples, which are produced in many regions of the humid–temperate latitudes (e.g. United States, Central Europe, New Zealand), perennial crops are of relatively minor importance.

Fig. 7.27 Rye (*Secale cereale*)

Fig. 7.28 Potato (*Solanum tuberosum*), shoot with leaves and flower

In many areas agricultural production in these regions is no longer directed toward provisioning the population but is, instead, oriented toward the market. This is one reason the importance of different crops has changed markedly in recent decades. In Germany, approximately half of the total agricultural land is currently used for the production of wheat, barley, and maize. Comparison with 1925 shows that at that time approximately half the agricultural land was used for rye, oats, and potatoes (Table 7.3).

Mostly intensive **grassland management** is primarily practiced in those areas which are less suitable for crop production because of climatic reasons. These

Fig. 7.29 Sugar beet (*Beta vulgaris*)

Table 7.3 Percentage of production area for important crops in Germany in the years 1925 and 2000 (Based on Piorr and Lehmann 2004)

	% of production area	
Crop	1925	2000
Wheat	7.2	22.5
Rye	21.4	7.1
Barley	6.4	18.6
Oats	16.4	2.5
Maize	2.0	12.8
Potato	13.6	2.6
Winter rapeseed	0.9	7.9
Sugar beet	2.0	4.3
Fodder beet	5.2	0.1
Grain legumes	3.2	1.5
Clover/grass mixtures	9.4	1.9
Alfalfa	3.1	0.3
Other	9.4	17.9

include most of the coastal regions and the mountainous regions. Another important type of land use is the forestry sector. In Germany, approximately 50% of the total land is used for agriculture and approximately 30% is covered by forest. In Japan, forest cover amounts to two-thirds of the total land area.

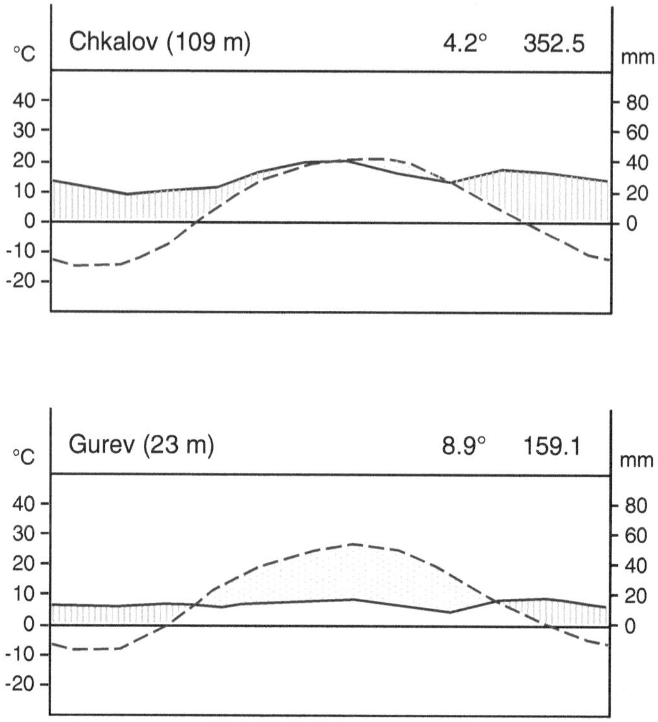

Fig. 7.30 Climographs from the dry–temperate latitudes: Orenburg (Chkalov) in Russia and Gurev in Kazakhstan

7.2.3.2 Dry–Temperate Latitudes

The largest areas of this climate zone are found in the continental regions of Asia and North America. In the southern hemisphere, the largest area lies in Argentina. On the basis of temperature and precipitation conditions, this climate zone can be subdivided into **steppes**, **desert steppes**, and **deserts**.

Climate and Vegetation

In the continental steppe regions of Asia and North America, average temperatures in the winter months are usually below −5°C. In the warmest summer months, average temperatures of more than 18°C are reached. Annual precipitation is in the range 300–500 mm and falls irregularly but throughout the year. A short arid phase in late summer is typical (Fig. 7.30, Orenburg). The steppes, which are called prairie in North America and pampa in South America, consist of grasslands in which no trees grow, in contrast with the savannas. Exceptions are small scattered forest plots, which are found in the transition zone between the humid temperate latitudes and between the steppes and the boreal climate zone.

Fig. 7.31 Wheat (*Triticum aestivum*)

In the continental desert steppe regions (e.g. in Kazakhstan) the difference in temperature between the winter and summer months is even greater than in the steppe regions. Annual precipitation is 100–200 mm and falls primarily in the winter months. In summer, arid conditions predominate (Fig. 7.30, Gurev). The vegetation consists primarily of tussock grasses.

The deserts of the dry–temperate latitudes are found in Central Asia (e.g. Karakum, Gobi, Takla-Makan), North America (Mojave), and South America (Atacama). They are characterized by hot, mostly dry summers and precipitation-poor, cold winters.

Soils

The characteristic soils of the steppe zones of Asia and North America are the **chernozems**, most of which have developed on loess. Their organic upper soil (Ah-horizon) can be 50–100 cm deep. The chernozems have a good nutrient supply, high CEC, and good water storage capacity; they are regarded as being among the most fertile agricultural soils. However, cultivation results in a substantial decline in humus content (cf. Sect. 3.6.2). In the somewhat drier regions (annual precipitation approx. 200–400 mm), the kastanozems have developed. They have similar properties to the Chernozems, but have a smaller organic horizon. Because of the relatively low water supply, such sites are primarily used as pasture land.

Land Use and Crops

The steppe regions are, worldwide, amongst the most important areas for production of wheat (Fig. 7.31) and account for most of the production areas in both the North American and Asian steppe zones. Other important crops are sunflower, sorghum

Fig. 7.32 Distribution of the boreal climate zone

(especially in the United States), maize, and, in the somewhat drier regions, barley and other cereal species. Soybean is also grown in the warmest regions.

When precipitation is below the agronomic dry limit (in the steppe regions approximately 300 mm annually), production of crops is usually only possible with irrigation. This is accompanied by the threat of salinization (Sect. 4.2.2). The dominant type of land use under these conditions is extensive grazing (usually with cattle or sheep), which leads to a risk of desertification (Sect. 6.2.1.2) when stocking densities are excessively high.

7.2.4 Boreal Climate Zone

Because of the global distribution of land masses, the boreal climate zone is only found in the northern hemisphere—in Asia and North America. On the western parts of the continents, the zone extends between 60 and 70° latitude north; on the eastern sides, the southern boundary is at approximately 50° latitude north (Fig. 7.32).

In the continental areas, average temperatures of the coldest months are often below −20°C; those of the warmest months are usually in the range 15–18°C. Average annual temperature is usually less than 0°C. Precipitation falls in all months of the year, but the largest amounts usually fall in summer. Total annual precipitation is approximately 300–600 mm (Fig. 7.33).

The natural vegetation of the boreal climate zone is primarily coniferous forests (**taiga**), in which spruces (*Picea*), pines (*Pinus*), or larches (*Larix*) usually predominate. In areas affected by oceans, birches (*Betula* species) are also common. The characteristic soils are the **podzols**, which often have a thick layer of raw humus,

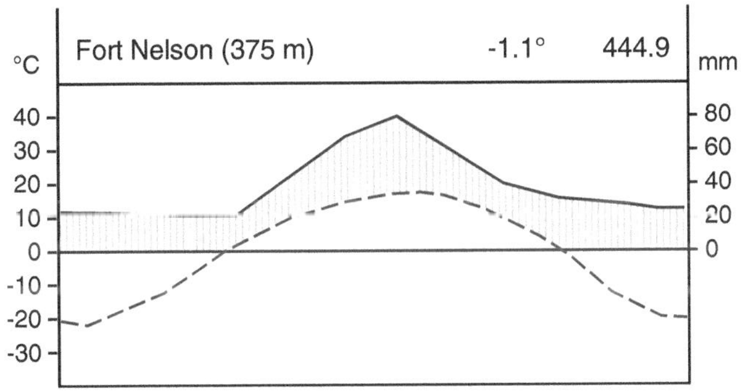

Fig. 7.33 Climograph of the boreal climate zone (Fort Nelson, Canada)

because of the limited rate of mineralization, and often very unfavourable properties for use overall. The short vegetation period of approximately 3–4 months is another reason agriculture is of no importance in this climate zone. The crop with the northernmost range is barley, which requires a vegetation period of at least 3 months. In Europe, the geographical boundary for the production of barley, rye, and potatoes is approximately the Arctic circle (66° N). The taiga is primarily used for commercial forestry.

7.2.5 Subpolar and Polar Climate Zones

The areas of the treeless arctic **tundra** and the **polar ice deserts** are found north of the Arctic forest line. With average annual temperatures of $-10°C$ at most and annual precipitation usually less than 300 mm, climate conditions in the subpolar and polar zones are more extreme than in the taiga. In the southern (subarctic) tundra, a closed vegetation cover exists, which is dominated by dwarf shrubs, grasses, sedges, or herbs, depending on the site conditions. A nomadic form of reindeer (*Rangifer tarandus*) herding is traditionally practiced in the Eurasian part of this zone, whereas hunting is more common in the North American tundra. In the coastal regions, fishing and hunting of marine mammals (whales and seals) predominates.

References

Jahn R (2000) Die Böden der Winterfeuchten Subtropen. Geographische Rundschau 52:28–33

Piorr H-P, Lehmann K (2004) Vielfalt im Wandel der Zeit – Historischer Überblick und Status Quo. In: Begemann F, Schröder S (eds) Schriften zu Genetischen Ressourcen 23: Produktvielfalt durch Ressourcenvielfalt – Potenziale genetischer Ressourcen. ZADI, Bonn, pp 1–15

Scholz U (1998) Die feuchten Tropen. Westermann Verlag, Braunschweig

Schultz J (2002) Die Ökozonen der Erde, 3rd edn. UTB, Eugen Ulmer Verlag, Stuttgart

Szinicz G, Martin K, Borberg C, Knaack A, Pressler D (2002) Agroecological studies in landuse systems around Ceinda, Baybay, Leyte, Philippines. In: Szinicz G, Göltenboth F (eds) Proceedings of the 9th international seminar and workshop on tropical ecology. Leyte State University, Leyte, pp 53–61

Walter H, Breckle S-W (1999) Vegetation und Klimazonen, 7th edn. UTB, Eugen Ulmer Verlag, Stuttgart

Walter H, Lieth H (1967) Klimadiagramm-Weltatlas. Gustav Fischer Verlag, Jena

Wezel A, Boecker R (1998) Fallow plant communities and site characteristics in semi-arid Niger, West Africa. J Arid Environ 40:269–280

Chapter 8
Agroecological Aspects of Global Change

The beginnings of agriculture over 10,000 years ago was a process closely related to the shortage of naturally-available human food resources. From this, the need to develop alternative food sources emerged and was achieved by production of crops (Sect. 2.1). Since then, the human diet has been almost exclusively based on agricultural products; these, however, have not always been available in sufficient quantities. Repeated famines characterized life in Europe in the Middle Ages and have also been a feature of later centuries. In the mid-nineteenth century, over one million people starved in Ireland after several years in which the potato harvest failed as a result of blight (caused by the fungus *Phytophtora infestans*). Measured by the development of the global population and the standard of living of, at least, the industrial nations, agriculture has, overall, proved to be a very successful strategy for securing and developing the livelihood of humans. Whereas 10,000 years ago there were only approximately 2–20 million people on earth (Cohen 1997), currently there are more than 7 billion.

8.1 Global Change

With increasing population density, it is not only the demand for foodstuffs that increases but also the demand for other resources, for example fuel and water. This development is characterized by diverse and sometimes irreversible changes in the global environment, which increasingly threaten the conditions necessary for maintenance of the livelihoods of humans and other organisms, and as a whole represent **global change**. The central aspects of global change are related to:

- the global population increase and the resulting direct effects on the livelihood of humans (e.g. food supply, place of residence, standard of living, social factors, and health), and
- anthropogenic effects on the abiotic and biotic environment (climate, soil, water, biodiversity) and thus on essential resources for humans and other species.

K. Martin and J. Sauerborn, *Agroecology*, DOI 10.1007/978-94-007-5917-6_8,

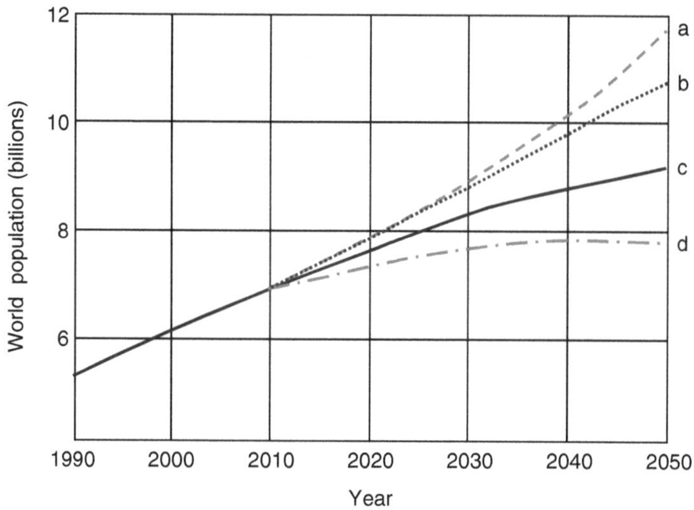

Fig. 8.1 Projected development of the world population by the year 2050 with (**a**) constant, (**b**) high, (**c**) medium, and (**d**) low fertility (Based on United Nations 2005)

8.1.1 World Population Growth

At the beginning of the nineteenth century the number of humans on Earth passed one billion; it reached three billion in 1960. In the subsequent 40 years, this number doubled. Overall, the world population nearly quadrupled in the twentieth century. The United Nations (UN) regularly provide predictions on the future development of the world population, which are based on different assumptions about fertility (average number of children per woman). The report from 2005 and the revised version from 2009 (United Nations 2005, 2009) contain four estimates of the development of the global population until 2050 (Fig. 8.1):

- According to the medium increase scenario, it is assumed that the average number of children per woman will decrease from 2.56 in 2005–2010 to 2.02 in 2045–2050. However, in the more developed regions, total fertility is projected to increase from 1.64 children per woman currently to 1.80 in 2045–2050. Overall, approximately 9.1 billion people would then be living on Earth.
- The low increase scenario is based on the assumption that the number of children per woman will decline by a further 0.5 compared with the medium scenario, which would result in a human world population of approximately 7.6 billion in the year 2050.
- In the high variant, in which fertility is assumed to remain mostly half a child higher than in the medium variant, the world population in 2050 would reach 10.5 billion persons.
- If the number of children per woman remains constant at current levels, in 2050 almost 11.7 billion people would be living on Earth.

According to these predictions the world population will increase, even if, as in the low and the medium increase scenarios, it is assumed that the number of children per woman will decline substantially.

There are large deviations from the global average in individual countries. Overall, the greatest differences are between industrialized nations and the least developed nations, which are primarily found in Africa, Latin America, and parts of Asia. The UN (United Nations 2005) provides the following details in relation to this:

- In the industrialized nations, the number of children per woman is approximately 1.6 on average. In 35 of the 148 countries that are regarded as developing nations, the number is above 5.0.
- In 51, mainly developed, countries, the number of people in 2050 is expected to be lower than it was in 2005. In contrast, in several other countries, for example Afghanistan, Niger, and Uganda, the 2005 population will probably more than triple by 2050.
- The three countries that will have the highest share of world population growth until 2050 are India, Pakistan, and Nigeria. These countries are among eight that will make up approximately half of the expected growth worldwide.

8.1.2 Development of Agricultural Production

A consequence of the development of agriculture was the emergence of permanent settlements (villages and cities), which was associated with a continuous increase in human population density (Sect. 2.3). This also led to an increase in the demand for food, which had to be met by agricultural production. Viewed globally, the quantity of food produced has been able to keep up with population growth until today. However, this does not mean that all people are adequately supplied. In many countries of Africa, Asia, and Latin America, the increase in food production has lagged behind population growth. With other factors, for example social and political inequality, wars, and natural disasters (e.g. droughts, floods), this accounts for the fact that today approximately one billion people worldwide are undernourished or malnourished.

To provide for the growing world population, the quantity of food produced globally must be increased substantially in the coming decades, primarily to meet the needs of the countries that are not, or will not be, in the position to produce sufficient food for their own populations.

Estimates of developments necessary in the near future are listed below (FAO 2003, 2006b, 2009; UNEP 2009; Eickhout et al. 2006; Pelletier and Tyedmers 2010):

- Annual world cereal production, which at the end of the 1990s amounted to 1.9 billion tons on average, must be increased by approximately 50% to 2.8 billion tons by 2030. Global production of vegetable oils, which are predominantly produced from soybeans, oil palm, rapeseed, and sunflowers, must increase by

Fig. 8.2 Observed and expected increase in global crop production (Based on Eickhout et al. 2006)

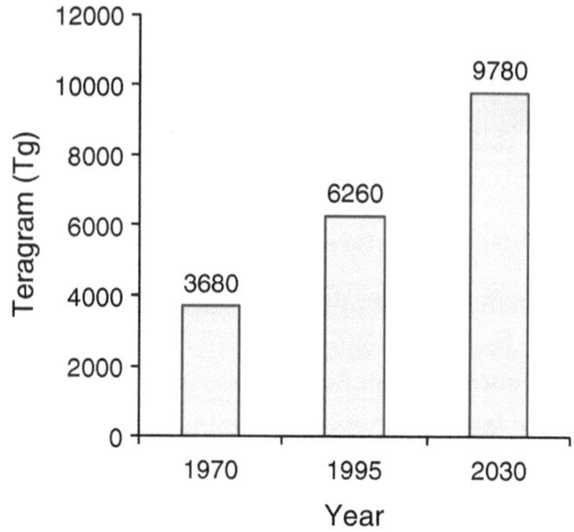

Fig. 8.3 Observed and expected increase in global meat production (Based on Eickhout et al. 2006)

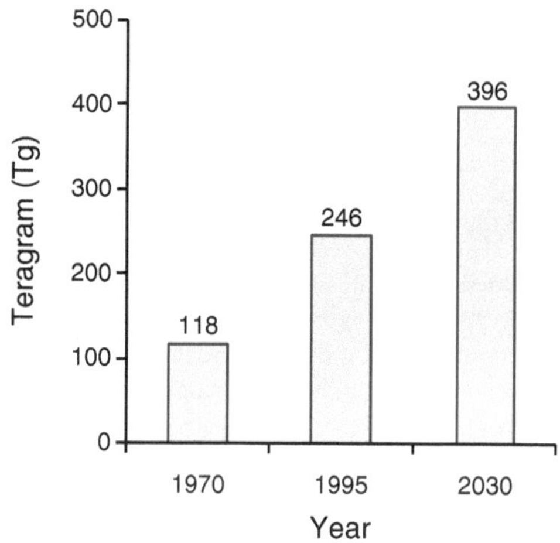

more than 108% by the year 2030. Overall global crop production is expected to grow by more than half by 2030 compared with production in 1995 (Fig. 8.2).

- Projections for 2030 indicate that approximately 80% of the increase in total crop production will result from higher yields per area and approximately 20% from agricultural land expansion. Expected food and feed demand will require a substantial increase of global food production of 70% by 2050.
- Global meat production has more than doubled between 1970 and 1995 and is expected to grow by more than 60% by 2030 compared with production in 1995 (Fig. 8.3). More than one-third of global crop production is currently needed for animal feed and the production of meat. If the trend in meat production continues, 88% of all biomass harvested from agricultural land will be needed for livestock feed production in 2050.

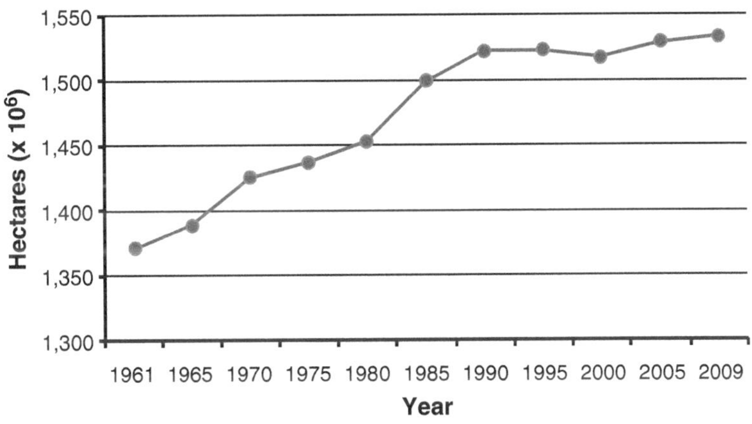

Fig. 8.4 Development of the world agricultural land area between 1961 and 2009 (FAOSTAT data)

These predictions are based on the assumption that in the year 2030 approximately 8.3 billion people will live on Earth, which corresponds to the medium increase scenario of the predictions from the United Nations (cf. Sect. 8.1.1).

Fundamentally, there are two possible means of increasing yields in agricultural production

- expansion of the areas of production, and
- intensification of production, i.e. increasing the yield per unit area.

8.1.2.1 Expansion of Production Areas

Until the middle of the twentieth century, the increasing demand for foodstuffs by the world's population was predominantly met by expansion of agricultural land. From the beginning of the 1960s, the increase in global production area slowed and has not increased significantly since the 1990s (Fig. 8.4).

Within this period, however, the world's population has more than doubled (from three to seven billion). Per capita agricultural land available has, therefore, approximately halved during this time. Agricultural expansion has had massive effects on ecosystems, biodiversity, carbon storage, and soil conditions. In fact, worldwide agriculture has already cleared or converted 70% of grassland, 50% of savanna, 45% of temperate deciduous forest, and 27% of tropical forest biome areas. Today, agriculture is mainly expanding in the tropics, where it is estimated that approximately 80% of new croplands are established by replacing forests (Foley et al. 2011). Furthermore, large areas of cropland have been abandoned, and production has diminished on others, because of soil erosion and salinization; compensation must also be made for this.

Fig. 8.5 Terraced fields on La Gomera (Canary Islands)

Nevertheless, on the basis of estimates by the FAO, significant potential still exists in unused areas suitable for production of crops. However, the amounts of such land vary substantially in different regions of the Earth. In South America and some regions of Africa south of the Sahara, only approximately 20% of the land area with potential for rainfed agriculture is currently used for production. In contrast, in South Asia, in Africa north of the Sahara, and in the Near East, approximately 90% of potentially usable areas are already used for cultivation of crops.

However, the possibility of developing more agricultural land will lead to other conflicts. More than half of the areas that might be considered for use are covered by forest or dedicated as nature reserves. Their transformation into agricultural land would be associated with the loss of natural ecosystems and, primarily in the tropics, with a substantial decline in biodiversity. Furthermore, many land resources could be productively cultivated only at high expense, because they are often located on nutrient-poor or otherwise unsuitable soils or on steep slopes. The latter can usually only be effectively utilized by terracing (Fig. 8.5).

8.1.2.2 Intensification of Production

Although in the second half of the twentieth century the world agricultural area increased at a much lower rate than the world's population, a global food crisis has not resulted thus far from this situation. The area required at the beginning of the 1960s to achieve the same yields as were needed at the end of the 1990s declined by 56%, on average, worldwide (FAO 2003). The halving of the per capita agricultural area in this period could thus be compensated.

This improvement in global food supply was initiated in the 1960s by the so-called **Green Revolution**. This term describes the process in agriculture that is primarily based on progress in breeding of the three most important cereal species (rice, wheat, and maize). High-yielding varieties of wheat and rice were developed, most notably at the International Maize and Wheat Improvement Center (CIMMYT) in Mexico and at the International Rice Research Institute (IRRI) in the Philippines. These differ from traditional varieties in being characterized by shorter stalks, which provide them with greater stability and with a higher proportion of grain in the overall biomass (Fig. 8.6). They often also have a significantly shorter growth period than the traditional varieties, which enables more harvest per unit of time. However, the increases in yield made possible by the high-yielding varieties cannot be achieved by use of these varieties alone, i.e., it is not sufficient simply to replace traditional varieties by high-yielding varieties in a traditional production system. Rather, different technical production measures, primarily high inputs of fertilizer and sufficient irrigation, are required to achieve the high yield. Because many of the high-yielding varieties are more susceptible to pests and diseases than traditional varieties, application of pesticides often also increases with their cultivation.

The high-yielding varieties of wheat and rice have widely replaced traditional varieties in Asia and Latin America. In the 1990s, the new seed was used on more than 80% of the production area of wheat and on more than 60% of the production area of rice in these regions (Evenson and Gollin 2003). Between 1970 and 1995, wheat production in Asia doubled whereas the population increased by 60% only in the same period.

In Africa south of the Sahara, the Green Revolution had only limited effects. There, increases in agricultural production during recent decades were achieved primarily by expansion of the agricultural area. This is essentially explained by the unfavourable conditions for the high-yielding production of crops in many of these regions. These include limited possibilities for irrigation in wet–dry climate zones and the limited availability of fertilizers, which would be needed in particularly large quantities for production of high-yielding varieties on the usually nutrient-poor soils found in these areas. However, progress has been made in breeding crop varieties (rice, maize, cassava, and others) since the 1990s which enable yield increases on marginal sites.

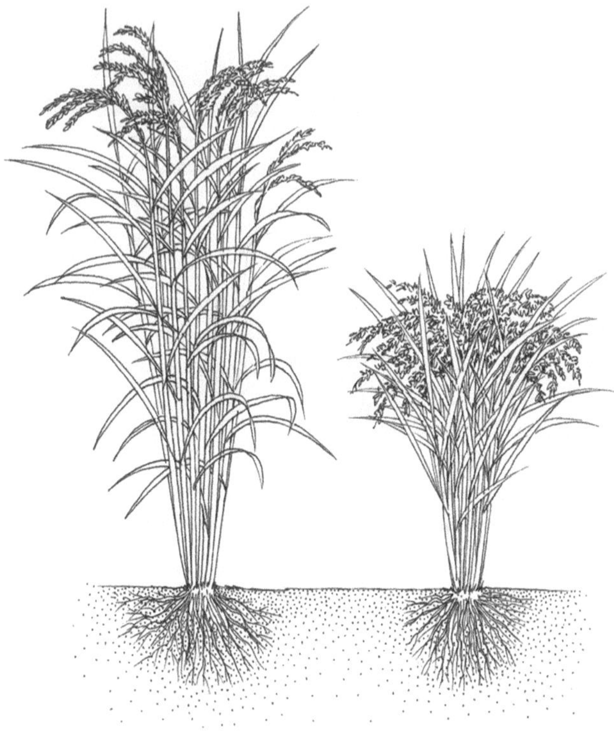

Fig. 8.6 Habitus of a traditional (*left*) and a high-yielding variety (*right*) of rice (*Oryza sativa*)

In relation to the Green Revolution, two aspects of the effects of agricultural production on the environment can be distinguished:

- The increasing of yields reduces the need to expand production areas, whereby natural ecosystems and other areas worthy of protection can be saved. Had the 1961 average world yield of cereal still prevailed, nearly 850 million hectares of additional land of the same quality would have been needed to equal the cereal harvest of 1999, to meet the food demand of the increased human population (Borlaug 2000). From this perspective, the Green Revolution made an important contribution to the conservation of natural resources and biodiversity.
- The increasing of yields by use of high-yielding varieties is possible only under intensive production conditions, which also has negative effects on the environment. As a result of irrigation, water scarcity is a threat in many regions, and is associated with lowering of the groundwater table, among other effects. Application of large amounts of fertilizer leads to eutrophication of water bodies and affects ecological processes in the soil. Between the 1960s and 1990s, the global quantity of fertilizer applied increased sevenfold (Tilman 1998). Additional problems related to intensive production arise from the consequences of pesticide and herbicide application, and from different forms of soil degradation, for example erosion, salinization and desertification.

8.1.2.3 Future Needs and Potentials

As a result of Green Revolution it was possible to largely avoid the world food crisis feared in the early 1960s. However, the "race between the plough and the stork" continues, because of the growing world population. The question of how the large demand for agricultural products can be met in the coming decades still remains. The FAO (2009) assumes that in the future, foodstuffs from plant production required by developing countries will have to be produced primarily by intensification of production, which means higher yields, better use of land with suitable species and varieties, and shorter fallow periods. However, in many countries it will not be possible to fully meet the demand for cereals with domestic production by the year 2030. These countries will, therefore, become increasingly dependent on imports from regions with high potential for productivity, for example the United States, the European Union, and elsewhere.

The environmental problems that result from conventional agriculture will continue to increase in the coming decades. To summarise, the major concerns include:

- The expansion of agricultural land into natural ecosystems, especially tropical forest, which leads to the destruction of globally valuable biological resources and ecosystem services. The main pressure on the remaining tropical forest comes from crops that contribute little or nothing to the world's total food supply, especially oil palm and sugarcane production (partly used for biofuels), rubber cultivation (for industrial purposes), and soybean production (largely for livestock feed).
- The increasing intensification of production is associated with an increasing demand for water and fertilizers and often also for pesticides and herbicides. The shortening of fallow periods reduces the natural regenerative capacity of the soil, whereby its usability for plant production is increasingly limited.
- The strongly increasing proportion of global crop production needed for livestock feedstuffs is a very inefficient contribution to global food security in terms of resource use, because it takes, on average, the equivalent of 3 kg grain to produce 1 kg meat. Furthermore, most meat and other livestock production is in industrial farming systems, which strongly affect the environment with animal waste products (especially nitrogen compounds).

Overall, this situation requires the development and application of production methods and technology that mitigate or eliminate the negative effects of agriculture on the environment yet fulfil high demands with regard to yields. In the future, contributions to this may, or should be, expected from the following approaches:

Sustainable Agriculture

The overall concept of sustainable agriculture comprises production methods that contribute to the conservation of natural resources, ecosystem services, and regulatory processes of agroecosystems. Sustainable agriculture includes a broad range of forms of production which are the different options to conventional agriculture (Sect. 2.3.6). These range from organic agriculture, in which nearly closed cycles in the overall system and few effects on the environment are the objective, to integrated agriculture, and to a wide range of special measures in which particular system components are the focus. For example, these include integrated plant protection, and different practices for soil conservation and the promotion and maintenance of soil fertility, for example conservation tillage and direct seeding (Box 4.2), crop rotations and the production of cover crops (Sect. 2.4.1.2).

In many regions of the tropics and subtropics, alternative production methods, for example no-tillage soil cultivation and organic agriculture, can produce equally high and, in the long-term, more stable yields than may be achieved by use of conventional methods. In developing countries, different forms of sustainable agriculture can make an important contribution to improvement of the food supply in the future, particularly in rural regions with marginal site conditions, in which small-scale and usually subsistence-oriented farms predominate. However, by the year 2030 more than half the population in developing countries will live in cities and not produce their own food.

Sustainable agriculture also includes the spatial and ecological recoupling of crop and livestock production in mixed farming systems by integrated management of livestock waste and the reduction of external energy inputs. Integrated crop–livestock systems could foster diverse cropping systems, including the use of perennial and legume forage which could be grown in selected areas of the landscape to achieve multiple environmental benefits. Integrated systems would inherently utilize animal manure, which enhances soil fertility and achieves C sequestration. Integration of crops and livestock could occur within a farm or among farms. In future, integrated crop–livestock systems will be especially important in industrialized countries (Russelle et al. 2007).

Productive and sustainable land use includes the use of crop varieties and livestock breeds best-adapted to the respective climates, soils, and landscapes.

New Technology

The negative effects of conventional agriculture provide a strong rationale for future technology and practices that can contribute to productivity and to sustainable intensification of agricultural systems.

Transgenic crops (GM crops). Great expectations with regard to global food security are dependent on the development and the application of genetically modified crops. As shown in Fig. 2.17, the production area has increased dramatically in recent years, but is essentially limited to varieties of soybean, maize, cotton, and rapeseed, which are resistant to specific herbicides and/or insects, pathogens, and/or viruses. Such modifications and applications, however, are not necessarily contributions to increased yields. They are, rather, a modern substitute for conventional herbicide or pesticide applications. Furthermore, these crop species and the properties of their transgenic varieties are of limited importance to improvement of the food supply of developing countries.

Until now, in addition to the agrochemical industry, it is primarily large operations supplying the world market with animal feed (soybean, maize) and plant raw materials (rapeseed, cotton) that have profited from the application of genetic engineering in plant breeding.

However, with approaches that are more closely oriented to the needs of developing countries, improvements in the food supply could also be achieved by means of genetic engineering. Such approaches are related primarily to the development of varieties that produce high yields on sites with unfavourable environmental conditions (e.g. salty or acidic soils, drought). On the basis of the estimates of the FAO (2003), the increases in global agricultural production necessary by 2030 can probably also be achieved without major advances in biotechnology.

Another approach in the genetic engineering of crop plants is modification of the photosynthetic efficiency to increase yields. This involves attempts to introduce a C_4 photosynthetic pathway into plants as an alternative to the standard C_3 pathway (Sect. 4.1.1). C_4 photosynthesis is found in drought-tolerant grasses, for example maize and sorghum, but not in wheat and rice. In comparison with C_3 crops, C_4 crops have higher yields, reduced water loss, and increased nitrogen use efficiency, particularly in hot and dry environments. Transfer of C_4 metabolism into rice could achieve an increase in yield of up to 50%; another gain could be a substantial increase in water use efficiency, a characteristic normally associated with C_4 plants (Hibberd et al. 2008). However, this is an ambitious objective that cannot be achieved in the near future. Other approaches to improve water use in crop production include the development and application of improved irrigation techniques (Sect. 4.2.1).

Precision agriculture. (or precision farming) uses techniques that apply water, nutrients, and pesticides only to the places and at the times when they are required in the field. This approach takes into account that conditions in a production area are usually not uniform, and that wide spatial variations in soil types can occur in fields in that area, depending on topography and parent material. Consequently, differences in water supply, nutrient availability, and potential leaching of fertilizers and pesticides occur which, in turn, lead to differences in crop yields. Precision farming is based on information on the exact location of the application areas. These data are recorded with the help of the global positioning system (GPS), geographical information systems (GIS), soil sensors, and yield monitors which, with GPS, enable geo-referenced records of yield to be collected during harvest. Overall, adoption of techniques and technology with greater precision for appropriate agroecosystem management can significantly reduce environmental effects by efficient input application and reduce the costs of inputs for the farmer.

Food use efficiency. Global food security in the coming decades is not a matter of production alone, but also of efficiency of use of the food products. It is estimated that, globally, 10–40% of the crop yields achieved in the field are lost after harvest for different reasons. In developing countries, a substantial proportion of grain yields are lost during storage, especially as a result of storage pests and mould fungi. For these reasons, approximately one-third of the rice harvest in Southeast Asia can be lost. Fresh products are often spoilt during transport or inappropriate storage. In India, 35–40% of fresh produce is estimated to be lost because neither wholesale nor retail outlets have cold storage (Godfray et al. 2010).

For other reasons, food losses in industrialized countries are as high as in developing countries. Whereas approximately 40% of food losses occur post-harvest and during processing in the latter, more than 40% of food losses occur during retail and consumption in the former. Food waste by consumers in industrialized countries (222 million tons) is almost as high as the total net food production of 230 million tons in sub-Saharan Africa (FAO 2011).

Part of the harvest of vegetables and fruits is destroyed solely because it does not fulfil the cosmetic standards of industry and consumers. Hence, many edible, yet only slightly blemished products are discarded during production and trade. In addition, many food products including bread, fruit, vegetables, and milk products are thrown away by wholesalers and retailers because they have not been sold. The largest proportion of food products, however, ends up as waste from private households. In such European countries as France, Germany, and the Netherlands, this accounts for more than 20% of purchased food products per private household. As shown by a study from Germany, private consumers account for 60% of all wasted food. The remaining proportion is contributed by traders, industry, and bulk consumers for example hospitals, the army, etc. (ISWA 2012; Fig. 8.7)

A further aspect of efficient food use is related to meat consumption, because of the expected dramatic increase in livestock and livestock feed production and the substantially higher energy requirements for meat compared with plant production. A global reduction in meat consumption, and thus the amount of plant production used as animal feed, could make a significant contribution to the basic supply of food for the world's population. Furthermore, diets rich in leguminous crops (including soybean) and other vegetable products provide well-balanced diets in terms of protein and are more healthy than those containing a high proportion of meat. Dietary changes in favour of more vegetable foods are therefore urgently required, especially in industrial countries.

Another important source of protein for many people, especially in coastal regions, is marine fish. According to the FAO (2012b), fish provides approximately three billion people with almost 20% of their intake of animal protein. Small-scale

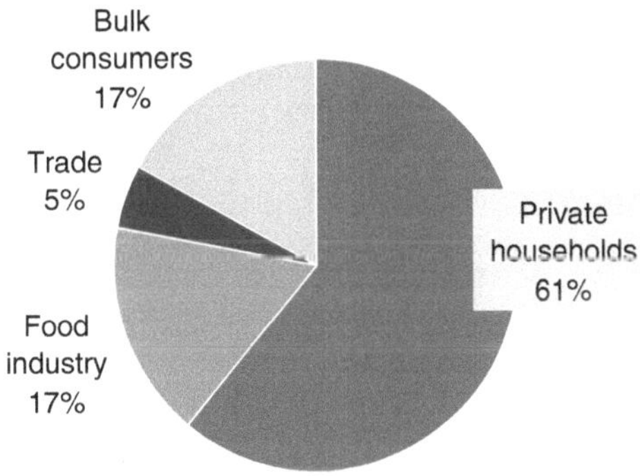

Fig. 8.7 Distribution of food wastes by sectors of production and consumption in Germany (Based on ISWA 2012)

fisheries employ more than 90% of the world's capture fishers. Since the 1950s, the total catch of fish from the world oceans has increased by nearly 400%, not only because of increased demand but also because of increasingly effective technology for detecting and recording shoals. As a consequence, most of the stocks of the top ten species, which account in total for approximately 30% of world marine capture fisheries production, are fully exploited. For example, of the seven principal tuna species, one-third were estimated to be overexploited and 37.5% were fully exploited in 2009 (FAO 2012a, b). Worm et al. (2006) project the global collapse of all taxa currently fished by the mid-twenty-first century, that is, to 100% in the year 2048. Sustainable fisheries methods that enable the regeneration of fish populations are therefore necessary to prevent this.

Production of marine fish in aquaculture systems (mainly in cages or in arms of the sea) has increased substantially in the last two decades and seems to be an alternative to the depletion of natural wild stocks. However, finfish farming requires large inputs of wild fish for feed and, therefore, even increases pressure on natural stocks and contributes to the collapse of fisheries stocks worldwide. In addition, the use of chemicals (antibiotics for medication and antifoulants) and water pollution as a result of feed wastage and faeces production have severe effects on the environment and tend to be most severe in areas with poor water exchange.

8.1.3 Climate Change

The average global temperature near the earth's surface has increased during the course of the twentieth century, especially over recent decades (Fig. 8.8). The total temperature increase from 1850–1899 to 2001–2005 is 0.76°C. Eleven of the twelve years in the period 1995–2006 rank among the 12 warmest years in the instrumental record of global surface temperature (since 1850). The temperature increase has been observed all over the globe and is greater at higher northern latitudes. Average Arctic temperatures at the top of the permafrost layer have generally increased since the 1980s by up to 3°C. Palaeoclimatic information

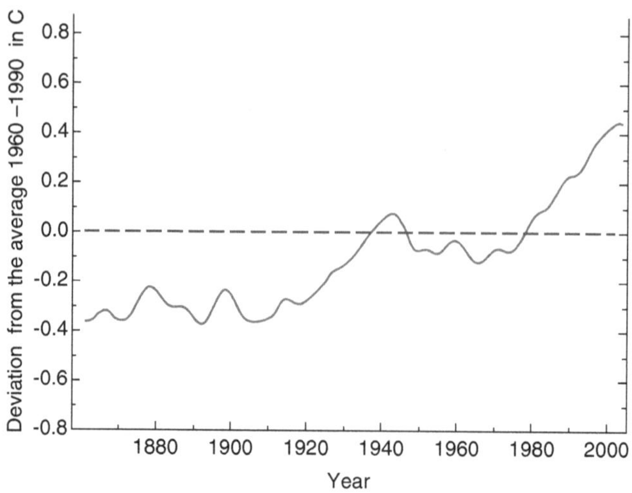

Fig. 8.8 Global averages of near-surface air temperature between 1861 and 2004 (Based on Met Office Hadley Centre 2005)

supports the interpretation that the warmth of the last half century is unusual in at least the previous 1,300 years.

These findings are some of the major conclusions regarding changes in global climate by the **Intergovernmental Panel on Climate Change (IPCC)**, as published in their fourth Assessment Report (IPCC 2007). The same report states that observed ecological changes as a consequence of this development in the last decades include:

- a rise in the average global sea level. Since 1993, the sea level has been rising at approximately 3 mm per year, significantly higher than the average during the previous half century. The two major causes of global sea level rise are thermal expansion of the oceans (water expands as it warms) and the loss of land-based ice sheets (especially of Greenland and Antarctica) because of increased melting.
- more intense and longer droughts over wider areas since the 1970s, particularly in the tropics and subtropics. Increased drying linked with higher temperatures and reduced precipitation has contributed to changes in drought.
- increased frequency of heavy precipitation over most land areas, consistent with warming and observed increases in atmospheric water vapour.
- widespread changes in extreme temperatures over the last 50 years. Cold days, cold nights, and frost have become less frequent whereas hot days, hot nights, and heat waves have become more frequent.

Together, these data leave no room for doubt that the current observed global warming in the course of the last century (and presumably far beyond that), is an exceptional process that is accompanied by further changes in the climate system.

8.1.3.1 Causes of Climate Change

In all phases of Earth's history the global climate has been subject to large and small changes. Development of the Earth's climate was, and is, basically defined by the incoming solar radiation, the composition and circulation of the atmosphere, the size and location of the continents and oceans, the polar icecaps, and the size and activity of the biosphere (Oschmann 2003).

History of the Earth's Climate

Approximately four billion years ago, the earth's atmosphere consisted of 95% CO_2 and had a very high water vapour content. At that time, the radiation energy of the sun was approximately 25–30% lower than today. After the initial cooling of the Earth's surface to less than 100°C, water formed and led to the emergence of an ocean approximately 3.5 billion years ago. As a result of an intense greenhouse effect (see below), high temperatures, estimated to be approximately 50°C, dominated on the Earth in this phase.

The presence of water initiated processes that led to a progressive decline in atmospheric CO_2 content. This happened as a result of chemical weathering and chemical and physical binding of CO_2 in the ocean. By these processes, carbonate rocks formed and were deposited on the cooling crust of the Earth. Primary production of the first organisms (bacteria, especially cyanobacteria), which have existed for at least 3.5 billion years, also contributed to the decline in the atmospheric CO_2 content. At the same time, enrichment of the atmosphere with oxygen began.

In connection with the continuing uptake of CO_2, the first great ice age began approximately 2.5 billion years ago. It persisted for approximately 200 million years, but was probably interrupted by numerous warmer periods. Four ice ages followed in the Palaeozoic era (550–250 million years ago). The Mesozoic era (250–65 million years ago) was a warm phase and was characterized by the existence of the dinosaurs. Until 50 million years ago, until the early Tertiary period, tropical temperatures prevailed worldwide. Subsequent gradual cooling resulted in the last major ice age period. In the last one million years approximately 20 major ice phases occurred, during which the northern hemisphere was covered with continental ice sheets several kilometres thick. They covered North America to the Great Lakes, and Scandinavia and parts of Central Europe, including the Alps. The ice ages were interrupted by warm periods, of which the last has been continuous since approximately 10,000 years ago.

The global decline in temperatures in the course of the Tertiary was accompanied by a series of processes which are also considered responsible for initiation of the last ice age. A central process, which began more than 190 million years ago, was the break up of the mega-continent Pangea, from which the modern continents were formed. As a result of continental drift, the landmasses gradually

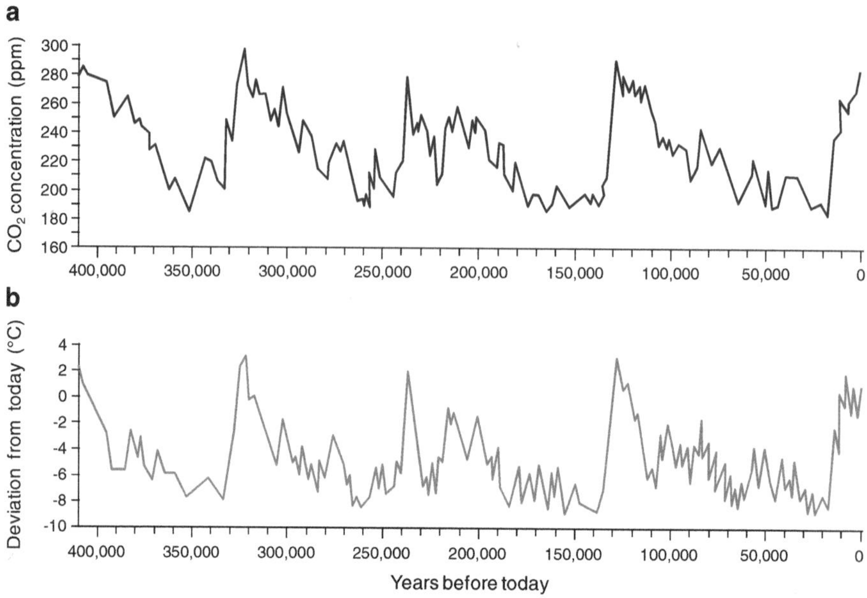

Fig. 8.9 (**a**) Atmospheric CO_2 concentrations and (**b**) air temperatures during the past 420,000 years. The data were extracted from ice cores from the Antarctic to a depth of 3,310 m. Determination of CO_2 concentration is based on air entrapped in the ice. Reconstruction of temperatures is based on the deuterium (heavy hydrogen) content of the ice (Based on Petit et al. 1999)

reached their current locations. This process was accompanied by horizontal and vertical changes in ocean currents, which prevented uniform heat transfer between the equatorial and higher latitudes (cf. Sect. 7.1). As a result, 35 million years ago Antarctica, which had drifted toward the South Pole, became covered in ice. The ice cover on the northern polar region has existed for the past three million years. Furthermore, drifting of the continents also resulted in changes in the atmospheric circulation, caused by the distribution and size of land masses and mountains. The latter were formed as a result of the collision of land masses or continental plates. For example, by this process the Alps, Himalayas, and Andes were formed. Another factor, which probably contributed to global cooling and to the last ice ages, was the reduction of the greenhouse effect (see below) as a result of the continuous decline in the atmospheric CO_2 concentration. In addition to chemical and physical processes, CO_2 was increasingly consumed by photosynthetic terrestrial plants and marine phytoplankton. The regular changes between cold and warm periods during the last ice age, and smaller temperature variations (Fig. 8.9) are also assumed to be a result of the so-called Milankovitch cycles (after the Serbian astrophysicist Milankovitch, who, in 1941, developed the theory that cyclical changes in the Earth's orbit around the sun affect the global climate). Cyclical variations in three characteristics of the Earth's orbit around the sun (namely eccentricity, axial tilt, and precession) occur on time scales of 100,000, 40,000, and 20,000 years, respectively. As a result of these processes, the distribution of the radiation energy of the sun on the Earth's surface is subject to changes, which in turn are related to global changes in climate. However, the Milankovitch cycles can only initiate ice ages when the polar regions have already cooled (Oschmann 2003).

The Natural Greenhouse Effect

Against the background of the climate history of the Earth, the question arises regarding the causes of the currently observed changes. The temperature conditions in the lower atmosphere are not determined by insolation alone, but also by the natural greenhouse effect. This effect is depends on the portion of the heat radiation which is emitted by the Earth's surface, absorbed in the atmosphere and re-emitted in the direction of the Earth's surface (counter radiation; cf. Sect. 4.1). Without the resulting additional warming effect, the average global air temperature near the surface would not be 15°C, which it currently is, but would instead be −18°C. This **natural greenhouse effect** is a result of the presence of specific atmospheric gases which contribute, in different amounts, to the 33°C difference in temperature. Water vapour is the most important, with a 62% share in the natural greenhouse effect, resulting in a temperature increase of 20.6°C. The second most important contribution to the natural greenhouse effect is CO_2, with a contribution of 22% (7.2°C). Other natural greenhouse gases include tropospheric ozone (O_3), nitrous oxide (N_2O), and methane (CH_4). The two most abundant gases in the atmosphere, N_2 and O_2, do not contribute to the greenhouse effect.

The Anthropogenic Greenhouse Effect

Enhancement of the natural greenhouse effect by gases that reach the atmosphere as a result of human activity makes up the **anthropogenic greenhouse effect**. Through this, the concentration of different greenhouse gases has increased substantially in comparison with the pre-industrial period (IPCC 2007):

- The global atmospheric concentration of CO_2 has increased from a pre-industrial value by 35%, from approximately 280 ppm (parts per million) to approximately 380 ppm in 2005. The atmospheric concentration of carbon dioxide in 2005 far exceeds the natural range over the last 650,000 years (180–300 ppm) as determined from ice cores. The annual growth in carbon dioxide concentration was larger during the period 1995–2005 (average: 1.9 ppm per year) than it has been since the beginning of continuous direct atmospheric measurements (1960–2005 average: 1.4 ppm per year) although there is year-to-year variability in growth. The global increases in carbon dioxide concentration are primarily because of anthropogenic fossil fuel combustion, with land-use change providing another significant but smaller contribution.
- The global atmospheric concentration of CH_4 has increased from the pre-industrial value by approximately 150%, from 715 to 1,774 ppb (parts per

billion) in 2005. The atmospheric concentration of methane in 2005 by far exceeds the natural range of the last 650,000 years (320–790 ppb) as determined from ice cores. It is very likely that the observed increase in methane concentration is a result of anthropogenic activity, predominantly use of fossil fuels, livestock production, and paddy rice production, but relative contributions from different source types are not well determined.
- The global atmospheric nitrous oxide (N_2O) concentration increased from a pre-industrial value of approximately 270–319 ppb in 2005. The growth rate has been approximately constant since 1980. More than a third of all nitrous oxide emissions are anthropogenic, primarily as a result of agriculture.

In different phases of the Earth's history there have been close correlations between the prevailing temperature conditions and the CO_2 content of the atmosphere. For example, data extracted from Antarctic ice cores (Fig. 8.9) have shown this to be true for the past 420,000 years. Therefore, it is an obvious assumption that there is a causal relationship between anthropogenic enhancement of the greenhouse effect and the currently observed increase in global temperatures.

On the basis of current knowledge, the IPCC (2007) came to the conclusion that the warming of the earth observed during recent decades is very likely to be a result of the increase in anthropogenic greenhouse gas concentrations. CO_2 is responsible for more than 60% of the anthropogenic contribution to the natural greenhouse effect.

8.1.3.2 Climate Projections for the Twenty-First Century

Development of the global climate during the next 100 years will be decisively determined by the quantity of anthropogenically released greenhouse gases; it therefore depends essentially on economic, social, technological and demographic changes within the human community. Under consideration of such factors, the IPCC developed 40 scenarios based on different assumptions, in which predictions were made about changes in the emission of greenhouse gases. These, in turn, serve as the bases of models for prediction of atmospheric greenhouse gas concentrations, and for future climate. Some of the overall predictions of the IPCC (2007) include:

- Continued greenhouse gas emissions at or above current rates would cause further warming and induce many changes in the global climate system during the twenty-first century that are *very likely* to be larger than those observed during the twentieth century
- For the next two decades, warming of approximately 0.2°C per decade is projected for a range of scenarios. Even if the concentrations of all greenhouse gases and aerosols had been kept constant at year 2000 levels, further warming of approximately 0.1°C per decade would be expected.
- Anthropogenic warming and sea level rise would continue for centuries, because of the time scales associated with climate processes and feedback, even if greenhouse gas concentrations were to be stabilized.

- Uptake of anthropogenic carbon since 1750 has led to the ocean becoming more acidic, with an average decrease in pH of 0.1 units. Increasing atmospheric CO_2 concentrations would lead to further acidification.

Ecological Effects of Climate Change

The changes in the global climate expected during the next 100 years will affect natural ecosystems, their goods and services, ecological cycles, and the living conditions of numerous species, including humans, in many ways. On the basis of the predictions of the IPCC (2007), the most important ecological effects of climate change will include:

Water cycle and water resources. The expected variability in precipitation, both periods between precipitation and the quantity of precipitation, will lead to changes in runoff and water availability. By the mid-century runoff is projected with high confidence to increase by 10–40% at higher latitudes and in some wet tropical areas, including populous areas in East and Southeast Asia, and decrease by 10–30% over some dry regions at mid-latitudes and in the dry tropics, because of decreases in rainfall and higher rates of evapotranspiration. There is also high confidence that many semi-arid areas (e.g. the Mediterranean Basin, western United States, southern Africa, and north-eastern Brazil) will suffer a decrease in water resources. Drought-affected areas are projected to increase in extent. Heavy precipitation, which is very likely to increase in frequency, will augment flood risk.

On a regional scale, mountain snow pack, glaciers, and small ice caps are crucial to the availability of fresh water. Widespread mass losses from glaciers and reductions in snow cover over recent decades are projected to accelerate throughout the twenty-first century, reducing water availability, hydropower potential, and changing the seasonality of flows in regions supplied by meltwater from major mountain ranges.

Ecosystems and biodiversity. Climate-induced changes and losses in terrestrial and aquatic habitats will affect the presence and population density of animal and plant species. Approximately 20–30% of plant and animal species assessed so far are likely to be at increased risk of extinction if increases in global average temperature exceed approximately 2°C. Especially vulnerable are species that exist only in very limited habitats (e.g. islands, mountainous regions, wetlands). As a result of the increase in temperature, the boundaries of climate zones and related distribution ranges of many species will change. In the northern hemisphere, expansion into the north and to mountainous regions will occur. The rise in the ocean levels is a threat to coastal ecosystems, which include mangrove forests, tidal flats, deltas of rivers, and coral reefs. Progressive acidification of oceans because of increasing atmospheric carbon dioxide is expected to have negative effects on marine shell-forming organisms (e.g., corals) and their dependent species. The flooding or loss of low coastal regions also has economic and social consequences for the resident humans who will, in extreme cases, be forced to migrate.

Carbon cycle. For C_3 plants, which include approximately 95% of all plant species (Sect. 4.1.1.1), the current atmospheric concentration of CO_2 of approximately 380 ppm is below the optimum. With increasing atmospheric CO_2 content, their photosynthetic performance may increase. However, it is not to be expected that this CO_2 fertilization effect will lead to increased net ecosystem production in all vegetation zones in which C_3 plants dominate. This process is also affected by other factors that will change with climate change (temperature and moisture) and by nutrient availability. In many regions (e.g. boreal climate zone) global warming can lead to greater droughts, whereby plant growth will be limited. An increase in temperature leads to an increase in microbial decomposition of organic substances when sufficient soil moisture is present; increased release of CO_2 from soil stores may occur as a result of this, which will, in turn, contribute to global warming. For example, this may be expected from the thawing of permafrost soils in the Arctic tundra. Global warming reduces terrestrial and ocean uptake of atmospheric CO_2, and thus increases the proportion of anthropogenic emissions remaining in the atmosphere. Over the course of this century, net uptake of CO_2 by terrestrial ecosystems is likely to peak before mid-century and then weaken or even reverse.

Effects of Climate Change on Agriculture

The changes in climate expected in the future will affect agricultural production on different scales, depending on the physiological performance of crops, processes in agroecosystems, and global and regional yield.

Crop productivity. C_3 and C_4 plants (Sect. 4.1.1), each of which is represented by important crops, react differently to increases in CO_2 availability. For numerous C_3 species, for example wheat, rice, and soybean, yields can increase by approximately 30% by doubling of the natural CO_2 concentration. The stomata also react to an increased supply of CO_2 by reducing the size of their openings, whereby the water consumption by the plant is reduced. This lower transpiration can delay drying of the soil, which is particularly important for production of biomass during precipitation-poor periods.

For most C_3 plants the temperature optimum for CO_2 fixation is between 15 and 25°C. Within this range, an increased supply of CO_2 leads to increased photosynthetic performance. However, an increase in temperatures to more than 26°C results in a decline of yields for such species as rice, wheat, and soybeans.

An improved supply of CO_2 also affects the nutrient balance of C_3 plants. With increased photosynthesis and biomass production, the demand for nutrients, especially nitrogen, increases. In many cases it has been shown that plants subjected to an increase in CO_2 concentration have a lower nitrogen content and thus a higher C/N ratio in their tissues than those grown with the current atmospheric CO_2 supply. This results in changes in the nutritional value of the plant tissue in C_3

plants, which may include reduced protein content in grains of rice and wheat. In many C_3 plants, another reaction to an increase in CO_2 is translocation of photosynthetic products to the roots. Thus the root/shoot ratio is altered, which has a positive effect on the yields of root and tuber crops.

For C_4 plants, an increase in atmospheric CO_2 content is of little importance, because they are, as a result of their physiology, already able to better utilize the currently available supply than are C_3 plants (Sect. 4.1.1.2).

Agroecosystems. An increase in atmospheric CO_2 concentration, and the resulting changes in climate, will also affect a variety of processes in agroecosystems:

- Via higher demand for nutrients of C_3 plants as a consequence of the increasing CO_2 concentrations and increasing photosynthetic rate, supply deficits can occur in extensively managed systems that must be compensated by the application of fertilizers.
- Organic carbon stores in the soil will be affected in a variety of ways by higher CO_2 concentrations. When the nitrogen content of plant tissues decreases (compare above), the C/N ratio increases in harvest residues and in green manures. As a result, rates of decomposition and mineralization of plant residues in the soil will decrease, which is associated with enrichment of organic carbon compounds (Sect. 4.3.5.1). To some extent this process counteracts the increase in temperature which leads to an increased rate of decomposition of organic substances and thus can lead to carbon loss from the soil.
- Little is known about the question of whether and to what extent climate change will affect the incidence of pests and the respective yield losses in crops. The interactions between crops and pests are affected by the nitrogen content of plant tissues (Sect. 4.5.4.2). When the nitrogen content declines as a consequence of an increased uptake of CO_2 by the plant, the quality of the food for the phytophages is reduced. This can have negative effects on the development of populations of many phytophagous species and thus to a lower incidence of pests. In contrast, the expected global warming might have predominantly positive effects on pests. With higher temperatures, the development cycles of insects become shorter, particularly in regions outside the tropics. This, combined with longer vegetation periods, means pests are able to produce more generations per year than was previously possible. Furthermore, primarily in the temperate latitudes, survival of pests through the winter may increase because of the milder weather conditions.
- An increase in CO_2 concentration can effect the relationships between weeds and crops (Sect. 4.4.1), primarily because of the different reactions of C_3 and C_4 plants. When the crop plant is a C_3 species and the weed is a C_4 species, with increased CO_2 supply lower competition-related yield losses are to be expected compared with the current CO_2 concentration. This has also been shown experimentally (Ziska 2000). In contrast, in C_4 crops (e.g. maize, sorghum, sugarcane) that stand in competition with C_3 plants, the CO_2 fertilization effect can enhance the negative effects of weed competition on yields.

- Infestation by many phytopathogenic bacteria and fungi is substantially affected by climate and weather conditions. Mild and wet winters in the temperate latitudes increase the risk of the spread of plant diseases, while greater summer droughts have the opposite effect.

Production conditions and yields. Worldwide, the potential for food production is projected to increase with increasing local average temperature over a range of 1–3°C; above this it is projected to decrease (IPCC 2007).

However, there are large differences between latitudes and regions:

- Crop productivity is projected to increase slightly at mid to high latitudes for local mean temperature increases of up to 1–3°C, depending on the crop, and then decrease beyond that in some regions. A temperature shift northward beyond the current climatic boundaries in Europe, North America, and Asia will enable expansion of agricultural areas, more or less far depending on the change in temperature. In large parts of Central Europe, warming of 2°C combined with an increasing atmospheric CO_2 concentration would lead primarily to an increase in wheat yields and at least unchanging yields in other crops. A further increase in temperature will tend to produce more negative effects on the production conditions in this region, because the risk of drought-related yield losses will increase. In many regions of south and southeast Europe, climate change is projected to worsen conditions (high temperatures and drought). There, even a relatively small increase in temperature with reduced water availability would have detrimental effects on crop production, even when the CO_2 fertilization effect is taken into account. For North America, moderate climate change is projected to increase aggregate yields of rainfed agriculture by 5–20% in the early decades of the century, but with important variability among regions. Major challenges are expected for crops near the warm end of their suitable range or which depend on highly utilized water resources.
- Temperature and CO_2 increases in different regions outside the tropics affect not only the distribution and growth of crops, but also enable many species of the agroecosystem to expand their range. As a result, pests, weeds, and diseases may appear in new regions. An example of this is the parasitic weed *Orobanche crenata*, which is currently mainly distributed in the Mediterranean area but which may expand northward with increasing temperatures. Its potential range expansion with a temperature increase of 3°C is shown in Fig. 8.10. In addition, the conditions for pests, weeds, and diseases that have been introduced from other regions and continents (Sect. 3.3.3) may improve and also affect productivity, species interactions, and the biodiversity of agroecosystems.
- At lower latitudes, especially in seasonally dry and tropical regions, crop productivity is projected to decrease for even small local temperature increases (1–2°C); this would increase the risk of famine. In the tropics, yields of several important crops will decrease with increasing warming, because these species react to increased temperatures not with an increase but rather with a decrease in their production performance. In addition, the greatest uncertainty in the water supply must be expected in the wet–dry tropics. The negative effects of climate change will be strongest in Africa, where the number of countries with insecure

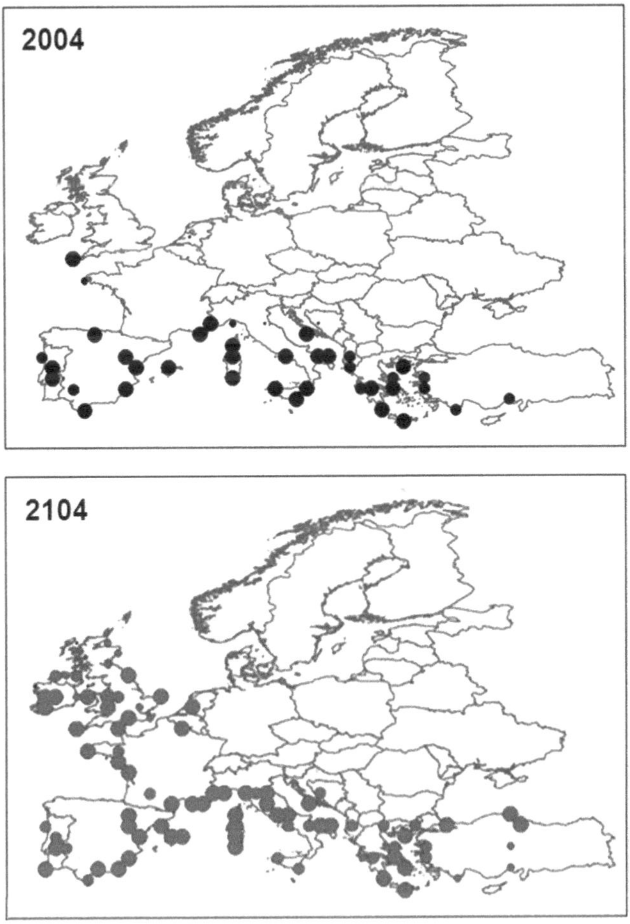

Fig. 8.10 Potential geographic range of *Orobanche crenata* in Europe under current climate conditions and under an assumed warming of 3°C in the next 100 years (based on the seedbank model by Grenz and Sauerborn 2007). The size of the symbols corresponds to the ecological suitability of the site, which was determined by use of the computer model CLIMEX (Sutherst et al. 1999)

food production is already high. By 2020, yields from rain-fed agriculture in some countries could be reduced by up to 50%. Agricultural production, including access to food, is projected to be severely threatened in many African countries. With a temperature increase of more than 2°C, it is expected that yields of rice and wheat will also decline in tropical and subtropical Asia. In many regions of Latin America the yield situation will also worsen with increasing temperatures.

In most regions of the world, an increase in the average global temperature of more than approximately 2°C will have negative affects on yields of the most important cereal species (wheat and rice) and probably also on other crops.

The effects of climate change on global and regional plant production are not only determined by the atmospheric CO_2 concentration and the temperature and precipitation conditions, but also by numerous additional factors relating to CO_2 concentration. These include the above described possible changes in soil properties, and the effects of pests, weeds, and diseases. As a result, predictions of the affects of climate change on yields in different regions of the earth are burdened with great uncertainties.

To some extent, the negative effects of climate change on plant production may be mitigated by adaptive measures, including selection of suitable species and varieties of crops and changes in planting and sowing dates and crop rotation. Loss of organic material from the soil, which could increase with increasing temperature, may be countered by changes in land use and adapted cultivation. However, it must be expected that the effects of such adaptations will be limited by increased variability in annual temperatures and precipitation. Modification of the production system might therefore prove suitable in 1 year and detrimental in another.

8.1.4 Outlook

The number of humans on Earth, their resource requirements, and anthropogenic release of greenhouse gases, will increase in the coming decades and will have substantial effects on large parts of the human population. These effects will include increasing threats as a result of environmental catastrophes, the threat of wars for resources, increasing numbers of hungry people, increasing poverty, and economic and environmental refugees. Fundamentally, the measures needed to mitigate or prevent these effects are known. They consist primarily of a reduction in the consumption of raw materials and other non-renewable products including natural and agricultural lands, whereby a simultaneous and substantial contribution would be made to resolving current and future environmental problems. Currently, the application of such measures on the global level is the greatest challenge to humanity and is connected with substantial difficulties. For example, according to IPCC estimates (2001), for successful application of measures to reduce greenhouse gas emissions, numerous technical, economic, political, cultural, social, behavioural, and/or institutional barriers which hinder total use of technological, economic, and social possibilities for reducing emissions must be overcome. This may also be true of the search for solutions to other problems, for example sustainable land use, agricultural productivity, and population growth, and is related to these. If and how the problems of global change can be solved by global cooperation, essentially depends on the willingness of nations to limit their use of resources so that future generations will face suitable living conditions.

There is enough on Earth for everybody's need, but not enough for everybody's greed (Mahatma Ghandi).

References

Borlaug N (2000) Ending world hunger. The promise of biotechnology and the threat of antiscience zealotry. Plant Physiol 124:487–490

Cohen JE (1997) Population economics, environment and culture: an introduction to human carrying capacity. J Appl Ecol 34:1325–1333

Eickhout B, Bouwman AF, van Zeijts H (2006) The role of nitrogen in world food production and environmental sustainability – agriculture. Ecosyst Environ 116:4–14

Evenson RE, Gollin D (2003) Assessing the impact of the green revolution, 1960 to 2000. Science 300:758–762

FAO (2003) World agriculture: towards 2015/2030 – An FAO perspective. Food and Agriculture Organization of the United Nations, Rome. Earthscan Publications, London, and http://www.fao.org/docrep/005/y4252e/y4252e00.htm

FAO (2006) World agriculture: towards 2030/2050. Global Perspective Studies Unit, Food and Agriculture Organization of the United Nations, Rome, http://www.fao.org/fileadmin/user_upload/esag/docs/Interim_report_AT2050web.pdf

FAO (2009) How to feed the world in 2050. In: Proceedings of the expert meeting on how to feed the world in 2050, 24–26 June 2009, Food and Agriculture Organization of the United Nations Headquarters, Rome. http://www.fao.org/docrep/012/ak542e/ak542e00.htm

FAO (2011) Global food losses and food waste. Food and Agriculture Organization of the United Nations, Rome, http://www.fao.org/docrep/014/mb060e/mb060e00.pdf

FAO (2012a) Biodiversity for a world without hunger. Food and Agriculture Organization of the United Nations, Rome, http://www.fao.org/biodiversity/components/plants/en/

FAO (2012b) The state of world fisheries and aquaculture 2012. Fisheries and Aquaculture Department of the Food and Agriculture Organization of the United Nations, Rome, http://www.fao.org/docrep/016/i2727e/i2727e.pdf

FAOSTAT. The statistical database of the Food and Agriculture Organization (FAO), Rome. http://www.faostat.fao.org/

Foley JA, Ramankutty N, Brauman KA, Cassidy ES, Gerber JS (2011) Solutions for a cultivated planet. Nature 478:337–342

Godfray HCJ, Beddington JR, Crute IR, Haddad L (2010) Food security: the challenge of feeding 9 billion people. Science 327:812–818

Grenz JH, Sauerborn J (2007) Mechanisms limiting the geographical range of the parasitic weed *Orobanche crenata*. Agric, Ecosyst Environ 122:275–281

Hibberd JM, Sheehy JE, Langdale JA (2008) Using C_4 photosynthesis to increase the yield of rice – rationale and feasibility. Curr Opin Plant Biol 11:228–231

IPCC (2001) Climate change 2001: synthesis report. Cambridge University Press, Cambridge, http://www.grida.no/publications/other/ipcc_tar/

IPCC (2007) Fourth assessment report: climate change 2007: synthesis report. http://www.ipcc.ch/publications_and_data/publications_ipcc_fourth_assessment_report_synthesis_report.htm

ISWA (2012) Ermittlung der Mengen weggeworfener Lebensmittel und Hauptursachen für die Entstehung von Lebensmittelabfällen in Deutschland. Zusammenfassung einer Studie der Universität Stuttgart, Institut für Siedlungswasserbau, Wassergüte- und Abfallwirtschaft (ISWA). http://www.iswa.uni-stuttgart.de/afw/aktuelles/tagungen/tagungen_download/03-12-Studie_Lebensmittel_Factsheet.pdf

Met Office Hadley Centre for Climate Prediction and Research (2005) http://www.metoffice.gov.uk/climate-change/resources/hadley

Oschmann W (2003) Vier Milliarden Jahre Klimageschichte im Überblick. In: Klimastatusbericht 2003. Deutscher Wetterdienst, Offenbach. pp 7–24

Pelletier N, Tyedmers P (2010) Forecasting potential global environmental costs of livestock production 2000–2050. Proc Natl Acad Sci 107:18371–18374

Petit JR, Jouzel J, Raynaud D, Barkov NI, Barnola J-M, Basile I, Benders M, Chappellaz J, Davis M, Delayque G, Delmotte M, Kotlyakov VM, Legrand M, Lipenkov VY, Lorius C, Pépin L, Ritz C,

Saltzman E, Stievenard M (1999) Climate and atmospheric history of the past 420,000 years from the Vostok ice core, Antarctica. Nature 399:429–436

Russelle MP, Entz MH, Franzluebbers AJ (2007) Reconsidering integrated crop-livestock systems in North America. Agron J 99:325–334

Sutherst RW, Maywald GF, Yonow T, Stevens PM (1999) CLIMEX – predicting the effects of climate on plants and animals. SCIRO Publishing, Collingwood

Tilman D (1998) The greening of the green revolution. Nature 396:211–212

UNEP (2009) The environmental food crisis – the environment's role in averting future food crises. A UNEP rapid response assessment. United Nations Environment Programme. http://www.unep.org/pdf/foodcrisis_lores.pdf

United Nations (2005) Population division of the Department of Economic and Social Affairs: world population prospects: the 2004 revision. Highlights. New York. www.un.org/esa/population/unpop.htm

United Nations (2009) Population division of the Department of Economic and Social Affairs: world population prospects: the 2008 revision. Highlights. New York. http://www.un.org/esa/population/publications/wpp2008/wpp2008_highlights.pdf

Worm B, Barbier EB, Beaumont N, Duffy JE et al (2006) Impacts of biodiversity loss on ocean ecosystem services. Science 314:787–790

Ziska LH (2000) The impact of elevated CO_2 on yield loss from a C_3 and C_4 weed in field-grown soybean. Glob Chang Biol 6:899–905

Index

K. Martin and J. Sauerborn, *Agroecology*, DOI 10.1007/978-94-007-5917-6,

FSC
www.fsc.org
MIX
Papier | Fördert
gute Waldnutzung
FSC® C083411

Kolibri 360 GmbH
Ferdinand-Jühlke-Straße 7
99095 Erfurt, Deutschland
produktsicherheit@kolibri360.de